AF903675

ALGEBRAIC GEOMETRY IN *East Asia*

Proceedings of the Symposium on

ALGEBRAIC GEOMETRY IN *East Asia*

Editors

Akira Ohbuchi

Tokushima University, Japan

Kazuhiro Konno

Sampei Usui

Osaka University, Japan

Atsushi Moriwaki

Kyoto University, Japan

Noboru Nakayama

RIMS, Japan

World Scientific

New Jersey • London • Singapore • Hong Kong

Published by

World Scientific Publishing Co. Pte. Ltd.
5 Toh Tuck Link, Singapore 596224
USA office: Suite 202, 1060 Main Street, River Edge, NJ 07661
UK office: 57 Shelton Street, Covent Garden, London WC2H 9HE

British Library Cataloguing-in-Publication Data
A catalogue record for this book is available from the British Library.

ALGEBRAIC GEOMETRY IN EAST ASIA
Proceedings of the Symposium on Algebraic Geometry in East Asia

Copyright © 2002 by World Scientific Publishing Co. Pte. Ltd.

All rights reserved. This book, or parts thereof, may not be reproduced in any form or by any means, electronic or mechanical, including photocopying, recording or any information storage and retrieval system now known or to be invented, without written permission from the Publisher.

For photocopying of material in this volume, please pay a copying fee through the Copyright Clearance Center, Inc., 222 Rosewood Drive, Danvers, MA 01923, USA. In this case permission to photocopy is not required from the publisher.

ISBN 981-238-265-8

Printed in Singapore by World Scientific Printers (S) Pte Ltd

Preface

This book is the proceedings of the conference "Algebraic Geometry in East Asia" which was held in International Institute for Advanced Studies (IIAS) (9-3 Kizugawadai, Kizu-cho, Soraku-gun, Kyoto 619-0225, Japan), during August 3 to August 10, 2001. The conference was partially supported by Grants-in-Aid for Scientific Researches ((A) (1) 11304001 by Sampei Usui and (B) (2) 13440008 by Atsushi Moriwaki).

Although many east Asian mathematicians now play a leading role in the international mathematical community, in modern times mathematics did not become a strength of east Asian scholarship until the early 20'th century. In particular, as a consequence of this relatively short history, the various east Asian mathematical communities, such as the algebraic geometers represented at this conference, have had less of a chance to meet and exchange ideas as their western counterparts. Accordingly, one of the primary goals of the conference was to facilitate such an exchange.

As the breadth of the topics covered in this proceedings demonstrate, the conference was indeed successful in assembling a wide spectrum of east Asian mathematicians, and gave them a welcome chance to discuss current state of algebraic geometry. It is the first time that such a conference has been held in algebraic geometry, and we hope that it is but the start of continuing tradition.

We wish to thank, first of all, the lecturers for their beautiful talks. We also wish to thank the participants for their cooperation and providing stimulating atmosphere.

During the conference, administrative staffs in IIAS as well as many graduate students from Osaka University and Kyoto University helped us. Without them, the conference would have been less successful than we had hoped for.

In particular, we wish to thank Ms. Yoshiko Kusaki and Ms. Minako Tanaka of the IIAS as well as the following graduate students from Osaka and Kyoto Universities: Mr. Masao Aoki, Mr. Takeshi Harui, Mr. Atsushi Ikeda, Mr. Michiaki Inaba, Mr. Tomokazu Kawahara, Mr. Hiraku Kawanoe, Mr. Shinya Kitagawa, Mr. Masaaki Murakami, Mr. Hiroto Nakayama, Ms. Noriko Tsuda and Mr. Daisuke Yanase.

Finally, we would like to thank Prof. Junjiro Kanamori, the Director of IIAS, for his dignified opening speech.

Organizers
Akira Ohbuchi(Chief)
Kazuhiro Konno
Atsushi Moriwaki
Noboru Nakayama
Sampei Usui

Contents

INTRODUCTION TO ARAKELOV GEOMETRY

SHU KAWAGUCHI, ATSUSHI MORIWAKI AND KAZUHIKO YAMAKI

CONTENT

Introduction

This note was first written in Japanese for intensive lectures of Arakelov geometry organized by Moriwaki from December 8 to December 10, 1998 at Kyoto University. These lectures were intended to give an overview of Arakelov geometry and a proof of Bogomolov's conjecture for general algebraic geometers. From that time, we have considered that this note should be available for not only Japanese but also a broader range of readers. We, however, had no chance to translate it into English. Fortunately, during the meeting "Algebraic Geometry in East Asia," Professor Usui recommended its translation. Here we would like to express hearty thanks for his proposal.

The final goal of this note is a generalization of conjectures of Bogomolov and Lang. For this purpose, in the first part, we introduce "Arithmetic Chow group" and "Arithmetic Riemann-Roch theorem," in which we do not give a rigorous proof for every result, but we believe that this is a good introduction of Arakelov geometry. In the middle part, we consider "Existence of a small section," "Adelic metric and admissible pairing" and "Arithmetic height function," in which several techniques of Arakelov geometry are used. In the final part, we give a proof of Bogomolov's conjecture and a generalization of conjectures of Bogomolov and Lang.

Here we would like to explain a generalization of conjectures of Bogomolov and Lang in the case of a curve and its Jacobian. Let K be a number field, X a geometrically irreducible projective curve of genus greater than or equal to 2 over K, and J the Jacobian of X. Let us fix an embedding $\iota : X(\overline{K}) \to J(\overline{K})$ and a Néron-Tate pairing

$$\langle\ ,\ \rangle : J(\overline{K}) \times J(\overline{K}) \to \mathbb{R}.$$

Let Γ be a subgroup of $J(\overline{K})$ with $\dim_{\mathbb{Q}} \Gamma \otimes \mathbb{Q} < \infty$. Let $(\Gamma \otimes \mathbb{R})^{\perp}$ be the orthogonal complement of $\Gamma \otimes \mathbb{R}$ in $J(\overline{K}) \otimes \mathbb{R}$ in terms of the Néron-Tate pairing. Let $\iota_{\Gamma} : X(\overline{K}) \to (\Gamma \otimes \mathbb{R})^{\perp}$ be the compositions of maps

$$X(\overline{K}) \xrightarrow{\ \iota\ } J(\overline{K}) \longrightarrow J(\overline{K}) \otimes \mathbb{R} \xrightarrow{\text{projection}} (\Gamma \otimes \mathbb{R})^{\perp}.$$

Then, a generalization of conjectures of Bogomolov and Lang says that the fiber of ι_{Γ} is finite and the image of ι_{Γ} is a discrete subset of $(\Gamma \otimes \mathbb{R})^{\perp}$ in terms of the metric arising from the Néron-Tate pairing. If we consider the case $\Gamma = J(K)$, then the first assertion is nothing more than Mordell's conjecture. Moreover, if we consider the case $\Gamma = \{0\}$, then we have Bogomolov's conjecture.

§1, §2, §3 and §7 were written by Kawaguchi, §4 by Yamaki, and §5 and §6 by Moriwaki. We hope that this note will be useful for anyone who wants to know Arakelov geometry.

1. Arithmetic Chow Group

1.1. **Introduction.** In the Arakelov geometry, one considers, roughly speaking,

- schemes over $\mathbb{Z}$ with "infinity," instead of algebraic varieties over a field,
- vector bundles with "metrics at infinity," instead of vector bundles.

Szpiro [34] wrote about the Arakelov geometry that

> Put metrics at infinity on vector bundles and you will have a geometric intuition of compact varieties to help you.

Let us see this analogy by comparing a compact Riemann surface with $\mathrm{Spec}(\mathbb{Z})$.

Let X be a compact Riemann surface.

(A) Let f be a nonzero rational function on X. Then we have

$$\mathrm{div}(f) = \sum_{p \in X} v_p(f) \cdot [p] \in \bigoplus_{p \in X} \mathbb{Z} \cdot [p] =: \mathrm{Div}(X).$$

It follows from the residue formula that

$$\deg(f) = \sum_{p \in X} v_p(f) = 0.$$

Set $\mathrm{Rat}(X) = \{\mathrm{div}(f) \mid f \in \mathbb{C}(X)^*\} \subset \mathrm{Div}(X)$, and $\mathrm{CH}^1(X) = \mathrm{Div}(X)/\mathrm{Rat}(X)$. Then the map $\deg : \mathrm{Div}(X) \to \mathbb{Z}$ induces the map $\deg : \mathrm{CH}^1(X) \to \mathbb{Z}$.

(B) Let L be a holomorphic line bundle over X, and s a nonzero rational section of L. Put

$$\mathrm{div}(s) = \sum_{p \in X} v_p(s) \cdot [p] \in \mathrm{CH}^1(X).$$

Then, as an element of $\mathrm{CH}^1(X)$, $\mathrm{div}(s)$ depends only on L, i.e., it is independent of the choice of s.

(C) Let $\mathrm{Pic}(X)$ be the set of isomorphism classes of holomorphic line bundles over X. Then, we have the isomorphism

$$c_1 : \mathrm{Pic}(X) \to \mathrm{CH}^1(X), \quad L \mapsto \mathrm{div}(s),$$

where s is any nonzero rational section of L. In particular, through this isomorphism, $\deg : \mathrm{Pic}(X) \to \mathbb{Z}$ is defined.

Next, we consider $\mathcal{X} = \mathrm{Spec}(\mathbb{Z})$.

(A') Since we consider a scheme at "infinity," let us set

$$\widehat{\mathrm{Div}}(\mathcal{X}) = \left(\bigoplus_{p:\mathrm{prime}} \mathbb{Z} \cdot [p] \right) \oplus \mathbb{R} \cdot [\infty].$$

Let $f \in \mathbb{Q}$ be a nonzero rational number. We set

$$v_\infty(f) = -\log |f|^2 \in \mathbb{R},$$

$$\widehat{\mathrm{div}}(f) = \sum_{p:\mathrm{prime}} v_p(f) \cdot [p] + v_\infty(f) \cdot [\infty] \in \widehat{\mathrm{Div}}(\mathcal{X}).$$

We define $\widehat{\deg} : \widehat{\mathrm{Div}}(\mathcal{X}) \to \mathbb{R}$ by

$$\sum_{p:\text{prime}} n_p \cdot [p] + a \cdot [\infty] \mapsto \sum_{p:\text{prime}} n_p \log p + \frac{1}{2} a.$$

It follows from the product formula that

$$\widehat{\deg}(\widehat{\mathrm{div}}(f)) = \sum_{p:\text{prime}} v_p(f) \log p + \frac{1}{2} v_\infty(f) = 0.$$

(In this case, the product formula is an obvious consequence of the prime factorization.) Set

$$\widehat{\mathrm{Rat}}(\mathcal{X}) = \{\widehat{\mathrm{div}}(f) \mid f \in \mathbb{Q} \setminus \{0\}\} \subset \widehat{\mathrm{Div}}(\mathcal{X})$$

and $\widehat{\mathrm{CH}}^1(\mathcal{X}) = \widehat{\mathrm{Div}}(\mathcal{X}) / \widehat{\mathrm{Rat}}(\mathcal{X})$. Then the map $\widehat{\deg} : \widehat{\mathrm{Div}}(\mathcal{X}) \to \mathbb{Z}$ induces the map $\widehat{\deg} : \widehat{\mathrm{CH}}^1(\mathcal{X}) \to \mathbb{Z}$.

(B') Let $\mathcal{L}$ be a line bundle over X. Since we consider a "metric at infinity," let us take a hermitian metric

$$h : \mathcal{L}_{\mathbb{C}} \times \mathcal{L}_{\mathbb{C}} \longrightarrow \mathbb{C}$$

on $\mathcal{L}_{\mathbb{C}} = \mathcal{L} \otimes_{\mathbb{Z}} \mathbb{C}$. We denote the pair $(\mathcal{L}, h)$ by $\overline{\mathcal{L}}$, and call it a hermitian line bundle.

Let s be a nonzero rational section of $\mathcal{L}$. Put

$$\widehat{\mathrm{div}}(s) = \sum_{p:\text{prime}} v_p(s) \cdot [p] + (-\log h(s_{\mathbb{C}}, s_{\mathbb{C}})) \cdot [\infty] \in \widehat{\mathrm{CH}}^1(X).$$

Then, as an element of $\widehat{\mathrm{CH}}^1(\mathcal{X})$, $\widehat{\mathrm{div}}(s)$ depends only on $\mathcal{L}$, i.e., it is independent of the choice of s.

(C') Two hermitian line bundles $\overline{\mathcal{L}}_1 = (\mathcal{L}_1, h_1)$ and $\overline{\mathcal{L}}_2 = (\mathcal{L}_2, h_2)$ are said to be isomorphic if there exist an isomorphism $\phi : \mathcal{L}_1 \to \mathcal{L}_2$ of line bundles such that the induced map $\phi_{\mathbb{C}} : (\mathcal{L}_{1\mathbb{C}}, h_1) \to (\mathcal{L}_{2\mathbb{C}}, h_2)$ is an isometry.

Let $\widehat{\mathrm{Pic}}(\mathcal{X})$ be the set of isomorphism classes of hermitian line bundles over $\mathcal{X}$. Then

$$\widehat{c}_1 : \widehat{\mathrm{Pic}}(\mathcal{X}) \to \widehat{\mathrm{CH}}^1(\mathcal{X}), \quad \overline{\mathcal{L}} \mapsto \widehat{\mathrm{div}}(s)$$

is an isomorphism, where s is any nonzero rational section of $\mathcal{L}$ (cf. Proposition 1.3.4). In particular, through this isomorphism, $\widehat{\deg} : \widehat{\mathrm{Pic}}(\mathcal{X}) \to \mathbb{Z}$ is defined.

To sum up, by adding "infinity" to $\mathrm{Spec}(\mathbb{Z})$ and considering hermitian line bundles over $\mathrm{Spec}(\mathbb{Z})$, one has the degree map $\widehat{\deg}$ for $\mathrm{Spec}(\mathbb{Z})$, similar to the degree map deg for a compact Riemann surface, in the sense that $\widehat{\deg}(\widehat{\mathrm{div}}(f)) = 0$ for $f \in \mathbb{Q} \setminus \{0\}$.

Such an analogy between the ring of integers of a number field and a compact Riemann surface have been noted by many mathematicians such as Hasse and Weil to name a few. Arakelov [2] generalized this analogy to a 2-dimensional case, and established the intersection theory on arithmetic surfaces, which corresponds to that on projective surfaces over $\mathbb{C}$. Then, Faltings proved in [5] among other things a Riemann-Roch theorem on arithmetic surfaces. Gillet and Soulé (cf. [9] [11], [3], [12], [13]) developed its higher dimensional theory, including arithmetic cycles and their intersections on arithmetic varieties, arithmetic Chern classes of hermitian vector bundles, and an arithmetic Riemann-Roch theorem. Such theory of arithmetic varieties is called the Arakelov geometry. We remark that [32] is a good reference of the Arakelov geometry and [33] is a good quick guide to it.

Let us consider a case of dimension ≥ 2. We set

$$\mathcal{X} = \operatorname{Proj}\left(\mathbb{Z}[X,Y,Z]/(Y^2Z = X^3 + XZ^2)\right).$$

This is an example of arithmetic surfaces (cf. §1.3). To consider "infinity" means to consider the compact Riemann surface

$$\mathcal{X}_{\mathbb{C}} = \operatorname{Proj}\left(\mathbb{C}[X,Y,Z]/(Y^2Z - X^3 - XZ^2)\right).$$

Moreover, to consider a line bundle with a "metric at infinity" means to consider a pair $\overline{\mathcal{L}} = (\mathcal{L}, h)$, where $\mathcal{L}$ is a line bundle over $\mathcal{X}$ and h is a hermitian metric on $\mathcal{L}_{\mathbb{C}}$.

Then, what is $\widehat{\operatorname{Div}}(\mathcal{X})$? The case $\mathcal{X} = \operatorname{Spec}(\mathbb{Z})$ kept in mind, it would be natural (undoubtedly with hindsight) to think that

$$(\operatorname{div}(s), -\log h(s_{\mathbb{C}}, s_{\mathbb{C}}))$$

becomes an "arithmetic divisor" on $\mathcal{X}$, where s is a nonzero section of $\mathcal{L}$.

In what follows, we will give the precise definitions of arithmetic varieties, arithmetic divisors, the arithmetic Chow groups on an arithmetic variety etc., due to Gillet and Soulé. In fact, an arithmetic divisor on an arithmetic variety $\mathcal{X}$ is a pair $(\mathcal{Z}, g)$ such that $\mathcal{Z}$ is a cycle on $\mathcal{X}$ and g is a "Green current" on $\mathcal{X}(\mathbb{C})$; And $(\operatorname{div}(s), -\log h(s_{\mathbb{C}}, s_{\mathbb{C}}))$ above is indeed an arithmetic divisor on $\mathcal{X}$. So, let us first define Green currents in the next subsection.

1.2. **Currents.** Let X be a d-dimensional compact complex manifold. Let $A^{p,q}(X)$ be the space of C^∞ differential forms of type (p,q) on X. We endow $A^{p,q}(X)$ with the compact C^∞ topology: Namely, a sequence $\{\eta_n\}$ converges to η_∞ in $A^{p,q}(X)$ if and only if (1) there exists a compact set K such that for any n the support of η_n is contained in K and (2) any order derivation of η_n uniformly converges to the corresponding derivation of η_∞.

Definition 1.2.1. We call a continuous linear functional

$$T : A^{d-p,d-q}(X) \to \mathbb{C}$$

a *current* of type (p,q) on X. Let $D^{p,q}(X)$ be the set of currents of type (p,q) on X.

Example 1.2.2. For $\omega \in A^{p,q}(X)$, set

$$[\omega] : A^{d-p,d-q}(X) \to \mathbb{C}, \quad \eta \mapsto \int_X \omega \wedge \eta.$$

Then, $[\omega] \in D^{p,q}(X)$. Via $[\cdot] : A^{p,q}(X) \hookrightarrow D^{p,q}(X), \omega \mapsto [\omega]$, $A^{p,q}(X)$ is regarded as a subspace of $D^{p,q}(X)$.

Example 1.2.3. Let ω be a differential form of type (p,q) on X with locally integrable coefficients. Then, in the same way as in Example 1.2.2, one obtains the current $[\omega] \in D^{p,q}(X)$.

Example 1.2.4. Let X be a non-singular projective variety over $\mathbb{C}$ and Y a subvariety of X of codimension p. Then, we have the Dirac type current $\delta_Y \in D^{p,p}(X)$ defined by

$$\delta_Y : A^{d-p,d-p}(X) \to \mathbb{C}, \quad \eta \mapsto \int_{Y^{ns}} \eta,$$

where Y^{ns} is the set of non-singular points of Y. Note that, if $\pi : \widetilde{Y} \to Y$ is a resolution of singularities of Y, then the equality $\delta_Y(\eta) = \int_{\widetilde{Y}} \pi^*(\eta)$ holds, and thus $\int_{Y^{ns}} \eta$ converges.

Example 1.2.5. More generally, let $Y = \sum_\alpha n_\alpha Y_\alpha$ $(n_\alpha \in \mathbb{Z})$ be a cycle of codimension p on X. Then, we have the current $\delta_Y \in D^{p,p}(X)$ defined by $\delta_Y = \sum_\alpha n_\alpha \delta_{Y_\alpha}$.

A current $T \in D^{p,p}(X)$ is said to be *real* if $T(\overline{\eta}) = \overline{T(\eta)}$ for any $\eta \in A^{p,p}(X)$. For example, δ_Y as above is a real current.

Let us define differential operators on $\bigoplus_{p,q} D^{p,q}(X)$. For $T \in D^{p,q}(X)$, we define $\partial T \in D^{p+1,q}(X)$ and $\overline{\partial} T \in D^{p,q+1}(X)$ by

$$\partial T(\eta) = (-1)^{p+q+1} T(\partial \eta) \qquad (\eta \in A^{d-(p+1),d-q}(X)),$$
$$\overline{\partial} T(\eta) = (-1)^{p+q+1} T(\overline{\partial} \eta) \qquad (\eta \in A^{d-p,d-(q+1)}(X)).$$

We see from the Stokes theorem that $[\partial\omega] = \partial[\omega]$ and $[\overline{\partial}\omega] = \overline{\partial}[\omega]$ for $\omega \in A^{p,q}(X)$. Moreover, we set

$$d = \partial + \overline{\partial},$$
$$d^c = \frac{1}{4\pi\sqrt{-1}}(\partial - \overline{\partial}).$$

Note that $dd^c = \frac{\sqrt{-1}}{2\pi}\partial\overline{\partial}$. For $\omega \in A^{p,q}(X)$, we similarly have $[d\omega] = d[\omega]$ and $[d^c\omega] = d^c[\omega]$.

The pull-back of differential forms induces the push-forward of currents. Indeed, let $\pi : X \to Y$ be a holomorphic map of compact complex manifolds and

g an element of $D^{p,q}(X)$. Then, the push-forward of g, which is an element of $D^{p+\dim X-\dim Y, q+\dim X-\dim Y}$ and is denoted by $\pi_*(g)$, is defined by $\pi_*(g)(\eta) = g(\pi^*\eta)$.

Definition 1.2.6 (Green Current). Let X be a compact Kähler manifold and $Z \subset X$ a cycle of codimension p. A *Green current* for Z is a current $g \in D^{p-1,p-1}(X)$ such that there exists $\omega \in A^{p,p}(X)$ with

$$dd^c(g) + \delta_Z = [\omega].$$

Let X be a compact Kähler manifold and L a line bundle over X. A C^∞-hermitian metric h on L is a C^∞-field of hermitian inner products in the fibers of L. Namely, for each $x \in X$,

$$h_x : L_x \times L_x \longrightarrow \mathbb{C}$$

is a hermitian inner product, and h_x is C^∞ with respect to x. We call $\overline{L} := (L, h)$ a *C^∞-hermitian line bundle* .

Example 1.2.7. Let X be a smooth projective variety over $\mathbb{C}$, $\overline{L} = (L, h)$ a C^∞-hermitian line bundle over X, and s a nonzero rational section of L. Then, since $-\log h(s,s)$ is locally integrable, $[-\log h(s,s)]$ defines a current in $D^{0,0}(X)$ (cf. Example 1.2.3). The following Poincaré-Lelong formula shows that $[-\log h(s,s)]$ is actually a Green current for $\operatorname{div}(s)$.

Theorem 1.2.8 (Poincaré-Lelong formula). *Let X be a smooth projective variety over $\mathbb{C}$, $\overline{L} = (L,h)$ a C^∞-hermitian line bundle over X, and s a nonzero rational section of L. Let $c_1(\overline{L}) \in A^{1,1}(X)$ be the first Chern form of $\overline{L}$. Then, the following formula holds in $D^{1,1}(X)$:*

(1.2.8.1) $$dd^c[-\log h(s,s)] + \delta_{\operatorname{div}(s)} = [c_1(\overline{L})].$$

Proof: Let d be the dimension of X.

Step 1 The assertion holds if the support of $\operatorname{div}(s)$ is a normal crossing divisor. Indeed, for any $p \in X$, one can take an open neighborhood U of p and local coordinates $z_1, \dots, z_d$ of U such that $\operatorname{Supp}(\operatorname{div}(s))$ is locally defined by $z_1 z_2 \cdots z_k = 0$. By the partition of unity and the linearity, it suffices to show that, for any $\eta \in A^{d-1,d-1}(U)$ with compact support,

$$\int_U \log|z_1|^2 \, dd^c\eta = \int_{z_1=0} \eta.$$

We will show this equality in the appendix (cf. Lemma A.1).

Step 2 We treat a general case. Set $D = \operatorname{div}(s)$. By Hironaka's theorem [17], there exists a proper morphism $\pi : \widetilde{X} \to X$ such that

(i) $\widetilde{X}$ is smooth,
(ii) $E = \pi^*(D)_{red}$ is a normal crossing divisor,

(iii) $\pi|_{\widetilde{X}\setminus \mathrm{Supp}(E)} : \widetilde{X} \setminus \mathrm{Supp}(E) \to X \setminus \mathrm{Supp}(D)$ is isomorphic.

On the other hand, we have

$$\int_X -\log h(s,s)\; dd^c\eta = \int_{\widetilde{X}} -\log \pi^* h(\pi^* s, \pi^* s)\; dd^c(\pi^*\eta)$$
$$\int_X c_1(\overline{L}) \wedge \eta = \int_{\widetilde{X}} c_1(\pi^*\overline{L}) \wedge \pi^*\eta.$$

We write $E = \overline{D} + E'$, where $\overline{D}$ is the strict transform of D. Then $\dim \pi_*(E') < \dim D$. Thus, we have

$$\int_{\mathrm{div}(\pi^* s)} \pi^*\eta = \int_{\overline{D}} \pi^*\eta + \int_{E'} \pi^*\eta = \int_D \eta = \int_{\mathrm{div}(s)} \eta.$$

Since $\mathrm{div}(\pi^* s)_{red}$ is a normal crossing divisor, by Step 1, we obtain the formula in a general case. □

Remark 1.2.9. In relation to the last part of the proof of Theorem 1.2.8, we remark that, for any morphism $\pi : X \to Y$ of compact complex manifolds and cycle Z of X, we have $\pi_*(\delta_Z) = \delta_{\pi_*(Z)}$.

In the rest of this subsection, we consider some basic properties of Green currents.

Lemma 1.2.10 (dd^c-lemma for currents)**.** *Let X be a compact Kähler manifold and η an element of $D^{p,p}(X)$. Assume that η is d-exact. Then, there exists $\gamma \in D^{p-1,p-1}(X)$ such that $\eta = dd^c\gamma$.*

For its proof, we refer to [14, p149], where the dd^c-lemma for C^∞ differential forms is proven. Since operators $\partial, \overline{\partial}^*, G_{\overline{\partial}}$ for C^∞ differential forms in [14, p149] are all extended to those for currents, the same argument goes for the dd^c-lemma for currents.

Proposition 1.2.11. *Let X be a compact Kähler manifold. Then, for any cycle Z of codimension p on X, there exists a Green current for Z.*

Proof: Take $\omega \in A^{p,p}(X)$ which represents Z in the cohomology class. Then, $[\omega] - \delta_Z$ is d-exact. By the dd^c-lemma, there exists a current $g \in D^{p-1,p-1}(X)$ with $[\omega] - \delta_Z = dd^c g$. □

Proposition 1.2.12. *Let X be a compact Kähler manifold and Z a cycle of codimension p on X. Let g_1 and g_2 be Green currents for Z. Then, there exist $\eta \in A^{p,p}(X)$, $u \in D^{p-2,p-1}(X)$ and $v \in D^{p-1,p-2}(X)$ such that*

$$g_1 - g_2 = [\eta] + \partial u + \overline{\partial} v.$$

Proof: For $i = 1, 2$, write $dd^c(g_i) + \delta_Y = [\omega_i]$ for some $\omega_i \in A^{p,p}(X)$. Then, we have $\frac{\sqrt{-1}}{2\pi}\partial\overline{\partial}(g_1 - g_2) = [\omega_1 - \omega_2]$. Then, the assertion follows from the following lemma. □

Lemma 1.2.13. *Let X be a compact Kähler manifold and g an element of $D^{p,q}(X)$. Assume that $\partial\overline{\partial}g$ is C^∞, i.e., $\partial\overline{\partial}g = [\omega]$ for some $\omega \in A^{p+1,q+1}(X)$. Then, there exist $\eta \in A^{p,q}(X)$, $u \in D^{p-1,q}(X)$ and $v \in D^{p,q-1}(X)$ such that*

$$g = [\eta] + \partial u + \overline{\partial} v.$$

Proof: By [14, p385], $d, \partial, \overline{\partial}$-cohomology of currents coincide with d, ∂, $\overline{\partial}$-cohomology of C^∞ differential forms.

Thus, if $\partial\overline{\partial}g = [\omega]$ for some $\omega \in A^{p+1,q+1}(X)$, then there exists a C^∞ differential form α with $\omega = \partial\alpha$. Since $\partial(\overline{\partial}g - [\alpha]) = 0$, there exist a C^∞ differential form β and a current g_1 such that $\overline{\partial}g - [\alpha] = [\beta] + \partial g_1$. Thus, $\partial\overline{\partial}g_1 = \overline{\partial}[\alpha+\beta] = [\overline{\partial}(\alpha+\beta)]$, where g_1 is a current of type $(p-1, q+1)$. By iterating this procedure, we get a current g_n of type $(p-n, q+n)$ and a C^∞ differential form α_n that satisfy $\overline{\partial}g_n = [\alpha_n] + \partial g_{n+1}$ $(n \geq 1)$.

Since $g_{n+1} = 0$ for $n \geq p$, we have $\overline{\partial}g_n = [\alpha_n]$. Since α_n is a C^∞ differential form, there exists a C^∞ differential form η_n with $g_n = [\eta_n] + \overline{\partial}v_n$. Then, since $\overline{\partial}(g_{n-1} + \partial v_n)$ $= [\alpha_{n-1}] + \partial g_n - \partial(g_n - [\eta_n]) = [\alpha_{n-1}] + \partial[\eta_n]$, there exists a C^∞ differential form η_{n-1} with $g_{n-1} = [\eta_{n-1}] + \partial u_{n-1} + \overline{\partial}v_{n-1}$. By iterating this procedure, we get $g = [\eta] + \partial u + \overline{\partial}v$ for some C^∞ differential form η. □

In Example 1.2.7, a C^∞-hermitian line bundle $\overline{L} = (L, h)$ and a nonzero rational section s of L determine a Green current for the divisor $\operatorname{div}(s)$. The next proposition shows that the converse also holds.

Proposition 1.2.14. *Let X be a smooth projective variety over $\mathbb{C}$ and D a divisor on X. Let s be a rational section of $\mathcal{O}_X(D)$ with $\operatorname{div}(s) = D$. Let g be a Green current for D. Then, there exist a C^∞-hermitian metric h over $\mathcal{O}_X(D)$ with*

$$g = [-\log h(s,s)].$$

Proof: Take any C^∞-hermitian metric h' on $\mathcal{O}_X(D)$. By Example 1.2.7, $[-\log h'(s,s)]$ is a Green current for D. Since D is a divisor, by Proposition 1.2.12, there exists a C^∞ function f with

$$g - [-\log h'(s,s)] = [f].$$

Set $h = \exp(-f)h'$. Then, h is a desired C^∞-hermitian metric over $\mathcal{O}_X(D)$. □

1.3. **Arithmetic Variety and Arithmetic Chow Group.** Following Gillet and Soulé, let us define arithmetic varieties, arithmetic cycles, arithmetic Chow groups, and the arithmetic first Chern class of C^∞-hermitian line bundles.

Definition 1.3.1 (arithmetic variety). An *arithmetic variety* is an integral scheme that is quasi-projective and flat over $\mathbb{Z}$.

Let $\mathcal{X}$ be a projective arithmetic variety. Assume that $\mathcal{X}_\mathbb{Q}$ is regular. Then $\mathcal{X}_\mathbb{C}$ is a compact complex manifold. Let $F_\infty : \mathcal{X}(\mathbb{C}) \to \mathcal{X}(\mathbb{C})$ be the complex conjugation. Set

$$\begin{aligned} D^{p,p}(\mathcal{X}) &= \{T \in D^{p,p}(\mathcal{X}(\mathbb{C})) \mid T \text{ is real, and } F_\infty^*(T) = (-1)^p T\} \\ A^{p,p}(\mathcal{X}) &= D^{p,p}(\mathcal{X}) \cap A^{p,p}(\mathcal{X}(\mathbb{C})) \\ \widetilde{A}^{p,p}(\mathcal{X}) &= A^{p,p}(\mathcal{X})/\{\operatorname{Image}(\partial) + \operatorname{Image}(\bar\partial)\}. \end{aligned}$$

Let $\mathcal{Z} \subset \mathcal{X}$ be a cycle of codimension p. A current $g \in D^{p-1,p-1}(\mathcal{X})$ is called a *Green current* for $\mathcal{Z}$ if there exists $\omega \in A^{p,p}(\mathcal{X})$ with

$$dd^c(g) + \delta_{\mathcal{Z}(\mathbb{C})} = [\omega].$$

A pair $(\mathcal{Z}, g)$ is called an *arithmetic cycle* of codimension p if $g \in D^{p-1,p-1}(\mathcal{X})$ is a Green current for $\mathcal{Z}$. Some examples of arithmetic cycles are the following:

Example 1.3.2. Let $\mathcal{Y} \subset \mathcal{X}$ be an integral closed subscheme of codimension $(p-1)$. Let $f \in k(\mathcal{Y})^*$ be a nonzero element of the functional field of $\mathcal{Y}$. Then, $(\operatorname{div}(f), [-\log|f|^2])$ is an arithmetic cycle of codimension p on $\mathcal{X}$. Here, $\operatorname{div}(f)$ is the divisor on $\mathcal{Y}$ associated with f and thus a cycle of codimension p on $\mathcal{X}$. And $[-\log|f|^2] \in D^{p-1,p-1}(\mathcal{X})$ is a current which sends η to $\int_{\mathcal{Y}(\mathbb{C})}(-\log|f|^2)\,\eta$. (Precisely speaking, for the integral, we need to consider a resolution of singularities of $\mathcal{Y}(\mathbb{C})$.)

Example 1.3.3. For $u \in D^{p-2,p-1}(X)$ and $v \in D^{p-1,p-2}(X)$, $(0, \partial u + \bar\partial v)$ is an arithmetic cycle of codimension p on $\mathcal{X}$.

We set

$$\widehat{Z}^p(\mathcal{X}) = \{\text{arithmetic cycle of codimension } p\}.$$

Let $\widehat{\operatorname{Rat}}^p(\mathcal{X})$ be the $\mathbb{Z}$-submodule of $\widehat{Z}^p(\mathcal{X})$ that is generated by two types of arithmetic cycles in Examples 1.3.2 and 1.3.3. Namely,

$$\widehat{\operatorname{Rat}}^p(\mathcal{X}) = \left\{ \left(\sum_i \operatorname{div}(f_i), \sum_i [-\log|f_i|^2] + \partial u + \bar\partial v \right) \;\middle|\; \begin{array}{c} \operatorname{codim}_\mathcal{X}(\mathcal{Y}_i) = p-1,\ f_i \in k(\mathcal{Y}_i)^* \\ u \in D^{p-2,p-1}(X), v \in D^{p-1,p-2}(X) \end{array} \right\}.$$

Then, the *arithmetic Chow group* of codimension p is defined by

$$\widehat{\operatorname{CH}}^p(\mathcal{X}) = \widehat{Z}^p(\mathcal{X}) / \widehat{\operatorname{Rat}}^p(\mathcal{X}).$$

We say that $\overline{\mathcal{L}} = (\mathcal{L}, h)$ is a C^∞-hermitian line bundle over $\mathcal{X}$ if $\mathcal{L}$ is a line bundle over $\mathcal{X}$ and h is a C^∞-metric on $\mathcal{L}_{\mathbb{C}}$ (cf. §1.2) that is invariant under the complex conjugation. Here, "invariant under the complex conjugation" means the following: For $x \in \mathcal{X}(\mathbb{C})$, let $\mathcal{L}_x \xrightarrow{F_\infty} \mathcal{L}_{\overline{x}}$ be the isomorphism induced by the complex conjugation F_∞; Then, h is said to be invariant under the complex conjugation if the equality $h_{\overline{x}}(F_\infty s, F_\infty t) = \overline{h_x(s,t)}$ holds for any $x \in \mathcal{X}(\mathbb{C})$ and $s, t \in \mathcal{L}_x$.

Two C^∞-hermitian line bundles $\overline{\mathcal{L}}_1 = (\mathcal{L}_1, h_1)$ and $\overline{\mathcal{L}}_2 = (\mathcal{L}_2, h_2)$ over $\mathcal{X}$ are said to be isomorphic if there exists an isomorphism $\phi : \mathcal{L}_1 \to \mathcal{L}_2$ of line bundles that induces an isometry $\phi_{\mathbb{C}} : (\mathcal{L}_{1\mathbb{C}}, h_1) \to (\mathcal{L}_{2\mathbb{C}}, h_2)$. Let $\widehat{\mathrm{Pic}}(\mathcal{X})$ denote the isomorphism classes of C^∞-hermitian line bundles over $\mathcal{X}$. Then, by the tensor product, $\widehat{\mathrm{Pic}}(\mathcal{X})$ becomes an abelian group.

Let $\overline{\mathcal{L}} = (\mathcal{L}, h)$ be a C^∞-hermitian line bundle. Let s be a nonzero rational section of $\mathcal{L}$. Then, we have

$$(\mathrm{div}(s), [-\log h(s_{\mathbb{C}}, s_{\mathbb{C}})]) \in \widehat{\mathrm{CH}}^1(\mathcal{X}).$$

As an element of $\widehat{\mathrm{CH}}^1(\mathcal{X})$, $(\mathrm{div}(s), [-\log h(s_{\mathbb{C}}, s_{\mathbb{C}})])$ does not depend on the choice of s. We denote this element by $\widehat{c}_1(\overline{\mathcal{L}})$ and call it the *arithmetic first Chern class* of $\overline{\mathcal{L}}$. As we quoted in §1.1, $\widehat{c}_1$ induces an isomorphism:

Proposition 1.3.4.

$$\widehat{c}_1 : \widehat{\mathrm{Pic}}(\mathcal{X}) \to \widehat{\mathrm{CH}}^1(\mathcal{X})$$

is isomorphic.

Proof: It is easy to check that $\widehat{c}_1$ is a group homomorphism. the injectivity of $\widehat{c}_1$ follows from $\widehat{\mathrm{Rat}}^1(\mathcal{X}) = \{(\mathrm{div}(f), [-\log|f|^2]) \mid f \in k(\mathcal{X})^*\}$, while surjectivity follows from Proposition 1.2.14. □

Remark 1.3.5. More generally, C^∞-hermitian vector bundles $\overline{\mathcal{E}}$ and arithmetic Chern classes $\widehat{c}_i(\overline{\mathcal{E}})$ are defined in §2.3.

The p-th Chow group of $\mathcal{X}$ is given by

$$\mathrm{CH}^p(\mathcal{X}) = \frac{\{\text{cycle of codimension } p\}}{\langle \mathrm{div}(f) \mid f \in k(\mathcal{Y})^*, \mathrm{codim}_{\mathcal{X}} \mathcal{Y} = p-1 \rangle}.$$

The following proposition relates the p-th Chow group and the p-th arithmetic Chow group.

Proposition 1.3.6. *There exists an exact sequence*

$$\widetilde{A}^{p-1,p-1}(\mathcal{X}) \xrightarrow{a} \widehat{\mathrm{CH}}^p(\mathcal{X}) \xrightarrow{z} \mathrm{CH}^p(\mathcal{X}) \longrightarrow 0,$$

where $z(\mathcal{Z}, g) = \mathcal{Z}$ *and* $a(\eta) = (0, \eta)$.

Proof: The surjectivity of z follows from Proposition 1.2.11. Take $(0, g) \in \widehat{\mathrm{CH}}^p(\mathcal{X})$. Then, by Proposition 1.2.12, $g = [\eta] + \partial u + \overline{\partial} v$ for some $\eta \in A^{p-1,p-1}(\mathcal{X})$, $u \in D^{p-2,p-1}(\mathcal{X})$, and $v \in D^{p-1,p-2}(\mathcal{X})$. Thus, $a(\eta) = (0, \eta) = (0, g) \in \widehat{\mathrm{CH}}^p(\mathcal{X})$. □

For $(\mathcal{Z}, g) \in \widehat{\mathrm{CH}}^p(\mathcal{X})$, let ω be an element of $A^{p,p}(\mathcal{X})$ with $dd^c(g) + \delta_{\mathcal{Z}(\mathbb{C})} = [\omega]$. Then, we have the map

$$\omega : \widehat{\mathrm{CH}}^p(\mathcal{X}) \to A^{p,p}(\mathcal{X}), \qquad (\mathcal{Z}, g) \mapsto \omega.$$

The Poincaré-Lelong formula says that, for a C^∞-hermitian line bundle $\overline{\mathcal{L}} = (\mathcal{L}, h)$ over $\mathcal{X}$, $\omega(\widehat{c}_1(\overline{\mathcal{L}})) = c_1((\mathcal{L}_{\mathbb{C}}, h))$.

1.4. **Arithmetic Intersection Theory.** One of the important properties of arithmetic Chow groups is that they carry a natural ring structure (when tensored by $\mathbb{Q}$). Namely, the following theorem holds.

Theorem 1.4.1 ([11]). *Let $\mathcal{X}$ be a regular projective arithmetic variety over $\mathcal{X}$. Then, for any $p, q \geq 0$, there exists a commutative associative bilinear pairing*

$$\widehat{\mathrm{CH}}^p(\mathcal{X}) \times \widehat{\mathrm{CH}}^q(\mathcal{X}) \longrightarrow \widehat{\mathrm{CH}}^{p+q}(\mathcal{X}) \otimes_{\mathbb{Z}} \mathbb{Q}.$$

For its proof, we refer to [11, Theorem 4.2.3], [32, III, Theorem 2]. We remark that [11, Theorem 4.2.3] is a stronger assertion. For example, it suffices that $\mathcal{X}$ is quasi-projective.

In the last of this subsection, we will discuss the intersection of an arithmetic cycle by the arithmetic first Chern class of a C^∞-hermitian line bundle. This type of intersection has many applications and is defined comparatively easily (and does not require the regularity of $\mathcal{X}$).

Roughly speaking, the intersection of $(\mathcal{Y}, g_{\mathcal{Y}}) \in \widehat{Z}^p(\mathcal{X})$ and $(\mathcal{Z}, g_{\mathcal{Z}}) \in \widehat{Z}^q(\mathcal{X})$ is given by

$$(\mathcal{Y}, g_{\mathcal{Y}}) \cdot (\mathcal{Z}, g_{\mathcal{Z}}) = (\mathcal{Y} \cap \mathcal{Z}, g_{\mathcal{Y}} * g_{\mathcal{Z}}), \tag{1.4.2.0}$$

where $\mathcal{Y} \cap \mathcal{Z}$ is the "intersection" of the cycles $\mathcal{Y}$ and $\mathcal{Z}$, and $g_{\mathcal{Y}} * g_{\mathcal{Z}}$ is the "Green current" of $\mathcal{Y} \cap \mathcal{Z}$ given by

$$g_{\mathcal{Y}} * g_{\mathcal{Z}} = \delta_{\mathcal{Y}(\mathbb{C})}\, g_{\mathcal{Z}} + g_{\mathcal{Y}}\, \omega(\mathcal{Z}, g_{\mathcal{Z}}).$$

However, there are two difficulties to rigorously define the intersection:

(A) Since Chow's moving lemma is not known, it is difficult to define $\mathcal{Y} \cap \mathcal{Z}$ in general.

(B) Since $g_{\mathcal{Z}}$ is a current, it is difficult to define $\delta_{\mathcal{Y}(\mathbb{C})}\, g_{\mathcal{Z}}$ in general.

Regarding (A), Gillet and Soulé [9] used the algebraic K-theory and defined $\mathcal{Y} \cap \mathcal{Z}$ in $\widehat{\mathrm{CH}}^{p+q}(\mathcal{X}) \otimes_{\mathbb{Z}} \mathbb{Q}$. (Here $\otimes \mathbb{Q}$ appears.) Regarding (B), they showed in [11] that there exists a Green current $g_{\mathcal{Z}}$ for $\mathcal{Z}$ such that $g_{\mathcal{Z}}$ is induced by an L^1-form which is C^∞ on $\mathcal{X}(\mathbb{C}) \setminus \mathrm{Supp}(\mathcal{Z}(\mathbb{C}))$ and is "of logarithmic type" along $\mathcal{Z}(\mathbb{C})$. Then the current $\delta_{\mathcal{Y}(\mathbb{C})} g_{\mathcal{Z}}$ is defined by $\delta_{\mathcal{Y}(\mathbb{C})} g_{\mathcal{Z}}(\eta) = \int_{\mathcal{Y}(\mathbb{C})} g_{\mathcal{Z}} \wedge \eta$. (Precisely speaking, one uses a resolution of singularities of $\mathcal{Z}(\mathbb{C})$ to define $\delta_{\mathcal{Y}(\mathbb{C})} g_{\mathcal{Z}}$.)

The following theorem states the pull-back and the push-forward of arithmetic Chow groups.

Theorem 1.4.3 ([11]). (i) *Let $\pi : \mathcal{X} \to \mathcal{Y}$ be a morphism of regular arithmetic varieties. Then, one has a pull-back*

$$\pi^* : \widehat{\mathrm{CH}}^p(\mathcal{Y}) \to \widehat{\mathrm{CH}}^p(\mathcal{X}) \otimes_{\mathbb{Z}} \mathbb{Q},$$

which is compatible with a, z and ω. Moreover, $\pi^*(xy) = \pi^*(x)\pi^*(y)$.

(ii) *Suppose further that π is proper and $\pi_{\mathbb{C}} : \mathcal{X}(\mathbb{C}) \to \mathcal{Y}(\mathbb{C})$ is smooth. Then, one has a push-forward*

$$\pi_* : \widehat{\mathrm{CH}}^p(\mathcal{X}) \longrightarrow \widehat{\mathrm{CH}}^{p-(\dim \mathcal{X} - \dim \mathcal{Y})}(\mathcal{X})$$

which is compatible with a, z and ω.

Proof: For the proof of (i), we refer to [11, 4.4.3]. For (ii), let $(\mathcal{Z}, g)$ be an element of $\widehat{Z}^p(\mathcal{X})$. By linearity, it suffices to define $\pi_*(\mathcal{Z}, g)$ when $\mathcal{Z}$ is reduced and irreducible. Let $\pi(\mathcal{Z})$ denote the image of $\mathcal{Z}$ by π as a set. Put

$$\pi_*(\mathcal{Z}) = \begin{cases} 0 & (\text{if } \dim \pi(\mathcal{Z}) < \dim \mathcal{Z}) \\ \deg(\mathcal{Z} \to \pi(\mathcal{Z}))\ [\pi(\mathcal{Z})] & (\text{if } \dim \pi(\mathcal{Z}) = \dim \mathcal{Z}). \end{cases}$$

Let $\pi_{\mathbb{C}*} g$ be the push-forward of the current g (cf. §1.2). We define the push-forward by $\pi_*(\mathcal{Z}, g) := (\pi_*(\mathcal{Z}), \pi_{\mathbb{C}*} g)$. We need to check that $\pi_{\mathbb{C}*} g$ is a Green current for $\pi_*(\mathcal{Z})$. Let ω be a C^∞ differential form with $dd^c(g) + \delta_{\mathcal{Z}(\mathbb{C})} = [\omega]$. Then, we have $dd^c(\pi_{\mathbb{C}*} g) + \delta_{\pi(\mathcal{Z})(\mathbb{C})} = [\pi_{\mathbb{C}*}(\omega)]$ (cf. Remark 1.2.9). Here, $\pi_{\mathbb{C}*}(\omega)$ is the integration along the fiber of π. Since $\pi_{\mathbb{C}} : \mathcal{X}(\mathbb{C}) \to \mathcal{Y}(\mathbb{C})$ is smooth, $\pi_{\mathbb{C}*}(\omega)$ is a C^∞ differential form. Thus, $\pi_{\mathbb{C}*} g$ is a Green current of $\pi_*(\mathcal{Z})$. □

In the remainder of this subsection, let us consider the intersection of an arithmetic cycle by the arithmetic first Chern class of a C^∞-hermitian line bundle. Here, let $\mathcal{X}$ be a projective arithmetic variety (not necessarily regular) such that $\mathcal{X}_{\mathbb{Q}}$ is regular.

Let $\overline{\mathcal{L}} = (\mathcal{L}, h)$ be a C^∞-hermitian line bundle and $(\mathcal{Y}, g_{\mathcal{Y}})$ an arithmetic cycle of codimension p on $\mathcal{X}$. Recall that $\widehat{c}_1(\overline{\mathcal{L}}) \in \widehat{\mathrm{CH}}^1(\mathcal{X})$ is represented by $(\mathrm{div}(s), [-\log h(s_{\mathbb{C}}, s_{\mathbb{C}})])$ for any nonzero section s of $\mathcal{L}$. In this case, regarding (A) as above, we have only to take a suitable s. Moreover, regarding (B), $-\log h(s_{\mathbb{C}}, s_{\mathbb{C}})$ is by definition a current induced from L^1-form that is of logarithmic type along $\mathrm{div}(s)$. Thus, we can define comparatively easily the intersection $\widehat{c}_1(\overline{\mathcal{L}}) \cdot (\mathrm{div}(s), [-\log h(s_{\mathbb{C}}, s_{\mathbb{C}})])$. Namely, when $\mathcal{Y}$ is reduced and irreducible, we take a nonzero rational section s of $\mathcal{L}$ with $\mathcal{Y} \not\subset \mathrm{Supp}(\mathrm{div}(s))$ and set

$$(\mathcal{Y}, g_{\mathcal{Y}}) \cdot \widehat{c}_1(\overline{\mathcal{L}}) = \left(\mathrm{div}(s|_{\mathcal{Y}}), [-\log(h|_{\mathcal{Y}})(s|_{\mathcal{Y}} \otimes \mathbb{C}, s|_{\mathcal{Y}} \otimes \mathbb{C})] + c_1(\overline{\mathcal{L}}) \wedge g_{\mathcal{Y}}\right),$$

where $[-\log(h|_{\mathcal{Y}})(s|_{\mathcal{Y}} \otimes \mathbb{C}, s|_{\mathcal{Y}} \otimes \mathbb{C})]$ is the current of type (p,p) on $\mathcal{X}(\mathbb{C})$ given by

$$[-\log(h_{\mathcal{Y}})(s|_{\mathcal{Y}} \otimes \mathbb{C}, s|_{\mathcal{Y}} \otimes \mathbb{C})](\eta) = \int_{\mathcal{Y}(\mathbb{C})} -\log(h_{\mathcal{Y}})(s|_{\mathcal{Y}} \otimes \mathbb{C}, s|_{\mathcal{Y}} \otimes \mathbb{C})\eta.$$

For a general $\mathcal{Y}$, we extend the map by linearity. Thus, we get

$$\widehat{c}_1(\overline{\mathcal{L}}) : \widehat{\mathrm{CH}}^p(\mathcal{X}) \longrightarrow \widehat{\mathrm{CH}}^{p+1}(\mathcal{X}) \qquad \alpha \mapsto \alpha \cdot \widehat{c}_1(\overline{\mathcal{L}}). \tag{1.4.4.0}$$

(This morphism is extended in §1.5.)

1.5. **Arithmetic Chow Group and Push-forward of Arithmetic Cycles.** In the previous subsection, the push-forward of arithmetic cycles is defined under the assumption that $\pi_{\mathbb{C}}$ is smooth. In this subsection, for later use, we define arithmetic D-cycles and their push-forward without the assumption of smoothness, and see a projection formula for them. We refer to [18] for details.

Let $\mathcal{X}$ be a projective arithmetic variety such that $\mathcal{X}_{\mathbb{Q}}$ is regular. We say that a pair $(\mathcal{Z}, g)$ is an *arithmetic D-cycle* of codimension p if $\mathcal{Z}$ is a cycle of codimension p on $\mathcal{X}$ and g is an element of $D^{p-1,p-1}(\mathcal{X})$. Let $\widehat{\mathrm{CH}}^p_D(\mathcal{X})$ denote the set of arithmetic D-cycles of codimension p. Set

$$\widehat{\mathrm{CH}}^p_D(\mathcal{X}) = \widehat{Z}^p_D(\mathcal{X}) / \widehat{\mathrm{Rat}}^p(\mathcal{X})$$

and call it the *arithmetic D-Chow group* of codimension p.

Let $\overline{\mathcal{L}} = (\mathcal{L}, h)$ be a C^∞-hermitian line bundle over $\mathcal{X}$. Then, similar to (1.4.4), we can define a map

$$\widehat{c}_1(\overline{\mathcal{L}}) : \widehat{\mathrm{CH}}^p_D(\mathcal{X}) \longrightarrow \widehat{\mathrm{CH}}^{p+1}_D(\mathcal{X}) \qquad \alpha \mapsto \alpha \cdot \widehat{c}_1(\overline{\mathcal{L}})$$

as follows. Let $(\mathcal{Y}, g_{\mathcal{Y}})$ be an element of $\widehat{\mathrm{CH}}^p_D(\mathcal{X})$. First, suppose $\mathcal{Y}$ is reduced and irreducible. For a nonzero rational section s of $L|_{\mathcal{Y}}$, put

$$(\mathcal{Y}, g_{\mathcal{Y}}) \cdot \widehat{c}_1(\overline{\mathcal{L}}) = \left(\mathrm{div}(s), [-\log(h|_{\mathcal{Y}})(s_{\mathbb{C}}, s_{\mathbb{C}})] + c_1(\overline{\mathcal{L}}) \wedge g_{\mathcal{Y}}\right),$$

which, as an element of $\widehat{\mathrm{CH}}^{p+1}_D(\mathcal{X})$, is independent of the choice of s. For non-reduced or reducible $\mathcal{Y}$, we extend the map by linearity.

Let $\pi : \mathcal{X} \to \mathcal{Y}$ be a morphism of projective arithmetic varieties such that $\mathcal{X}_{\mathbb{Q}}$ and $\mathcal{Y}_{\mathbb{Q}}$ are regular. For $(\mathcal{Z}, g) \in \widehat{Z}^p_D(\mathcal{X})$, we define the *push-forward* by $(\pi_* Z, \pi_{\mathbb{C}*} g)$. This gives rise to a map

$$\pi_* : \widehat{\mathrm{CH}}^p_D(\mathcal{X}) \longrightarrow \widehat{\mathrm{CH}}^{p-(\dim \mathcal{X} - \dim \mathcal{Y})}_D.$$

Then the following projection formula holds.

Proposition 1.5.1. *Let $\pi : \mathcal{X} \to \mathcal{Y}$ be a morphism of projective arithmetic varieties such that $\mathcal{X}_{\mathbb{Q}}$ and $\mathcal{Y}_{\mathbb{Q}}$ are regular. Then, for any C^∞-hermitian line bundle $\overline{\mathcal{L}} = (\mathcal{L}, h)$ and $z \in \widehat{\mathrm{CH}}^p_D(\mathcal{X})$, we have*

$$\pi_*(\widehat{c}_1(\pi^*(\mathcal{L}), \pi^*h) \cdot z) = \widehat{c}_1(\mathcal{L}, h) \cdot \pi_*(z).$$

Proof: Let $(\mathcal{Z}, g)$ represent z. To prove the formula, we may assume that $\mathcal{Z}$ is reduced and irreducible. Set $\mathcal{T} = \pi(\mathcal{Z})$ and $\phi = \pi|_{\mathcal{Z}} : \mathcal{Z} \to \mathcal{T}$. Let s be a nonzero section of $\mathcal{L}|_{\mathcal{T}}$. Then, $\phi^*(s)$ is an element of $\pi^*(\mathcal{L})|_{\mathcal{Z}} = \phi^*(\mathcal{L}|_{\mathcal{T}})$. Thus, $\widehat{c}_1(\pi^*(\mathcal{L}), \pi^*h) \cdot z$ is represented by

$$(\mathrm{div}(\phi^*(s)), [-\log \phi^*(h|_{\mathcal{T}})(\phi^*(s), \phi^*(s))] + c_1(\pi^*(\mathcal{L}), \pi^*h) \wedge g),$$

where $[-\log \phi^*(h|_{\mathcal{T}})(\phi^*(s), \phi^*(s))]$ is the current given by

$$[-\log \phi^*(h|_{\mathcal{T}})(\phi^*(s), \phi^*(s))](\eta) = \int_{\mathcal{Z}(\mathbb{C})} (-\log \phi^*(h|_{\mathcal{T}})(\phi^*(s), \phi^*(s)))\,\eta.$$

Set

$$\deg(\phi) = \begin{cases} 0 & (\text{if } \dim \mathcal{T} < \dim \mathcal{Z}) \\ \deg(\mathcal{Z} \to \mathcal{T}) & (\text{if } \dim \mathcal{T} = \dim \mathcal{Z}). \end{cases}$$

Then, for any C^∞ differential form η on $\mathcal{Y}(\mathbb{C})$, we get

$$\begin{aligned} \int_{\mathcal{Z}(\mathbb{C})} (-\log \phi^*(h|_{\mathcal{T}})(\phi^*(s), \phi^*(s)))\,\pi^*(\eta) &= \int_{\mathcal{Z}(\mathbb{C})} \phi^*((-\log(h|_{\mathcal{T}})(s,s))\,\eta) \\ &= \deg(\phi) \int_{\mathcal{T}(\mathbb{C})} (-\log(h|_{\mathcal{T}})(s,s))\,\eta. \end{aligned}$$

Thus, we have

$$\pi_*[-\log \phi^*(h|_{\mathcal{T}})(\phi^*(s), \phi^*(s))] = \deg(\phi)[-\log(h|_{\mathcal{T}})(s,s)].$$

Then, the formula follows from

$$\begin{aligned} \pi_*(\widehat{c}_1(\pi^*(\mathcal{L}), \pi^*h) \cdot z) &= (\deg(\phi)\,\mathrm{div}(s), \deg(\phi)[-\log(h|_{\mathcal{T}})(s,s)] + c_1(\mathcal{L}, h) \wedge \pi_*(g)) \\ &= \widehat{c}_1(\mathcal{L}, h) \cdot (\deg(\phi)T, \pi_*(g)) = \widehat{c}_1(\mathcal{L}, h) \cdot \pi_*(z). \end{aligned}$$

□

1.6. **Height of Arithmetic Variety.** Let $\mathcal{X}$ be a projective arithmetic variety of dimension $(d+1)$.

Assume first that $\mathcal{X}_{\mathbb{Q}}$ is regular. The degree map $\widehat{\deg}$ is defined by

$$\widehat{\deg} : \widehat{\mathrm{CH}}^{d+1}(\mathcal{X}) \longrightarrow \mathbb{R}, \qquad \left(\sum_{i=1}^k n_i P_i, g\right) \mapsto \sum_{i=1}^k n_i \log \#k(P_i) + \frac{1}{2} \int_{\mathcal{X}(\mathbb{C})} g,$$

where P_i is a closed point of $\mathcal{X}$ and g is an element of $D^{d,d}(\mathcal{X})$. Note that, if $X = \operatorname{Spec}(\mathbb{Z})$, then $\widehat{\deg}$ coincides with the one in §1.1. Let $\overline{\mathcal{L}} = (\mathcal{L}, h)$ be a C^∞-hermitian line bundle. Then, we obtain the real number $\widehat{\deg}(\widehat{c}_1(\overline{\mathcal{L}})^{d+1})$.

We will see that $\widehat{\deg}(\widehat{c}_1(\overline{\mathcal{L}})^{d+1})$ is also defined when $\mathcal{X}_{\mathbb{Q}}$ is not necessarily regular. So, let $\mathcal{X}$ be a projective arithmetic variety of dimension $(d+1)$ such that $\mathcal{X}_{\mathbb{Q}}$ is not necessarily regular.

A pair $\overline{\mathcal{L}} = (\mathcal{L}, h)$ is called a continuous hermitian line bundle over $\mathcal{X}$ if $\mathcal{L}$ is a line bundle over $\mathcal{X}$, $h_x : L_x \times L_x \longrightarrow \mathbb{C}$ is a hermitian inner product on L_x for any $x \in \mathcal{X}(\mathbb{C})$, and $h = \{h_x\}_{x \in \mathcal{X}(\mathbb{C})}$ is continuous with respect x and invariant under the complex conjugation. (Here, we consider continuity in terms of the underling topology of the analytic space $\mathcal{X}(\mathbb{C})$.)

A continuous hermitian line bundle $\overline{\mathcal{L}} = (\mathcal{L}, h)$ over $\mathcal{X}$ is said to be C^∞ if, for any complex manifold M and analytic map $\mu : M \to \mathcal{X}(\mathbb{C})$, $(\mu^*(\mathcal{L}_{\mathbb{C}}), \mu^*(h))$ is C^∞. When $\mathcal{X}_{\mathbb{Q}}$ is regular, this definition coincides with the one in §1.3. In the rest of this subsection, we assume that $\overline{\mathcal{L}}$ is C^∞.

Let $\pi : \widetilde{\mathcal{X}} \to \mathcal{X}$ be a birational morphism of projective arithmetic varieties such that $\widetilde{\mathcal{X}}_{\mathbb{Q}}$ is regular. Such $\pi : \widetilde{\mathcal{X}} \to \mathcal{X}$ is called a *generic resolution of singularities* of $\mathcal{X}$ (cf. [37, §1]). By Hironaka's theorem [17], a generic resolution of singularities always exists. Since $\pi^*(\overline{\mathcal{L}}) = (\pi^*(\mathcal{L}), \pi^* h)$ is a C^∞-hermitian line bundle, the value $\widehat{\deg}(\widehat{c}_1(\pi^*(\overline{\mathcal{L}})^{d+1}))$ is defined. Put

$$\widehat{\deg}(\widehat{c}_1(\overline{\mathcal{L}})^{d+1}) = \widehat{\deg}(\widehat{c}_1(\pi^*(\overline{\mathcal{L}}))^{d+1}).$$

let us check that the value $\widehat{\deg}(\widehat{c}_1(\overline{\mathcal{L}})^{d+1})$ does not depend on $\pi : \widetilde{\mathcal{X}} \to \mathcal{X}$. Suppose $\pi_1 : \mathcal{X}_1 \to \mathcal{X}$ and $\pi_2 : \mathcal{X}_2 \to \mathcal{X}$ are two generic resolutions of singularities of $\mathcal{X}$. We take a generic resolution of singularities

$$g : \mathcal{X}_3 \longrightarrow (\text{main part of } \mathcal{X}_1 \times_{\mathcal{X}} \mathcal{X}_2).$$

For $i = 1, 2$, let p_i : (main part of $\mathcal{X}_1 \times_{\mathcal{X}} \mathcal{X}_2) \to \mathcal{X}_i$ be the projection to the i-th factor. Then, by the projection formula (Proposition 1.5.1), we find

$$\widehat{\deg}(\widehat{c}_1((g \circ p_i \circ \pi_i)^*(\overline{\mathcal{L}})^{d+1})) = \widehat{\deg}(\widehat{c}_1(\pi_i^*(\overline{\mathcal{L}})^{d+1})).$$

Thus, we have $\widehat{\deg}(\widehat{c}_1(\pi_1^*(\overline{\mathcal{L}})^{d+1})) = \widehat{\deg}(\widehat{c}_1(\pi_2^*(\overline{\mathcal{L}})^{d+1}))$.

Let us define the height of arithmetic varieties. Let $\mathcal{X}$ be a $(d+1)$-dimensional projective arithmetic variety and $\overline{\mathcal{L}} = (\mathcal{L}, h)$ a C^∞-hermitian line bundle over $\mathcal{X}$.

Let $\pi : \mathcal{X} \to \mathbb{Z}$ be the structural morphism. By the Stein factorization theorem, one can factor π into $g \circ f$ where $f : \mathcal{X} \to \Gamma$ is a projective morphism with connected fibers and $g : \Gamma \to \operatorname{Spec}(\mathbb{Z})$ is a finite map. We assume that there exists a number field K such that $\Gamma = \operatorname{Spec}(O_K)$ with O_K being the ring of integers of K. Note that this assumption is satisfied if X is normal.

Set

$$h_{\overline{\mathcal{L}}}(\mathcal{X}) = \frac{\widehat{\deg}(\widehat{c}_1(\overline{\mathcal{L}})^{d+1})}{(d+1)\deg_{\mathcal{L}_{\mathbb{Q}}}(\mathcal{X}_{\mathbb{Q}})},$$

where $\deg_{\mathcal{L}_{\mathbb{Q}}}(\mathcal{X}_{\mathbb{Q}}) = (\mathcal{L}_{\mathbb{Q}}^d)$. We call $h_{\overline{\mathcal{L}}}(\mathcal{X})$ the *height of* $\mathcal{X}$ *with respect to* $\overline{\mathcal{L}}$.

More generally, let $\mathcal{Y}$ be an integral closed subscheme of $\mathcal{X}$ such that $\dim \mathcal{Y}_{\mathbb{Q}} \geq 1$. Then, we define the *height of* $\mathcal{Y}$ *with respect to* $(\mathcal{X}, \overline{\mathcal{L}})$ by

$$h_{(\mathcal{X},\overline{\mathcal{L}})}(\mathcal{Y}) = \frac{\widehat{\deg}(\widehat{c}_1(\overline{\mathcal{L}}|_{\mathcal{Y}})^{\dim \mathcal{Y}_{\mathbb{Q}}+1})}{(\dim \mathcal{Y}_{\mathbb{Q}}+1)\deg_{\mathcal{L}_{\mathbb{Q}}}(\mathcal{Y}_{\mathbb{Q}})},$$

where $\deg_{\mathcal{L}_{\mathbb{Q}}}(\mathcal{Y}_{\mathbb{Q}}) = c_1(\mathcal{L}_{\mathbb{Q}})^{\dim \mathcal{Y}_{\mathbb{Q}}} \cap [\mathcal{Y}_{\mathbb{Q}}]$.

Also, let Y be a subvariety of $\mathcal{X}_{\overline{\mathbb{Q}}}$ with $\dim Y \geq 1$. Let $\mathcal{Y}$ be the Zariski closure of Y in $\mathcal{X}$. Then, we define the *height of* Y *with respect to* $(\mathcal{X}, \overline{\mathcal{L}})$ by

$$h_{(\mathcal{X},\overline{\mathcal{L}})}(Y) = h_{(\mathcal{X},\overline{\mathcal{L}})}(\mathcal{Y}).$$

Remark 1.6.1. The value $\dfrac{1}{[K:\mathbb{Q}]} h_{\overline{\mathcal{L}}}(\mathcal{X})$ is also often called the height of $\mathcal{X}$.

When $\dim Y = 0$, i.e., Y is a maximal point y, we define the height of y with respect to $(\mathcal{X}, \overline{\mathcal{L}})$ by

$$h_{(\mathcal{X},\overline{\mathcal{L}})}(y) = \frac{\widehat{\deg}\left(\overline{\mathcal{L}}|_{\Delta_y}\right)}{[K(y):K]},$$

where Δ_y is the Zariski closure of y in $\mathcal{X}$. Note that, if $\mu : \widetilde{\Delta_y} \to \Delta_y \subset \mathcal{X}$ is the normalization of Δ_y, then we have $\widehat{\deg}\left(\overline{\mathcal{L}}|_{\Delta_y}\right) = \widehat{\deg}\left(\mu^*(\overline{\mathcal{L}})\right)$.

The height of Y with respect to $(\mathcal{X}, \overline{\mathcal{L}})$ has played an important role in Ullmo and Zhang's proof of the Bogomolov conjecture. Indeed, roughly speaking, Theorem 5.5.1 says that

$$\begin{aligned}&\inf\{\text{the height of sufficiently general point of } \mathcal{X}_{\mathbb{Q}}\}\\ &\qquad\qquad \geq \text{the height of } \mathcal{X} \geq \inf\{\text{the height of point of } \mathcal{X}_{\mathbb{Q}}\}.\end{aligned}$$

Further, the equidistribution theorem (Theorem 6.1.1) is applied when these three terms coincide. To obtain a "nice" height of Y, in §4.1, we take various models $(\mathcal{X}, \overline{\mathcal{L}})$ and consider the intersection as a limit of models.

2. Arithmetic Riemann-Roch Theorem

2.1. **Characteristic Forms.** In this subsection, we review characteristic forms without proof. Let X be a complex manifold of dimension d. Let E be a holomorphic vector bundle of rank r over X. We denote by $A^0(X)$ (or by $C^\infty(X)$) the vector space of C^∞ functions on X. Moreover, we denote by $A^{p,q}(E)$ (or by $A^{p,q}(X, E)$) the space of C^∞ forms of type (p, q) on X with values in E. Set $A^n(E) = \bigoplus_{p+q=n} A^{p,q}(E)$.

A connection ∇ in E is a $\mathbb{C}$-linear map

$$\nabla : A^0(E) \longrightarrow A^1(E)$$

such that, for any $s \in A^0(E)$ and $f \in A^0(X)$,

$$\nabla(fs) = df \otimes s + f\nabla s.$$

According to the decomposition $A^1(E) = A^{1,0}(E) \oplus A^{0,1}(E)$, we have the decomposition $\nabla = \nabla^{1,0} \oplus \nabla^{0,1}$.

We extend a connection ∇ to the covariant exterior differential

$$\nabla : A^k(E) \longrightarrow A^{k+1}(E)$$

by the Leibnitz rule

$$\nabla(\theta \otimes s) = d\theta \otimes s + (-1)^k \theta \wedge \nabla s \qquad (\theta \in A^k(X), s \in A^0(E)).$$

The curvature of ∇ is defined to be

$$\nabla^2 : A^0(E) \to A^2(E).$$

Then, ∇^2 is $A^0(X)$-linear. Hence, ∇^2 is an element of $A^2(E \otimes E^\vee)$.

A C^∞-hermitian metric h on E is a C^∞-field of hermitian inner products in the fibers of E (cf. §1.2). Namely, for each $x \in X$,

$$h_x : E_x \times E_x \longrightarrow \mathbb{C}$$

is a hermitian inner product, and h_x is C^∞ with respect to x. We call $\overline{E} = (E, h)$ a *C^∞-hermitian vector bundle* .

A hermitian metric h determines a natural connection in E as follows.

Lemma 2.1.1. *Let $\overline{E} = (E, h)$ be a C^∞-hermitian vector bundle. Then, there exists a unique connection $\nabla_{\overline{E}}$ with the following properties:*

(i) $\nabla_{\overline{E}}^{0,1} = \overline{\partial}_E$, *where $\overline{\partial}_E$ is the Cauchy-Riemann operator;*

(ii) $\nabla_{\overline{E}}$ *is unitary, i.e.,*

$$d\, h(s, t) = h(\nabla_{\overline{E}} s, t) + h(s, \nabla_{\overline{E}} t).$$

Since $\nabla_{\overline{E}}^{0,2} = \overline{\partial}_E^2 = 0$ and by the unitarity $\nabla_{\overline{E}}^{2,0} = 0$, the curvature $\nabla_{\overline{E}}^2$ is a $(1,1)$-form with values in $E \otimes E^\vee$.

Let $\phi \in \mathbb{Q}[[T_1, \ldots, T_r]]$ be a symmetric formal power series in r-variables and $\phi^{(k)}$ the k-th part of ϕ. Let A be a commutative $\mathbb{Q}$-algebra and $M_n(A)$ the set of (n, n)-matrices with coefficients in A.

Then, there exists a unique map

$$\Phi^{(k)} : M_n(A) \longrightarrow A$$

such that (1) $\Phi^{(k)}(\operatorname{diag}(\lambda_1,\ldots,\lambda_n)) = \phi^{(k)}(\lambda_1,\ldots,\lambda_n)$, and (2) for any $P \in \mathrm{GL}_r(A)$ and $M \in M_r(A)$, $\Phi^{(k)}(PMP^{-1}) = \Phi^{(k)}(M)$.

For a C^∞-hermitian vector bundle $\overline{E} = (E,h)$ over X, we define the differential form $\phi(\overline{E})$ by

$$\phi(\overline{E}) = \sum_{p\geq 0} \Phi^{(p)}\left(-\frac{1}{2\pi\sqrt{-1}}\nabla^2_{\overline{E}}\right) \in \bigoplus_{p\geq 0} A^{p,p}(X).$$

The form $\phi(\overline{E})$ (denoted also by $\phi(E,h)$) is called the *characteristic form* of $\overline{E}$ associated to ϕ.

The form $\phi(E,h)$ is d-closed. If h' is another hermitian metric of E, then $\phi(E,h)-\phi(E,h')$ is d-exact. Thus, as a cohomology class, $\phi(\overline{E})$ does not depend on the choice of h, although, as a C^∞ differential form, $\phi(E,h)$ depends on h.

Some examples of ϕ are

- k-th Chern form: $c_k = \sum_{1\leq i_1<\cdots<i_k\leq r} T_{i_1}\cdots T_{i_k}$,
- total Chern form: $c = \sum_{k\geq 0} c_k$,
- Chern character form: $\mathrm{ch}(T_1,\ldots,T_r) = \sum_{i=1}^r \exp(T_i)$,
- Todd form: $\mathrm{td}(T_1,\ldots,T_r) = \displaystyle\prod_{i=1}^r \frac{T_i}{(1-\exp(-T_i))}$.

2.2. **The Bott-Chern Secondary Characteristic Class.** Let X be a complex manifold of dimension d. Let

$$\mathbf{E} : 0 \to S \to E \to Q \to 0$$

be an exact sequence of holomorphic vector bundles over X. Let h_S, h_E, h_Q be C^∞-hermitian metrics on S, E, Q, respectively. We set $\overline{\mathbf{E}} = (\mathbf{E}; h_S, h_E, h_Q)$ and write

$$\overline{\mathbf{E}} : 0 \to \overline{S} \to \overline{E} \to \overline{Q} \to 0.$$

Note that h_S need not be the induced metric of h_E, and h_Q need not be the quotient metric of h_E, either.

Let $\phi \in \mathbb{Q}[[T_1,\ldots,T_r]]$ be a symmetric formal power series. Since $\phi(\overline{E})$ and $\phi(\overline{S}\oplus\overline{Q})$ define the same element in the cohomology, $\phi(\overline{S}\oplus\overline{Q}) - \phi(\overline{E})$ is d-exact. Then, by the dd^c-lemma, there exist a form $\eta \in \bigoplus_{p\geq 0} A^{p,p}(X)$ with

$$\phi(\overline{S}\oplus\overline{Q}) - \phi(\overline{E}) = dd^c(\eta).$$

The theory of Bott-Chern secondary characteristic forms says that, modulo Image $\partial+$ Image $\overline{\partial}$, one can choose a good η.

We set

$$\widetilde{A}^{p,p}(X) = A^{p,p}(X)/(\operatorname{Image}\partial + \operatorname{Image}\overline{\partial}).$$

Theorem 2.2.1 (the Bott-Chern secondary characteristic form). *Let $\phi \in \mathbb{Q}[[T_1, \dots, T_r]]$ be a symmetric formal power series in r-variables. Then, for each complex manifold X and short exact sequence*

$$\overline{\mathbf{E}} : 0 \to \overline{S} \to \overline{E} \to \overline{Q} \to 0$$

of C^∞-hermitian vector bundles over X with $\operatorname{rk} E = r$, one can uniquely associate a form $\widetilde{\phi}(\overline{\mathbf{E}}) \in \oplus_{p \geq 0} \widetilde{A}^{p,p}(X)$ with the following three properties:

(i) $\phi(\overline{S} \oplus \overline{Q}) - \phi(\overline{E}) = dd^c(\widetilde{\phi}(\overline{\mathbf{E}}))$;

(ii) *For any holomorphic map $\pi : N \to M$ of complex manifolds, $\widetilde{\phi}(\pi^*(\overline{\mathbf{E}})) = \pi^* \widetilde{\phi}(\overline{\mathbf{E}})$;*

(iii) *If (E, h_E) splits, i.e., $(E, h_E) = (S \oplus Q, h_S \oplus h_Q)$, then $\widetilde{\phi}(\overline{\mathbf{E}}) = 0$.*

The form $\widetilde{\phi}(\overline{\mathbf{E}}) \in \widetilde{A}^{p,p}(X)$ is called the Bott-Chern secondary characteristic form of $\overline{\mathbf{E}}$.

For its proof, we refer to [12], [32, IV Theorem 2].

2.3. **Arithmetic characteristic class.** Let $\mathcal{X}$ be a regular arithmetic variety. Let $\mathcal{E}$ be a vector bundle over $\mathcal{X}$ and h a C^∞-hermitian metric of $\mathcal{E}_\mathbb{C}$ that is invariant under the complex conjugation. The pair $\overline{\mathcal{E}} := (\mathcal{E}, h)$ is called a *C^∞-hermitian vector bundle* over $\mathcal{X}$ (cf. §1.3).

Let $a : \widetilde{A}^{p-1,p-1}(\mathcal{X}) \longrightarrow \widehat{\mathrm{CH}}^p(\mathcal{X})$ and $\omega : \widehat{\mathrm{CH}}^p(\mathcal{X}) \to A^{p,p}(\mathcal{X})$ be the morphisms defined in §1.3.

Theorem 2.3.1 (Arithmetic Characteristic Class). *Let $\phi \in \mathbb{Q}[[T_1, \dots, T_r]]$ be a symmetric formal power series in r variables. Then, for each regular arithmetic variety $\mathcal{X}$ and C^∞-hermitian vector bundle $\overline{\mathcal{E}} = (\mathcal{E}, h)$ of rank r over $\mathcal{X}$, one can uniquely associate*

$$\widehat{\phi}(\overline{\mathcal{E}}) \in \bigoplus_{p \geq 0} \widehat{\mathrm{CH}}^p(\mathcal{X}) \otimes_\mathbb{Z} \mathbb{Q}$$

with following properties:

(i) *For any morphism $f : \mathcal{X} \to \mathcal{Y}$ of regular arithmetic varieties, $f^*(\widehat{\phi}(\overline{\mathcal{E}})) = \widehat{\phi}(f^*(\overline{\mathcal{E}}))$;*

(ii) *If $(\mathcal{E}, h) = (\mathcal{L}, h_1) \oplus \cdots \oplus (\mathcal{L}, h_r)$, then*

$$\widehat{\phi}(\overline{\mathcal{E}}) = \phi(\widehat{c}_1(\mathcal{L}_1, h_1), \dots, \widehat{c}_1(\mathcal{L}_r, h_r));$$

(iii) $\omega(\widehat{\phi}(\overline{\mathcal{E}})) = \phi(\overline{\mathcal{E}}_\mathbb{C})$;

(iv) *If $\overline{\mathbf{E}} : 0 \to \overline{\mathcal{S}} \to \overline{\mathcal{E}} \to \overline{\mathcal{Q}} \to 0$ is a short exact sequence of C^∞-hermitian vector bundles, then*

$$\widehat{\phi}(\overline{\mathcal{S}} \oplus \overline{\mathcal{Q}}) - \widehat{\phi}(\overline{\mathcal{E}}) = a(\widetilde{\phi}(\overline{\mathbf{E}}_\mathbb{C}));$$

(v) *Write* $\phi(T_1 + T, \ldots, T_r + T) = \sum_{i \geq 0} \phi_i(T_1, \ldots, T_r) T^i$*; Then, for any* C^∞*-hermitian line bundle,*

$$\widehat{\phi}(\overline{\mathcal{E}} \otimes \overline{\mathcal{L}}) = \sum_{i \geq 0} \widehat{\phi}_i(\overline{\mathcal{E}}) \cdot \widehat{c}_1(\overline{\mathcal{L}})^i.$$

The class $\widehat{\phi}(\overline{\mathcal{E}})$ *is called the* arithmetic characteristic class *of* $\overline{\mathcal{E}}$ *associated to* ϕ.

We refer to [12, §4.1] and [32, IV Theorem 3] for its proof. Notice that, taking ω in (iv) and using (iii) and $\omega \circ a = dd^c$, we get $\phi(\overline{\mathcal{S}}_{\mathbb{C}} \oplus \overline{\mathcal{Q}}_{\mathbb{C}}) - \phi(\overline{\mathcal{E}}_{\mathbb{C}}) = dd^c(\widetilde{\phi}(\overline{\mathbf{E}}_{\mathbb{C}}))$. This is one of the properties of the Bott-Chern secondary characteristic (cf. Theorem 2.2.1 (ii)). We remark that for a C^∞-hermitian line bundle $\overline{\mathcal{L}}$, (iii) becomes $\omega(\widehat{c}_1(\overline{\mathcal{L}})) = c_1(\overline{\mathcal{L}}_{\mathbb{C}})$. This is the Poincaré-Lelong formula.

By taking c_k, ch, and td as ϕ, we obtain the arithmetic k-th Chern class $\widehat{c}_k(\overline{\mathcal{E}})$, the arithmetic Chern character $\widehat{\mathrm{ch}}(\overline{\mathcal{E}})$, and the arithmetic Todd class $\widehat{\mathrm{td}}(\overline{\mathcal{E}})$, respectively.

2.4. **Analytic torsion and Quillen metric.** Let X be a compact Kähler manifold of dimension d with Kähler form Ω. Let E be a holomorphic vector bundle. The determinant of cohomology of E is the one-dimensional vector space

$$\lambda(E) = \bigotimes_{q=0}^{d} (\det H^q(X, E))^{(-1)^q}.$$

Let h be a C^∞-hermitian metric on E. In what follows, we will define two metrics (called the L^2-metric and the Quillen metric) on $\lambda(E)$.

Let $A^{0,q}(X, E)$ be the vector space of C^∞ sections of the vector bundle of $\bigwedge^q(T^{*(0,1)}X) \otimes E$. For $x \in X$, let $\langle\ \rangle_x$ be the hermitian product on $\bigwedge^q(T^{*(0,1)}X) \otimes E$ associated with ω and h. Then the normalized L^2-hermitian metric $(\ ,\ )_{L^2}$ is defined by

$$(s, t)_{L^2} = \frac{1}{(2\pi)^d} \int_X \langle s(x), t(x) \rangle_x \frac{\Omega^d}{d!} \qquad (s, t \in A^{0,q}(X, E)).$$

Let $\overline{\partial} : A^{0,q-1}(X, E) \to A^{0,q}(X, E)$ be the Dolbeault operator and $\overline{\partial}^* : A^{0,q}(X, E) \to A^{0,q-1}(X, E)$ its adjoint operator with respect to $(\ ,\ )_{L^2}$. Let $\Box_q = \overline{\partial}\,\overline{\partial}^* + \overline{\partial}^* \overline{\partial} : A^{0,q}(X, E) \to A^{0,q}(X, E)$ be the $\overline{\partial}$-Laplacian.

Set $\mathcal{H}^q(X, E) = \operatorname{Ker} \Box_q$. This is the vector space of harmonic $(0, q)$ forms with values in E. By Hodge theory, $\mathcal{H}^q(X, E)$ is isomorphic to $H^q(X, E)$. Since $\mathcal{H}^q(X, E)$ inherits the L^2-metric from that on $A^{0,q}(X, E)$, it induces a metric $h_{H^q(X,E)}$ on $H^q(X, E)$ by the above isomorphism. The hermitian metric on $\lambda(E)$, induced from $\mathcal{H}^q(X, E)$'s is denoted by h_{L^2} and called the L^2-metric.

Next, we will define the Quillen metric on $\lambda(E)$. Let $\sigma(\Box_q)$ be the set of eigenvalues of $\Box_q$ and $E^q(\lambda)$ the eigenspace of $A^{0,q}(X, E)$ with eigenvalue $\lambda \in \sigma(\Box_q)$. For

$s \in \mathbb{C}$ with $\mathrm{Re}(s) \gg 0$, set

$$\zeta_q(s) = \sum_{\lambda \in \sigma(\Box_q) \setminus \{0\}} \frac{\dim E^q(\lambda)}{\lambda^s}.$$

It is known that $\zeta_q(s)$ converges absolutely for $\mathrm{Re}(s) \gg 0$, extends meromorphically to the whole complex plane, and is regular at $s = 0$.

Definition 2.4.1 (Analytic torsion). The real number

$$T(E, h) = \sum_{q \geq 0} (-1)^{q+1} q \zeta_q'(0)$$

is called the *analytic torsion* of (E, h).

Remark 2.4.2. Originally, Ray and Singer [28] defined the analytic torsion as $\exp(\frac{1}{2}T(E, h))$.

Definition 2.4.3 (Quillen metric, [27]). The metric

$$h_Q = \exp T(E, h) \cdot h_{L^2}$$

on $\lambda(E)$ is called the *Quillen metric* of (E, h).

Here, we digress a little. Let (V_q, h_q) $(0 \leq q \leq n)$ be finite dimensional hermitian vector spaces. Suppose a complex of vector spaces

$$0 \longrightarrow V_0 \xrightarrow{d} V_1 \xrightarrow{d} \cdots \xrightarrow{d} V_n \longrightarrow 0$$

is given. Set $H_q = \mathrm{Ker}(d : V_q \to V_{q+1}) / \mathrm{Image}(d : V_{q-1} \to V_q)$ and $\lambda(V) = \bigotimes_{q=0}^{n} (\det H_q)^{(-1)^q}$. As a toy case, we will see what metric on $\lambda(V)$ corresponds to the L^2-metric and the Quillen metric.

Let $d^* : V_q \to V_{q-1}$ be the adjoint operator of $d : V_{q-1} \to V_q$ with respect to h_{q-1} and h_q. Define the self-adjoint operator $\Box_q$ by $dd^* + d^*d : V_q \to V_q$. For $\lambda \geq 0$, set $E_q(\lambda) = \{v \in V_q \mid \Box_q(v) = \lambda v\}$. Notice that, except for finitely many λ, $E_q(\lambda) = 0$. For $\lambda = 0$, we denote $E_q(0)$ by $\mathcal{H}_q$. Put

$$\zeta_q(s) = \sum_{\lambda > 0} \frac{\dim E_q(\lambda)}{\lambda^s} \quad \text{and} \quad T = \sum_{q=0}^{n} (-1)^{q+1} q \zeta_q'(0).$$

Lemma 2.4.4. (i) *Assume $\lambda > 0$. Then, $d(E_{q-1}(\lambda)) \subset E_q(\lambda)$ and*

$$0 \longrightarrow E_0(\lambda) \xrightarrow{d} E_1(\lambda) \xrightarrow{d} \cdots \xrightarrow{d} E_n(\lambda) \longrightarrow 0$$

is exact. Thus, canonically, $H_q \simeq \mathcal{H}_q$.

(ii) *The vector space $\bigotimes_{q=0}^{n} (\det \mathcal{H}_q)^{(-1)^q}$ is canonically isomorphic to $\bigotimes_{q=0}^{n} (\det V_q)^{(-1)^q}$. Thus, $\lambda(V)$, $\bigotimes_{q=0}^{n} (\det \mathcal{H}_q)^{(-1)^q}$ and $\bigotimes_{q=0}^{n} (\det V_q)^{(-1)^q}$ are canonically isomorphic to each other by* (i).

(iii) *By* (ii), *h_q's induce a metric on* $\bigotimes_{q=0}^n (\det V_q)^{(-1)^q} \simeq \lambda(V)$. *We denote this metric on* $\lambda(V)$ *by* h_Q. *On the other hand, since* $\mathcal{H}_q \subset V_q$ *inherits the metric* h_q *of* V_q, *it induces a metric on* $\bigotimes_{q=0}^n (\det \mathcal{H}_q)^{(-1)^q} \simeq \lambda(V)$ *by* (i). *We denote this metric* $\lambda(V)$ *by* h_{L^2}. *Then, we have*

$$h_Q = \exp(T) \cdot h_{L^2}.$$

¿From Lemma 2.4.4, one might say that the Quillen metric is, as it were, the metric on $\lambda(E)$ that is induced from the L^2-metric on "$\bigotimes_{q=0}^d \left(\det A^{0,q}(X,E)\right)^{(-1)^q}$". (Of course, since $A^{0,q}(X,E)$ is infinite dimensional, "$\det A^{0,q}(X,E)$" is not defined in an ordinary sense.)

Let X be a noetherian scheme and F a coherent sheaf on X. Then, one can define a line bundle $\det F$ over X. When X is regular, $\det F$ is given as below. (We refer to [20] for details.) Let T be the torsion part of F. Set

$$D = \sum_{P \in X, \mathrm{depth}(P)=1} \mathrm{length}(T_P) \cdot \overline{\{P\}}$$

and $\det(T) = \mathcal{O}_X(D)$. Moreover, set $\det(F/T) = \left(\bigwedge^{\mathrm{rk}(F)}(F/T)\right)^{**}$. Then, we have $\det F = \det(T) \otimes \det(F/T)$.

Let $f : \mathcal{X} \to \mathcal{Y}$ be a projective morphism of regular arithmetic varieties such that $f_{\mathbb{C}} : \mathcal{X}_{\mathbb{C}} \to \mathcal{Y}_{\mathbb{C}}$ is smooth. Let $\mathcal{E}$ be a vector bundle on $\mathcal{X}$. Then, the *determinant line bundle* over $\mathcal{Y}$ is defined by

$$\det \mathbf{R}f_*(\mathcal{E}) := \bigotimes_{q \geq 0} (\det R^q f_*(\mathcal{E}))^{(-1)^q}.$$

Let $\overline{\mathcal{E}} = (\mathcal{E}, h)$ be a C^∞-hermitian vector bundle over $\mathcal{X}$. We fix a Kähler metric h_f on the relative tangent bundle $T_{\mathcal{X}_{\mathbb{C}}/\mathcal{Y}_{\mathbb{C}}} := \mathrm{Ker}(T\mathcal{X}_{\mathbb{C}} \to f^*(T\mathcal{Y}_{\mathbb{C}}))$. Namely, $h_f|_{X_y}$ is Kähler for any fiber X_y and $\{h_f|_{X_y}\}_{y \in \mathcal{Y}(\mathbb{C})}$ is C^∞ with respect to y. We further assume that h_f is invariant under the complex conjugation.

Since

$$\det \mathbf{R}f_*(\mathcal{E})_y = \bigotimes_{q \geq 0} (\det H^q(X_y, E_y))^{(-1)^q} \qquad (y \in \mathcal{Y}(\mathbb{C})),$$

the L^2-metric $h_{L^2,y}$ and the Quillen metric $h_{Q,y}$ are defined on $\bigotimes_{q \geq 0} (\det H^q(X_y, E_y))^{(-1)^q}$. Thus, they induce the L^2-metric $h_{L^2} = \{h_{L^2,y}\}_{y \in \mathcal{Y}(\mathbb{C})}$ and the Quillen metric $h_Q = \{h_{Q,y}\}_{y \in \mathcal{Y}(\mathbb{C})}$ on $\det \mathbf{R}f_*(\mathcal{E})$, respectively.

Bismut, Gillet and Soulé proved the following theorem:

Theorem 2.4.5 ([3]). *The Quillen metric* h_Q *on* $\det \mathbf{R}f_*(\mathcal{E})$ *is* C^∞ *with respect to* $y \in \mathcal{Y}(\mathbb{C})$.

For its proof, we refer to [3, Corollary 3.9], [32, VI].

2.5. **Arithmetic Riemann-Roch Theorem.** Let $f : \mathcal{X} \to \mathcal{Y}$ be a projective morphism of regular arithmetic varieties such that $f_{\mathbb{C}} : \mathcal{X}_{\mathbb{C}} \to \mathcal{Y}_{\mathbb{C}}$ is smooth.

Let $T_{\mathcal{X}/\mathcal{Y}} = \operatorname{Ker}(T\mathcal{X} \to f^*(T\mathcal{Y}))$ be the relative tangent sheaf. We fix a Kähler metric h_f on $T_{\mathcal{X}(\mathbb{C})/\mathcal{Y}(\mathbb{C})}$ such that h_f is invariant under the complex conjugation. Then, the arithmetic Todd class $\widehat{\operatorname{td}}(\overline{T_{\mathcal{X}/\mathcal{Y}}}) \in \bigoplus_{p\geq 0} \widehat{\operatorname{CH}}^p(\mathcal{X})$ is defined, although the coherent sheaf $T_{\mathcal{X}/\mathcal{Y}}$ is not in general a vector bundle (cf. [32, VIII, 1.1]).

Let $R(T)$ be the Gillet-Soulé power series:

$$R(T) = \sum_{m:\text{odd}, m\geq 1} \left(2\zeta'(m) + \left(1 + \frac{1}{2} + \cdots + \frac{1}{m}\right)\zeta(-m)\right)\frac{T^m}{m!},$$

where $\zeta(s)$ is the Riemann zeta function. Set $R(T_1, \ldots, T_r) = R(T_1) + \cdots + R(T_r) \in \mathbb{R}[[T_1, \cdots, T_r]]$. Then, for any complex manifold X and rank r C^∞-hermitian vector bundle $\overline{E} = (E, h)$, the characteristic form $R(\overline{E}) \in \bigoplus_{p\geq 0} A^{p,p}(X)$ is defined as in §2.1. (In §2.1, $\mathbb{Q}$-coefficient power series are treated, but the same argument goes for $\mathbb{R}$-coefficient power series.)

Set

$$\widehat{\operatorname{td}}^R(\overline{T_{\mathcal{X}/\mathcal{Y}}}) = \widehat{\operatorname{td}}(\overline{T_{\mathcal{X}/\mathcal{Y}}}) \cdot \left(1 - a\left(R(\overline{T_{\mathcal{X}_{\mathbb{C}}/\mathcal{Y}_{\mathbb{C}}}})\right)\right) \in \bigoplus_{p\geq 0} \widehat{\operatorname{CH}}^p(\mathcal{X}) \otimes_{\mathbb{Z}} \mathbb{Q}.$$

Theorem 2.5.1 (Arithmetic Riemann-Roch Theorem). *Let $f : \mathcal{X} \to \mathcal{Y}$ be a projective morphism of regular arithmetic varieties such that $f_{\mathbb{C}} : \mathcal{X}_{\mathbb{C}} \to \mathcal{Y}_{\mathbb{C}}$ is smooth. Let $T_{\mathcal{X}/\mathcal{Y}}$ be the relative tangent sheaf. Fix a Kähler metric h_f on $T_{\mathcal{X}(\mathbb{C})/\mathcal{Y}(\mathbb{C})}$ such that h_f is invariant under the complex conjugation.*

Then, for any C^∞-hermitian vector bundle $\overline{\mathcal{E}} = (\mathcal{E}, h)$, the following formula holds in $\widehat{\operatorname{CH}}^1(\mathcal{Y}) \otimes_{\mathbb{Z}} \mathbb{Q}$:

$$\widehat{c}_1(\det \mathbf{R}f_*(\mathcal{E}), h_Q) = f_*\left(\widehat{\operatorname{ch}}(\overline{\mathcal{E}}) \cdot \widehat{\operatorname{td}}^R(\overline{T_{\mathcal{X}/\mathcal{Y}}})\right)^{(1)}. \tag{2.5.1}$$

For its proof, we refer to [13, Theorem 7], [32, VIII]. We remark that [13, Theorem 7] is a stronger assertion. Let $\mathcal{X}$ and $\mathcal{Y}$ be arithmetic varieties (that need not be regular) such that $\mathcal{X}_{\mathbb{Q}}$ and $\mathcal{Y}_{\mathbb{Q}}$ are regular. Let $f : \mathcal{X} \to \mathcal{Y}$ be a projective morphism such that $f_{\mathbb{C}} : \mathcal{X}_{\mathbb{C}} \to \mathcal{Y}_{\mathbb{C}}$ is smooth. Then, if (i) f is a locally complete intersection, or (ii) $\mathcal{Y}$ is regular, then (2.5.1) holds.

The equality (2.5.1) is the degree one part of a Riemann-Roch formula. For regular arithmetic varieties, Faltings ([6]) proved an arithmetic Riemann-Roch theorem which includes the higher degree parts.

3. Existence of a Small Section

For a hermitian line bundle $(\mathcal{L}, h)$, we set $\|\cdot\| = \sqrt{h(\cdot,\cdot)}$. In the sequel, we often denote $(\mathcal{L}, h)$ by $(\mathcal{L}, \|\cdot\|)$.

Apart from the arithmetic Riemann-Roch theorem (Theorem 2.5.1), the asymptotic behavior of analytic torsion (Proposition 3.5.3), and Minkowski's convex body theorem, this section is written in a self-contained way.

3.1. **Small Section.** Let X be a projective variety over $\mathbb{C}$ of dimension d, and L and N be line bundles over X. By the Hilbert-Samuel theorem (a weak form of the Riemann-Roch theorem), we have

$$\chi(X, L^{\otimes n} \otimes N) = \frac{\deg(L^d)}{n!} n^d + O(n^{d-1}).$$

(See, for example, [19] for details.) Assume further that L is nef and big. Then, by the Hilbert-Samuel theorem and the Kodaira vanishing theorem, $L^{\otimes n}$ has a nonzero section for large n.

Following [13] and [37], we will consider its arithmetic version (cf. Theorem 3.3.1 and Corollary 3.3.2).

Definition 3.1.1 (Small Section). Let $\mathcal{X}$ be an arithmetic variety, and $\overline{\mathcal{L}} = (\mathcal{L}, \|\cdot\|)$ be a C^∞-hermitian line bundle over $\mathcal{X}$. We say that $s \in H^0(\mathcal{X}, \mathcal{L})$ is a *small section* if $\|s\|_{\sup} := \sup\{\|s\|(x) \mid x \in \mathcal{X}(\mathbb{C})\} < 1$.

Remark 3.1.2. A small section of a C^∞-hermitian line bundle over an arithmetic variety is an arithmetic counterpart of a section of a line bundle over an algebraic variety.

A section $s \in H^0(\mathcal{X}, \mathcal{L})$ with $\|s\|_{\sup} \leq 1$ (or $\|s\|_{L^2} < 1$, or $\|s\|_{L^2} \leq 1$ in §3.2) is also often called a small section. In this note, however, by a small section is meant a section $s \in H^0(\mathcal{X}, \mathcal{L})$ with $\|s\|_{\sup} < 1$.

3.2. **Arithmetic Euler Characteristic.** Let $(V, \|\cdot\|)$ be a real normed space of dimension b. Let Γ be a lattice of V, i.e., Γ is a discrete $\mathbb{Z}$-submodule of V which generates V over $\mathbb{R}$.

For a $\mathbb{Z}$-basis $v_1, \ldots, v_b$ of Γ, we define the fundamental domain V/Γ by

$$V/\Gamma = \{v = \lambda_1 v_1 + \cdots + \lambda_b v_b \mid 0 \leq \lambda_i \leq 1 \ (1 \leq i \leq b)\}.$$

Moreover, let $B(V) = \{x \in V \mid \|x\| \leq 1\}$ be the closed unit ball in V.

For an isomorphism $\phi : V \xrightarrow{\sim} \mathbb{R}^b$ of vector spaces, let μ_ϕ denote the Lebesgue measure on V induced by ϕ. Set

$$\mathrm{vol}_{\|\cdot\|}(\Gamma) = \frac{\mu_\phi(V/\Gamma)}{\mu_\phi(B(V))}.$$

The value $\mathrm{vol}_{\|\cdot\|}(\Gamma)$ is independent of the choice of $v_1, \ldots, v_b$ and ϕ. We set

$$\chi_{\|\cdot\|}(\Gamma) = -\log \mathrm{vol}_{\|\cdot\|}(\Gamma).$$

We will state here Minkowski's convex body theorem.

Definition 3.2.1 (successive minima). Let $(V, \|\cdot\|)$ be a real normed space of dimension b and Γ a lattice of V. Then, there exist linearly independent vectors $v_1, \ldots, v_b \in V$, and positive numbers $\lambda_1(\Gamma), \ldots, \lambda_b(\Gamma)$ with the following properties:

(i) $\|v_i\| = \lambda_i(\Gamma)$ $(1 \le i \le b)$;

(ii) $\lambda_1(\Gamma) \le \lambda_2(\Gamma) \le \cdots \le \lambda_b(\Gamma)$;

(iii) For any b' with $1 \le b' \le b$, and for any $v \in \Gamma$ that is linearly independent of $v_1, \ldots, v_{b'-1}$, we have $\|v\| \ge \lambda_{b'}(\Gamma)$.

$\lambda_1(\Gamma), \ldots, \lambda_b(\Gamma)$ depends only on $(V, \|\cdot\|)$ and Γ. We call $\lambda_1(\Gamma), \ldots, \lambda_b(\Gamma)$ the successive minima of Γ in $B(V)$.

Theorem 3.2.2 (Minkowski's convex body theorem). *Let $(V, \|\cdot\|)$ be a real normed space of dimension b and Γ a lattice of V. Then, the following inequalities hold:*

$$\frac{2^b}{b!} \operatorname{vol}_{\|\cdot\|}(\Gamma) \le \lambda_1(\Gamma) \cdot \cdots \cdot \lambda_b(\Gamma) \le 2^b \operatorname{vol}_{\|\cdot\|}(\Gamma).$$

Its proof is found, for example, in [15].

Let $\mathcal{X}$ be a projective arithmetic variety and $\overline{\mathcal{L}} = (\mathcal{L}, \|\cdot\|)$ a C^∞-hermitian line bundle over $\mathcal{X}$. Assume that $\mathcal{X}$ is regular. Then, by [16, Theorem 8.8 (b)], $H^0(\mathcal{X}, \mathcal{L})$ is a torsion-free finite $\mathbb{Z}$-module.

Set $V = H^0(\mathcal{X}, \mathcal{L}) \otimes_{\mathbb{Z}} \mathbb{R}$. Let Γ be the image of the natural inclusion $H^0(\mathcal{X}, \mathcal{L}) \longrightarrow H^0(\mathcal{X}, \mathcal{L}) \otimes_{\mathbb{Z}} \mathbb{R}$. Then Γ is a lattice of V. (From now on, we will identify $H^0(\mathcal{X}, \mathcal{L})$ with Γ.)

Let us define norms on V. We fix a normalized volume element dx of $\mathcal{X}(\mathbb{C})$. As in §1.3, let $F_\infty : \mathcal{X}(\mathbb{C}) \to \mathcal{X}(\mathbb{C})$ be the complex conjugation. Since

$$\begin{aligned} V &= H^0(\mathcal{X}, \mathcal{L}) \otimes_{\mathbb{Z}} \mathbb{R} \\ &= \text{"the } F_\infty\text{-invariant space of } H^0(\mathcal{X}, \mathcal{L}) \otimes_{\mathbb{Z}} \mathbb{C}\text{"} \\ &= \text{"the } F_\infty\text{-invariant space of } H^0(\mathcal{X}_{\mathbb{C}}, \mathcal{L}_{\mathbb{C}})\text{"} \subset H^0(\mathcal{X}_{\mathbb{C}}, \mathcal{L}_{\mathbb{C}}), \end{aligned}$$

V is an $\mathbb{R}$-vector subspace of $H^0(\mathcal{X}_{\mathbb{C}}, \mathcal{L}_{\mathbb{C}})$.

(i) We define the L^p-norm $(1 \le p < \infty)$ on V by

$$\|s\|_{L^p} = \left\{ \int_{\mathcal{X}(\mathbb{C})} \|s\|^p(x) dx \right\}^{\frac{1}{p}} \qquad (s \in V).$$

In the right-hand side, s is regarded as an element of $H^0(\mathcal{X}_{\mathbb{C}}, \mathcal{L}_{\mathbb{C}})$ via the inclusion $V \subset H^0(\mathcal{X}_{\mathbb{C}}, \mathcal{L}_{\mathbb{C}})$.

(ii) We define the sup-norm by

$$\|s\|_{\sup} = \sup_{x \in \mathcal{X}(\mathbb{C})} \|s\|(x).$$

We define $\chi_{L^p}(\mathcal{X}, \overline{\mathcal{L}})$ (resp. $\chi_{\sup}(\mathcal{X}, \overline{\mathcal{L}})$) by $\chi_{L^p}(\mathcal{X}, \overline{\mathcal{L}}) = \chi_{\|\cdot\|_{L^p}}(H^0(\mathcal{X}, \mathcal{L}))$ (resp. $\chi_{\sup}(\mathcal{X}, \overline{\mathcal{L}}) = \chi_{\|\cdot\|_{\sup}}(H^0(\mathcal{X}, \mathcal{L}))$), and call it the *arithmetic Euler characteristic* with respect to the L^p-norm (resp. the sup-norm).

Notice that the value $\chi_{L^p}(\mathcal{X}, \overline{\mathcal{L}})$ depends on the choice of volume element dx of $\mathcal{X}_{\mathbb{C}}$. However, when $c_1(\overline{\mathcal{L}})$ is everywhere positive, $dx := \dfrac{c_1(\overline{\mathcal{L}})^{\wedge d}}{(\mathcal{L}_{\mathbb{C}}^{\dim \mathcal{X}_{\mathbb{C}}})}$ gives a natural volume element.

Let us compute the arithmetic Euler characteristic in the case of $\mathcal{X} = \operatorname{Spec}(\mathbb{Z})$. Let $\overline{\mathcal{L}} = (\mathcal{L}, h)$ be a hermitian line bundle over $\mathcal{X} = \operatorname{Spec}(\mathbb{Z})$, and e a $\mathbb{Z}$-basis of $\mathcal{L}$. In this case, since $X_{\mathbb{C}}$ is one point, the L^p-norm and the sup-norm are the same (and thus we will omit the suffix). Then, we get

$$\chi(\mathcal{X}, \overline{\mathcal{L}}) = -\log \frac{\sqrt{h(e,e)}}{2} = -\frac{1}{2}\log h(e,e) + \log 2.$$

Set $\overline{\mathcal{O}_{\mathcal{X}}}^{can} = (\mathcal{O}_{\mathcal{X}}, |\cdot|)$, where $|\cdot|$ is the absolute value. Then, $\chi(\mathcal{X}, \overline{\mathcal{O}_{\mathcal{X}}}^{can}) = \log 2$. On the other hand, by §1.1, we have $-\frac{1}{2}\log h(e,e) = \widehat{\deg}(\overline{\mathcal{L}})$. Substituting these to the above equality, we find

$$\chi(\mathcal{X}, \overline{\mathcal{L}}) - \chi(\mathcal{X}, \overline{\mathcal{O}_{\mathcal{X}}}^{can}) = \widehat{\deg}(\overline{\mathcal{L}}). \tag{3.2.2}$$

More generally, a similar equality holds when $\mathcal{X}$ is the spectrum of the ring of integers of a number field. On the analogy of the Riemann-Roch theorem for compact Riemann surfaces, $\chi(\mathcal{X}, \overline{\mathcal{L}})$ corresponds to the Euler characteristic of a line bundle over a compact Riemann surface.

3.3. **Arithmetic Hilbert-Samuel Theorem and the Existence of a Small Section.** Let M be a reduced analytic space. Let $\overline{L} = (L, h)$ be a continuous hermitian line bundle over M. As in §1.6, $\overline{L}$ is said to be C^∞ if, for any complex manifold N and analytic morphism $\phi : N \to M$, $\phi^*(\overline{L})$ is C^∞. Similarly, a continuous function f on M is said to be C^∞ if $\phi^*(f)$ is C^∞.

Assume that $\overline{L}$ is C^∞. Then, $c_1(\overline{L})$ is said to be *semipositive* if, for any complex manifold N and analytic morphism $\phi : N \to M$, the first Chern form $c_1(\phi^*(\overline{L}))$ is semipositive. Moreover, $c_1(\overline{L})$ is said to be *positive* if, for any $x \in M$ and real-valued function f defined over a neighborhood of x, there exists a $\lambda_0 > 0$ with the following property: for any λ with $|\lambda| \le \lambda_0$, $\lambda dd^c(f) + c_1(\overline{L})$ is semipositive in some neighborhood of x. When M is a manifold, semipositivity (resp. positivity) coincides with that in the ordinary sense.

Let $f : \mathcal{X} \to \operatorname{Spec}(\mathbb{Z})$ be a projective arithmetic variety. Let $\overline{\mathcal{L}} = (\mathcal{L}, h)$ be a C^∞-hermitian line bundle over $\mathcal{X}$. We define $\overline{\mathcal{L}}$ to be *vertically nef* if it satisfies the following conditions:

(i) For any one-dimensional integral scheme C such that $f(C)$ is zero-dimensional, $\deg(\mathcal{L}|_C) \geq 0$ (Nef along the fibers);
(ii) $c_1(\overline{\mathcal{L}})$ is semipositive.

Moreover, we define $\overline{\mathcal{L}}$ to be *vertically ample* if it satisfies the following conditions:

(i) $\mathcal{L}$ is f-ample;
(ii) $c_1(\overline{\mathcal{L}})$ is positive.

The purpose of this section is to prove the following theorem and its corollary.

Theorem 3.3.1 (Arithmetic Hilbert-Samuel Theorem). *Let $f : \mathcal{X} \to \operatorname{Spec}(\mathbb{Z})$ be a projective arithmetic variety. Let $\overline{\mathcal{L}}$ and $\overline{\mathcal{N}}$ be C^∞-hermitian line bundles over $\mathcal{X}$. Assume the following:*

(i) *$\mathcal{L}_{\mathbb{Q}}$ is ample over $\mathcal{X}_{\mathbb{Q}}$;*
(ii) *$\overline{\mathcal{L}}$ is vertically nef.*

Then, we have

$$\chi_{\sup}(\mathcal{X}, \overline{\mathcal{L}}^{\otimes n} \otimes \overline{\mathcal{N}}) = \frac{\widehat{\deg}(\widehat{c}_1(\overline{\mathcal{L}})^{d+1})}{(d+1)!} n^{d+1} + o(n^{d+1}).$$

Corollary 3.3.2 (Existence of a small section). *Let $\mathcal{X}$ be a projective arithmetic variety and $\overline{\mathcal{L}}$ a C^∞-hermitian line bundle over $\mathcal{X}$. Assume the following:*

(i) *$\mathcal{L}_{\mathbb{Q}}$ is ample $\mathcal{X}_{\mathbb{Q}}$;*
(ii) *$\overline{\mathcal{L}}$ is vertically nef;*
(iii) $\widehat{\deg}(\widehat{c}_1(\overline{\mathcal{L}})^{d+1}) > 0$

Then, for any sufficiently large n, $\overline{\mathcal{L}}^{\otimes n}$ has a nonzero small section.

In order to prove Theorem 3.3.1, we first prove the following theorem.

Theorem 3.3.3. *Let $f : \mathcal{X} \to \operatorname{Spec}(\mathbb{Z})$ be a projective arithmetic variety. Let $\overline{\mathcal{L}}$ and $\overline{\mathcal{N}}$ be C^∞-hermitian line bundles over $\mathcal{X}$. Assume the following:*

(i) *$\mathcal{X}_{\mathbb{Q}}$ is regular;*
(ii) *$\overline{\mathcal{L}}$ is vertically ample.*

Then, we have

$$\chi_{\sup}(\mathcal{X}, \overline{\mathcal{L}}^{\otimes n} \otimes \overline{\mathcal{N}}) = \frac{\widehat{\deg}(\widehat{c}_1(\overline{\mathcal{L}})^{d+1})}{(d+1)!} n^{d+1} + O(n^d \log n).$$

Remark 3.3.4. For the proof of Theorem 3.3.3, we use the arithmetic Riemann-Roch theorem. We remark that, Abbes and Bouche [1] gave a comparatively short proof of Theorem 3.3.3 (in the case $\mathcal{N} = \mathcal{O}_{\mathcal{X}}$) that does not use the arithmetic Riemann-Roch theorem.

Following [32], we will prove Theorem 3.3.3 in four steps:

Step 1: We compare $\chi_{\sup}(\mathcal{X}, \overline{\mathcal{L}}^{\otimes n} \otimes \overline{\mathcal{N}})$ and $\chi_{L^2}(\mathcal{X}, \overline{\mathcal{L}}^{\otimes n} \otimes \overline{\mathcal{N}})$ (§3.4);

Step 2: We relate $\chi_{L^2}(\mathcal{X}, \overline{\mathcal{L}}^{\otimes n} \otimes \overline{\mathcal{N}})$ to $\widehat{\deg}(\widehat{c}_1(\det \mathbf{R} f_*(\mathcal{L}^{\otimes n} \otimes \mathcal{N}), h_Q))$ (§3.5);

Step 3: We state the asymptotic behavior of analytic torsions, which appear as the the difference of L^2-metrics and Quillen metrics in Step 2 (§3.5);

Step 4: We relate $\widehat{\deg}(\widehat{c}_1(\det \mathbf{R} f_*(\mathcal{L}^{\otimes n} \otimes \mathcal{N}), h_Q))$ and $\dfrac{\widehat{\deg}(\widehat{c}_1(\mathcal{L})^{d+1})}{(d+1)!} n^{d+1}$, using the arithmetic Riemann-Roch theorem (§3.5).

In §3.6, following [37], we will prove Theorem 3.3.1, by taking a generic resolution of singularities of $\mathcal{X}$ and using Theorem 3.3.3.

In the sequel, we will see that Corollary 3.3.2 follows from Theorem 3.3.1 and Minkowski's convex body theorem.

Proof of Corollary 3.3.2 (Assuming Theorem 3.3.1): Set $\Gamma_n = H^0(\mathcal{X}, \mathcal{L}^{\otimes n})$ and $V_n = H^0(\mathcal{X}, \mathcal{L}^{\otimes n}) \otimes_{\mathbb{Z}} \mathbb{R}$. We endow V_n with the sup-norm induced by $\overline{\mathcal{L}}^{\otimes n}$. By Minkowski's convex body theorem (Theorem 3.2.2), we find

$$(\dim V_n) \log \lambda_1(\Gamma_n) \leq (\dim V_n) \log 2 + \log \mathrm{vol}_{\sup}(\Gamma_n). \tag{3.3.4}$$

On the other hand, by Theorem 3.3.1, we get

$$\chi_{\sup}(\Gamma_n) = \frac{\widehat{\deg}(\widehat{c}_1(\overline{\mathcal{L}})^{d+1})}{(d+1)!} n^{d+1} + o(n^{d+1}).$$

Since $\mathcal{L}_{\mathbb{Q}}$ is ample, we have, for sufficiently large n,

$$\dim V_n = \frac{\deg(\mathcal{L}_{\mathbb{Q}}^d)}{d!} n^d + O(n^{d-1}).$$

Thus, we obtain

$$\begin{aligned}
&(\dim V_n) \log 2 + \log \mathrm{vol}_{\sup}(\Gamma_n) \\
=&(\dim V_n) \log 2 - \chi_{\sup}(\Gamma_n) \\
=&\left\{ \frac{\deg(\mathcal{L}_{\mathbb{Q}}^d)}{d!} n^d + O(n^{d-1}) \right\} \log 2 - \left\{ \frac{\widehat{\deg}(\widehat{c}_1(\overline{\mathcal{L}})^{d+1})}{(d+1)!} n^{d+1} + o(n^{d+1}) \right\} \\
=& - \frac{\widehat{\deg}(\widehat{c}_1(\overline{\mathcal{L}})^{d+1})}{(d+1)!} n^{d+1} + o(n^{d+1}).
\end{aligned}$$

Hence, if $\widehat{\deg}(\widehat{c}_1(\overline{\mathcal{L}})^{d+1}) > 0$, then

$$(\dim V_n) \log 2 + \log \mathrm{vol}_{\sup}(\Gamma_n) \longrightarrow -\infty \quad (n \to \infty). \tag{3.3.4}$$

It follows from (3.3.4) and (3.3.4) that, for sufficiently large n,

$$\log \lambda_1(\Gamma_n) < 0. \tag{3.3.4}$$

Since, by definition, $\lambda_1(\Gamma_n) = \min\{\|s\|_{\sup} \mid s \in \Gamma_n,\ s \neq 0\}$, (3.3.4) shows that there exists a nonzero element $s \in \Gamma_n$ with $\|s\|_{\sup} < 1$. Therefore, for sufficiently large n, $\mathcal{L}^{\otimes n}$ has a nonzero small section. □

3.4. **The comparison of L^p-norm and sup-norm.** Let $U \subset \mathbb{C}$ be an open unit disk. Let $\phi(z)$ be a holomorphic function on U that extends continuously to $\overline{U}$. Then, for any $p > 0$, we have

$$\iint_{|z|<1} |\phi(z)|^p dxdy = \int_0^1 \left(\int_0^{2\pi} |\phi(re^{i\theta})|^p d\theta \right) rdr$$
$$\geq \int_0^1 (2\pi |\phi(0)|^p)\, rdr = \pi |\phi(0)|^p,$$

where we used the subharmonicity of $|\phi(z)|^p$ in the second line. This inequality bounds from above the value of ϕ at the origin by the L^p-metric of ϕ.

For global sections of line bundles (and their powers) over compact complex manifolds, the following proposition gives a comparison of the L^p-metric and the sup-metric. Its proof, using the subharmonicity of some functions similar to the above estimate, is due to Gromov (cf. [10], [23]).

Lemma 3.4.1 (L^p-sup comparison). *Let X be a compact Kähler manifold, $\overline{L} = (L, h_L)$, $\overline{N} = (N, h_N)$ C^∞-hermitian line bundles over X, and dx a volume form on X.*

Then, for any p, $1 \leq p < \infty$, there exist positive constants C_1, C_2 such that, for any positive integer n and section $s \in H^0(X, L^{\otimes n} \otimes N)$, we have

$$C_1 \|s\|_{L^p} \leq \|s\|_{\sup} \leq C_2 n^{\frac{2d}{p}} \|s\|_{L^p},$$

where $\|s\|_{\sup}$ and $\|s\|_{L^p}$ are respectively defined by

$$\|s\|_{\sup} = \sup_{x \in X} \sqrt{h_L^n \cdot h_N(s,s)},$$
$$\|s\|_{L^p} = \left(\int_X (h_L^n \cdot h_N(s,s))^{\frac{p}{2}}\, dx \right)^{\frac{1}{p}}.$$

Proof: Since

$$\|s\|_{L^p}^p = \int_X (h_L^n \cdot h_N(s,s))^{\frac{p}{2}}\, dx \leq \left(\int_X dx \right) \|s\|_{\sup}^p,$$

we have the left-hand side inequality with $C_1 = (\int_X dx)^{-1/p}$.

In order to show the right-hand-side inequality, for each $x \in X$, take a neighborhood U_x of x and coordinate functions $(\phi_x^1, \ldots, \phi_x^d)$ such that L and N are locally trivial over U_x. Let e_x and f_x be respectively local frames of L and N. Put $l_x = h_L(e_x, e_x)$ and $n_x = h_N(f_x, f_x)$. Then, l_x and n_x are C^∞ positive functions over U_x. ¿From now on, thorough $(\phi_x^1, \dots, \phi_x^d) : U_x \longrightarrow \mathbb{C}^d$, we identify U_x with its image and regard U_x as an open subset of $\mathbb{C}^d$.

For $a = (a_1, \dots, a_d), b = (b_1, \dots, b_d) \in U_x$, put

$$r_x(a, b) = |a_1 - b_1| + \cdots + |a_d - b_d|.$$

For $a = (a_1, \dots, a_d) \in U_x$, $R > 0$, put

$$B_x(a, R) = \{z = (\zeta_1, \dots, \zeta_d) \in \mathbb{C}^d \mid |z_i - a_i| < R \quad (1 \le i \le d)\}.$$

Then, one can show that there exist finite open coverings $W_1, \dots, W_k$ of X and positive numbers R and K which satisfies the following conditions:

(i) For any i, $1 \le i \le k$, there exists $x_i \in X$ with $W_i \subset U_{x_i}$;
(ii) For any $a \in W_i$, $B_{x_i}(a, R) \subset U_{x_i}$.
(iii) For any $a \in W_i$ and $\zeta \in B_{x_i}(a, R)$, $l_{x_i}(\zeta) \ge l_{x_i}(a) - K r_{x_i}(a, \zeta) \ge 0$.

Let s be an element of $H^0(X, L^{\otimes n} \otimes N)$. Take $a \in X$ such that $\|s\|$ attains the maximum at a, i.e., $\|s\|_{\sup} = \|s\|(a)$. Take W_i with $a \in W_i$. We write

$$s = g e_{x_i}^n f_{x_i},$$

where e_{x_i} and f_{x_i} are respectively local frame of L and N over U_{x_i}, and g is a holomorphic function over U_{x_i}. For simplicity, we omit the suffix i (and x_i) in the following. Since $|g|^p$ is subharmonic, we have

$$\begin{aligned}
\|s\|_{L^p}^p &= \int_X h_L^n \cdot h_N(s,s)^{\frac{p}{2}} dx \\
&\ge \int_{B(a,R)} |g(\zeta)|^p l(\zeta)^{\frac{np}{2}} n(\zeta)^{\frac{p}{2}} dx \\
&\ge C_3 \int_{B(a,R)} |g(\zeta)|^p \left(l(a) - K r(a,\zeta)\right)^{\frac{np}{2}} d\xi_1 d\eta_1 \cdots d\xi_d d\eta_d \quad (\zeta_i = \xi_i + \sqrt{-1}\eta_i) \\
&\ge C_4 |g(a)|^p \int_0^R \cdots \int_0^R r_1 \cdots r_d \left(l(a) - K(r_1 + \cdots r_d)\right)^{\frac{np}{2}} dr_1 \cdots dr_d \\
&\ge n^{-2d} C_5 |g(a)|^p l(a)^{\frac{np}{2}} n(a)^{\frac{p}{2}}
\end{aligned}$$

Thus, putting $C_2 = C_5^{\frac{1}{p}}$, we have

$$\|s\|_{\sup} \le C_2 n^{\frac{2d}{p}} \|s\|_{L^p}.$$

□

Set $e_n = \dim H^0(X, L^{\otimes n} \otimes N)$. Then, by the lemma above, for $l \in \det H^0(X, L^{\otimes n} \otimes N)$, one has $C_1^{e_n} \|l\|_{L^2} \le \|l\|_{\sup} \le (C_2 n^d)^{e_n} \|l\|_{L^2}$. Since $e_n = O(n^d)$, one has

$$\log \|l\|_{L^2} = \log \|l\|_{\sup} + O(n^d \log n). \tag{3.4.1}$$

3.5. **Proof of Theorem 3.3.3.** In this subsection, we will prove Theorem 3.3.3.

First, let us relate $\chi_{L^2}(\mathcal{X}, \overline{\mathcal{L}}^{\otimes n} \otimes \overline{\mathcal{N}})$ to $\widehat{\deg}(\widehat{c}_1(\det \mathbf{R}f_*(\mathcal{L}^{\otimes n} \otimes \mathcal{N}), h_Q))$. In §1.1, $\widehat{\deg}$ is defined for hermitian line bundles over $\operatorname{Spec}(\mathbb{Z})$. We will extend $\widehat{\deg}$ for coherent sheaves H over $\mathbb{Z}$ with a hermitian metric on $H_{\mathbb{C}}$. So, let H be a finite $\mathbb{Z}$-module of rank r. Let h a hermitian metric on $(H/H_{tor}) \otimes_{\mathbb{Z}} \mathbb{C}$, where H_{tor} is the torsion part of H. Set $\overline{H} = (H, h)$. Take a $\mathbb{Z}$-basis $e_1, \ldots, e_r$ of H/H_{tor} and define $\widehat{\deg}(\overline{H}) \in \mathbb{R}$ by

$$\log \# H_{tor} - \frac{1}{2} \log \det(h(e_i, e_j))_{i,j}.$$

Indeed, this number does not depend on the choice of $\mathbb{Z}$-basis $e_1, \ldots, e_r$. When $\overline{H} = (H, h)$ is a hermitian line bundle over $\operatorname{Spec}(\mathbb{Z})$, $\widehat{\deg}(\overline{H})$ coincides with the one in §1.1. Note that one has

$$\widehat{\deg}(\det H, \det h) = \widehat{\deg}(H, h). \tag{3.5.1}$$

Lemma 3.5.2. *Let $f : \mathcal{X} \to \operatorname{Spec}(\mathbb{Z})$ be a projective arithmetic variety such that $\mathcal{X}_{\mathbb{Q}}$ is regular. Let $\overline{\mathcal{E}}$ be a C^∞-hermitian vector bundle over $\mathcal{X}$. Let $\operatorname{vol}(B_{\operatorname{rk} H^q(\mathcal{X}, \mathcal{E})}) \in \mathbb{R}$ denote the volume of the unit ball in the Euclidean space $\mathbb{R}^{\operatorname{rk} H^q(\mathcal{X}, \mathcal{E})}$. Then, we have*

$$\begin{aligned} &\widehat{\deg}(\widehat{c}_1(\det \mathbf{R}f_*(\mathcal{E}), h_Q)) \\ &= \sum_{q \geq 0} (-1)^q \left\{ \log \# H^q(\mathcal{X}, \mathcal{E})_{tor} + \chi_{h_{L^2}}(H^q(\mathcal{X}, \mathcal{E})) + \log \operatorname{vol}(B_{\operatorname{rk} H^q(\mathcal{X}, \mathcal{E})}) \right\} + T(\overline{\mathcal{E}}_{\mathbb{C}}), \end{aligned}$$

where $T(\overline{\mathcal{E}}_{\mathbb{C}})$ is the analytic torsion of $\overline{\mathcal{E}}_{\mathbb{C}}$ (cf. Definition 2.4.1)

Proof: Recall that $\det \mathbf{R}f_*(\mathcal{E}) = \bigotimes_{q \geq 0} \det H^q(\mathcal{X}, \mathcal{E})^{(-1)^q}$. By (3.5.1), if $\{e_i^q\}_i$ is a $\mathbb{Z}$-basis of $H^q(\mathcal{X}, \mathcal{E})$, then we have

$$\begin{aligned} \widehat{\deg}(\widehat{c}_1(\det \mathbf{R}f_*(\mathcal{E}), h_Q)^{(-1)^q}) - T(\overline{\mathcal{E}}_{\mathbb{C}}) &= \widehat{\deg} \left(\widehat{c}_1 \left(\bigotimes_{q \geq 0} \det H^q(\mathcal{X}, \mathcal{E}), h_{L^2} \right)^{(-1)^q} \right) \\ &= \sum_{q \geq 0} (-1)^q \widehat{\deg} \left(H^q(\mathcal{X}, \mathcal{E}), h_{L^2} \right) \\ &= \sum_{q \geq 0} (-1)^q \{ \log \# H_{tor} \\ &\qquad - \frac{1}{2} \log \det(h_{L^2}(e_i^q, e_j^q))_{i,j} \bigg\}. \end{aligned}$$

On the other hand, we have

$$\chi_{h_{L^2}}(H^q(\mathcal{X}, \mathcal{E})) = -\frac{1}{2} \log \det(h_{L^2}(e_i^q, e_j^q))_{i,j} + \log \operatorname{vol}(B_{\operatorname{rk} H^q(\mathcal{X}, \mathcal{E})}).$$

Combining these two equalities, we get the assertion. □

Bismut and Vasserot proved the following estimate of the analytic torsion of $L^{\otimes n} \otimes N$ when n goes to infinity. We refer to [4] for its proof.

Proposition 3.5.3 ([4]). *Let X be a compact Kähler manifold of dimension d. Let $\overline{L}$ and $\overline{N}$ be C^∞-hermitian line bundles over X. Assume that $c_1(\overline{L})$ is everywhere positive. Then, one has*

$$T(L^{\otimes n} \otimes N) = O(n^d \log n).$$

Next, using the arithmetic Riemann-Roch theorem, let us relate $\widehat{\deg}(\widehat{c}_1(\det \mathbf{R}f_*(\mathcal{L}^{\otimes n} \otimes \mathcal{N}), h_Q))$ and $\dfrac{\widehat{\deg}(\widehat{c}_1(\mathcal{L})^{d+1})}{(d+1)!}n^{d+1}$. Let $f : \mathcal{X} \to \mathrm{Spec}(\mathbb{Z})$ be a projective arithmetic variety such that $\mathcal{X}_{\mathbb{Q}}$ is regular. Let $\overline{\mathcal{L}}$ and $\overline{\mathcal{N}}$ be C^∞-hermitian vector bundles over $\mathcal{X}$.

By the arithmetic Riemann-Roch theorem (Theorem 2.5.1), we have

$$\widehat{c}_1(\det \mathbf{R}f_*(\mathcal{L}^{\otimes n} \otimes \mathcal{N}), h_Q) = f_*\left(\widehat{\mathrm{ch}}(\overline{\mathcal{L}}^{\otimes n} \otimes \overline{\mathcal{N}})) \cdot \widehat{\mathrm{td}}^R(\overline{T_{X/Y}})\right)^{(1)}.$$

Since we have

$$\begin{aligned}&\left(\widehat{\mathrm{ch}}(\overline{\mathcal{L}}^{\otimes n} \otimes \overline{\mathcal{N}})) \cdot \widehat{\mathrm{td}}^R(\overline{T_{X/Y}})\right)^{(d+1)} \\ &\qquad = \left(\exp(\widehat{c}_1(\overline{\mathcal{L}}^{\otimes n} \otimes \overline{\mathcal{N}})) \cdot \widehat{\mathrm{td}}^R(\overline{T_{X/Y}})\right)^{(d+1)} = \frac{\widehat{c}_1(\overline{\mathcal{L}})^{d+1}}{(d+1)!}n^{d+1} + O(n^d),\end{aligned}$$

we obtain

$$\widehat{c}_1(\det \mathbf{R}f_*(\mathcal{L}^{\otimes n} \otimes \mathcal{N}), h_Q) = \frac{\widehat{c}_1(\overline{\mathcal{L}})^{d+1}}{(d+1)!}n^{d+1} + O(n^d). \tag{3.5.4}$$

Finally, let us prove Theorem 3.3.3. ¿From what we have seen, we find for $n \to \infty$

$$\begin{aligned}\chi_{\sup}(\mathcal{X}, \overline{\mathcal{L}}^{\otimes n} \otimes \overline{\mathcal{N}}) &= \chi_{L^2}(\mathcal{X}, \overline{\mathcal{L}}^{\otimes n} \otimes \overline{\mathcal{N}}) + O(n^d \log n) \\ &= \widehat{\deg}(\widehat{c}_1(\det \mathbf{R}f_*(\mathcal{L}^{\otimes n} \otimes \mathcal{N}), h_Q)) - T(\mathcal{L}_{\mathbb{C}}^{\otimes n} \otimes \mathcal{N}_{\mathbb{C}}) + O(n^d \log n) \\ &= \widehat{\deg}(\widehat{c}_1(\det \mathbf{R}f_*(\mathcal{L}^{\otimes n} \otimes \mathcal{N}), h_Q)) + O(n^d \log n) \\ &= \frac{\widehat{\deg}(\widehat{c}_1(\overline{\mathcal{L}})^{d+1})}{(d+1)!}n^{d+1} + O(n^d \log n).\end{aligned}$$

Indeed, the first equality follows from (3.4.1). On the other hand, since $\mathcal{L}$ is f-ample, $H^q(\mathcal{X}, \mathcal{L}^{\otimes n} \otimes \mathcal{N}) = 0$ for $n \gg 1$. Also, we have $\log \mathrm{vol}(B_{\mathrm{rk}\, H^0(\mathcal{X}, \mathcal{L}^{\otimes n} \otimes \mathcal{N})}) = O(n^d)$. Applying Lemma 3.5.2, we get the second equality. The third equality follows from the estimate of analytic torsions (Proposition 3.5.3), while the fourth follows from (3.5.4).

3.6. **Proof of the arithemtic Hilbert-Samuel theorem.** In this subsection, following [37], we will prove the arithemtic Hilbert-Samuel theorem (Theorem 3.3.1).

First, we need some lemmas.

Lemma 3.6.1. *Let $(V, \|\cdot\|)$ be a real normed space of dimension b and Γ a lattice of V. As in §3.2, let $\lambda_1(\Gamma) \leq \cdots \leq \lambda_b(\Gamma)$ be the successive minima of Γ.*

Let $\Gamma' \subset \Gamma$ be a $\mathbb{Z}$-submodule of rank b' and V' the vector space generated by Γ' over $\mathbb{R}$. We give the norm on V' induced from that on V. Then, we have

$$\chi_{\|\cdot\|}(\Gamma) - \chi_{\|\cdot\|}(\Gamma') \geq -\log(b!) - (b-b')\log\left(\frac{1}{2}\lambda_b(\Gamma)\right).$$

Proof: Let $\lambda_1(\Gamma') \leq \cdots \leq \lambda_{b'}(\Gamma')$ be the successive minima of Γ'. Then, by definition, we have $\lambda_i(\Gamma) \leq \lambda_i(\Gamma')$ $(1 \leq i \leq b')$.

On the other hand, Minkowski's convex body theorem says that

$$\frac{2^b}{b!}\operatorname{vol}_{\|\cdot\|}(\Gamma) \leq \lambda_1(\Gamma)\cdots\lambda_b(\Gamma) \leq 2^b \operatorname{vol}_{\|\cdot\|}(\Gamma).$$

Thus, we obtain

$$\begin{aligned}\operatorname{vol}_{\|\cdot\|}(\Gamma) &\leq \frac{b!}{2^b}\lambda_1(\Gamma)\cdots\lambda_b(\Gamma)\\ &\leq \frac{b!}{2^b}\lambda_1(\Gamma')\cdots\lambda_{b'}(\Gamma')\cdot\lambda_b(\Gamma)^{b-b'} \leq b!\operatorname{vol}_{\|\cdot\|}(\Gamma')\left(\frac{\lambda_b(\Gamma)}{2}\right)^{b-b'}.\end{aligned}$$

Taking $-\log$ on both sides, we get the assertion. □

Let us recall "m-regular" here. Let X be a projective variety over a field k and L a very ample line bundle over X. A coherent sheaf F on X is said to be m-regular if $H^q(X, F \otimes L^{\otimes(m-q)}) = 0$ holds for every $q > 0$. If F is m-regular, then, for any $n \geq m$, (1) F is n-regular and (2) $H^0(X, F \otimes L^{\otimes n}) \otimes H^0(X, L) \to H^0(X, F \otimes L^{\otimes(n+1)})$ is surjective (cf. [19, II §1 Proposition 1]).

Lemma 3.6.2. *Let k be a field and $\pi : Y \to X$ a morphism of projective varieties over k. Let L be an ample line bundle over X and M an ample line bundle on Y. Then,*

$$\bigoplus_{a,b\geq 0} H^0(Y, \pi^*(L)^{\otimes a} \otimes M^{\otimes b})$$

is a finitely generated k-algebra.

Proof: **Step 1:** Suppose L and M are very ample. Set $d = \dim Y$. Let b_0 be a sufficiently large integer. Then, for any j, $0 \le j \le d$ and q, $1 \le q \le d$,

$$H^q\left(Y, \pi^*(L)^{\otimes j} \otimes M^{\otimes(b_0-q)}\right) = 0. \tag{3.6.2.1}$$

This equality shows that, for fixed j, $\pi^*(L)^{\otimes j}$ is b_0-regular. Then, for any $b \ge b_0$,

$$H^q\left(Y, \pi^*(L)^{\otimes j} \otimes M^{\otimes(b-q)}\right) = 0 \quad (q > 0).$$

Thus, for any $b \ge b_0$,

$$H^q\left(X, L^{\otimes(d-q)} \otimes \pi_*(M^{\otimes b})\right) = H^q\left(Y, \pi^*(L)^{\otimes(d-q)} \otimes M^{\otimes b}\right) = 0 \quad (q > 0).$$

This equality shows that $\pi_*(M^{\otimes b})$ is d-regular. Then, by [19, II §1 Proposition 1], for any $a \ge d$, we have

$$H^0\left(X, L^{\otimes a} \otimes \pi_*(M^{\otimes b})\right) \otimes H^0(X, L) \twoheadrightarrow H^0\left(X, L^{\otimes(a+1)} \otimes \pi_*(M)^{\otimes b}\right),$$

and hence, we have

$$H^0\left(Y, \pi^*(L)^{\otimes a} \otimes M^{\otimes b}\right) \otimes H^0(Y, \pi^*(L)) \twoheadrightarrow H^0\left(X, \pi^*(L)^{\otimes(a+1)} \otimes M^{\otimes b}\right).$$

In particular, for any $a \ge d$, we have

(3.6.2.2)

$$H^0\left(Y, \pi^*(L)^{\otimes d} \otimes M^{\otimes b}\right) \otimes H^0\left(Y, \pi^*(L)^{\otimes(a-d)}\right) \twoheadrightarrow H^0\left(X, \pi^*(L)^{\otimes a} \otimes M^{\otimes b}\right).$$

On the other hand, substituting $j = d$ in (3.6.2.1), we get

$$H^q\left(Y, \pi^*(L)^{\otimes d} \otimes M^{\otimes(b_0-q)}\right) = 0 \quad (q > 0).$$

This equality shows that $\pi^*(L)^{\otimes d}$ is b_0-regular. Then, for any $b \ge b_0$,

(3.6.2.3)

$$H^0\left(Y, \pi^*(L)^{\otimes d} \otimes M^{\otimes b_0}\right) \otimes H^0\left(Y, M^{\otimes(b-b_0)}\right) \twoheadrightarrow H^0\left(X, \pi^*(L)^{\otimes d} \otimes M^{\otimes b}\right).$$

Combining (3.6.2.2) and (3.6.2.3), we find that, for any $a \ge d$ and $b \ge b_0$,

$$H^0\left(X, \pi^*(L)^{\otimes d} \otimes M^{\otimes b_0}\right) \otimes H^0\left(Y, \pi^*(L)^{\otimes(a-d)}\right) \otimes H^0\left(Y, M^{\otimes(b-b_0)}\right)$$
$$\twoheadrightarrow H^0\left(X, \pi^*(L)^{\otimes a} \otimes M^{\otimes b}\right).$$

This gives the assertion.

Step 2: Take A and B such that $L^{\otimes A}$ and $M^{\otimes B}$ are both very ample. By a similar argument in Step 1, for each $0 \le i < A$ and $0 \le j < B$,

$$\bigoplus_{a,b \ge 0} H^0\left(Y, \pi^*(L^{\otimes A})^{\otimes a} \otimes (M^{\otimes B})^{\otimes b} \otimes \pi^*(L^{\otimes i}) \otimes M^{\otimes j}\right)$$

is a finitely generated k-algebra. Then,

$$\bigoplus_{a,b\geq 0} H^0(Y, \pi^*(L)^{\otimes a} \otimes M^{\otimes b})$$
$$= \bigoplus_{0\leq i<A, 0\leq j<B} \bigoplus_{a,b\geq 0} H^0\left(Y, \pi^*(L^{\otimes A})^{\otimes a} \otimes (M^{\otimes B})^{\otimes b} \otimes \pi^*(L^{\otimes i}) \otimes M^{\otimes j}\right)$$

is also a finitely generated k-algebra. □

Let $\mathcal{X}$ be a projective arithmetic variety, and $\overline{\mathcal{L}} = (\mathcal{L}, \|\cdot\|)$ a C^∞-hermitian line bundle over $\mathcal{X}$. We give the sup-norm on $H^0(\mathcal{X}, \mathcal{L}) \otimes_{\mathbb{Z}} \mathbb{R}$. As in §3.2, let $\lambda_i(H^0(\mathcal{X}, \mathcal{L}))$ be the successive minima of $H^0(\mathcal{X}, \mathcal{L})$ for $1 \leq i \leq \operatorname{rk} H^0(\mathcal{X}, \mathcal{L})$. We denote $\lambda_{\operatorname{rk} H^0(\mathcal{X},\mathcal{L})}(H^0(\mathcal{X}, \mathcal{L}))$ by $\lambda_{\max}(H^0(\mathcal{X}, \mathcal{L})$.

Lemma 3.6.3. *Let $\pi : \mathcal{Y} \to \mathcal{X}$ be a morphism of projective arithmetic varieties. Let $\overline{\mathcal{L}}$ and $\overline{\mathcal{N}}$ be C^∞-hermitian line bundles over $\mathcal{X}$, and $\overline{\mathcal{M}}$ a C^∞-hermitian line bundle over $\mathcal{Y}$. Assume that $\overline{\mathcal{L}}_{\mathbb{Q}}$ and $\overline{\mathcal{M}}_{\mathbb{Q}}$ are ample over $\mathcal{X}_{\mathbb{Q}}$ and $\mathcal{Y}_{\mathbb{Q}}$, respectively. Then, there exists a constant C such that for any integer $a \geq 0$ and $b \geq 0$,*

$$\lambda_{\max}(H^0(\mathcal{X}, \pi^*(\mathcal{L})^{\otimes a} \otimes \mathcal{M}^{\otimes b} \otimes \overline{\mathcal{N}})) \leq C^{a+b}.$$

Proof: For simplicity, we assume $\overline{\mathcal{N}} = (\mathcal{O}_{\mathcal{X}}, |\cdot|)$. By Lemma 3.6.2,

$$\bigoplus_{a,b\geq 0} H^0(\mathcal{Y}_{\mathbb{Q}}, \pi^*(\mathcal{L}_{\mathbb{Q}})^{\otimes a} \otimes \mathcal{M}_{\mathbb{Q}}^{\otimes b})$$

is a finitely generated $\mathbb{Q}$-algebra. Moreover, from its proof, there exist positive integers a_0, b_0 and elements $s_1, \ldots, s_k \in \bigoplus_{a,b\geq 0} H^0(\mathcal{Y}_{\mathbb{Q}}, \pi^*(\mathcal{L}_{\mathbb{Q}})^{\otimes a} \otimes \mathcal{M}_{\mathbb{Q}}^{\otimes b})$ with the following properties: (1) s_i is an element of $H^0(\mathcal{Y}_{\mathbb{Q}}, \pi^*(\mathcal{L}_{\mathbb{Q}})^{\otimes a_i} \otimes \mathcal{M}_{\mathbb{Q}}^{\otimes b_i})$ for some $(a_i, b_i) \neq 0$; (2) If a, b satisfy $a \geq a_0$ or $b \geq b_0$, then $H^0(\mathcal{Y}_{\mathbb{Q}}, \pi^*(\mathcal{L}_{\mathbb{Q}})^{\otimes a} \otimes \mathcal{M}_{\mathbb{Q}}^{\otimes b})$ is generated by $s_1, \ldots, s_k$ over $\mathbb{Q}$.

Moreover, we can take s_i so that s_i is an element of $H^0(\mathcal{Y}, \pi^*(\mathcal{L}^{\otimes a}) \otimes \mathcal{M}^{\otimes b})$. Set $C' = \max_{1\leq i\leq k} \|s_i\|_{\sup}$.

For integers a, b with $a \geq a_0$ or $b \geq b_0$, consider

$$\left\{ l_\alpha = \prod_i s_i^{\alpha_i} \mid \alpha_i \geq 0, \sum_i \alpha_i a_i = a, \sum_i \alpha_i b_i = b \right\}.$$

This set generates $H^0(\mathcal{Y}_{\mathbb{Q}}, \pi^*(\mathcal{L}_{\mathbb{Q}})^{\otimes a} \otimes \mathcal{M}_{\mathbb{Q}}^{\otimes b})$ over $\mathbb{Q}$. Thus, we have

$$\lambda_{\max}(H^0(\mathcal{X}, \pi^*(\mathcal{L})^{\otimes a} \otimes \mathcal{M}^{\otimes b} \otimes \overline{\mathcal{N}})) \leq \|l_\alpha\|_{\sup} \leq C'^{\sum_i \alpha_i} \leq C'^{\sum_i \alpha_i(a_i+b_i)} = C'^{a+b}$$

Now, take $C \geq C'$ such that $\lambda_{\max}(H^0(\mathcal{X}, \pi^*(\mathcal{L})^{\otimes a} \otimes \mathcal{M}^{\otimes b} \otimes \overline{\mathcal{N}})) \leq C^{a+b}$ holds for $a < a_0$ and $b < b_0$. □

Lemma 3.6.4. *Let k be a field and $\pi : Y \to X$ a morphism of projective varieties over k. Let L be an ample line bundle over X and M an ample line bundle on Y. Let H be a line bundle over Y. Then, for any sufficiently large n, sufficiently large N, and i, $0 \le i \le N-1$, we have*

$$H^q(Y, \pi^*(L)^{\otimes(Nn+i)} \otimes M^{\otimes n} \otimes H) = 0 \quad (q > 0).$$

Proof: It suffices to show that

$$H^q(Y, \pi^*(L)^{\otimes N} \otimes M^{\otimes n} \otimes H) = 0 \quad (q > 0)$$

for sufficiently large n and N. We may assume that L and M are very ample. Set $d = \dim Y$. Take a sufficiently large n_0. Then, for any j, $0 \le j \le d$ and q, $1 \le q \le d$, we have

$$H^q\left(Y, \pi^*(L)^{\otimes j} \otimes M^{\otimes(n_0-q)} \otimes H\right) = 0.$$

Thus, $\pi^*(L)^{\otimes j}$ is n_0-regular for each j. Then, for any $n \ge n_0$,

$$H^q\left(Y, \pi^*(L)^{\otimes j} \otimes M^{\otimes(n-q)} \otimes H\right) = 0 \quad (q > 0),$$

and hence, for any $n \ge n_0$,

$$H^q\left(X, L^{\otimes(d-q)} \otimes \pi_*(M^{\otimes n} \otimes H)\right) = H^q\left(Y, \pi^*(L)^{\otimes(d-q)} \otimes M^{\otimes n} \otimes H\right) = 0 \quad (q > 0).$$

This shows that $\pi_*(M^{\otimes n} \otimes H)$ is d-regular. Thus, for any $N \ge d$,

$$H^q\left(X, L^{\otimes(N-q)} \otimes \pi_*(M^{\otimes n} \otimes H)\right) = 0 \quad (q > 0).$$

Then, for any $N \ge d$ and $n \ge n_0$,

$$H^q\left(Y, \pi^*(L)^{\otimes(N-q)} \otimes M^{\otimes n} \otimes H\right) = 0 \quad (q > 0).$$

□

Proof of Theorem 3.3.1: For simplicity, we assume $\mathcal{N} = \mathcal{O}_{\mathcal{X}}$.

(i) We fix a generic resolution of singularities of $\mathcal{X}$, which we denote by $\pi : \widetilde{\mathcal{X}} \to \mathcal{X}$ (cf. §1.6). Also, we fix a C^∞-hermitian line bundle $\overline{\mathcal{M}}$ over $\widetilde{\mathcal{X}}$ such that $\mathcal{M}$ is very ample and the first Chern form $c_1(\overline{\mathcal{M}}_{\mathbb{C}})$ is everywhere positive. Moreover, we fix a nonzero section s_1 of $H^0(\widetilde{\mathcal{X}}, \mathcal{M})$. Let C_1' be the sup-norm of s_1. Set $C_1 = \max(C_1', 1)$.

(ii) We fix a sufficiently large a such that

$$H^0\left(\widetilde{\mathcal{X}}_{\mathbb{Q}}, \pi^*(\mathcal{L}_{\mathbb{Q}})^{\otimes a} \otimes \mathcal{M}_{\mathbb{Q}}^{-1}\right) = H^0\left(\mathcal{X}_{\mathbb{Q}}, \mathcal{L}_{\mathbb{Q}}^{\otimes a} \otimes \pi_*(\mathcal{M}_{\mathbb{Q}}^{-1})\right) \neq 0.$$

Since

$$H^0\left(\widetilde{\mathcal{X}}_{\mathbb{Q}}, \pi^*(\mathcal{L}_{\mathbb{Q}})^{\otimes a} \otimes \mathcal{M}_{\mathbb{Q}}^{-1}\right) = H^0\left(\widetilde{\mathcal{X}}, \pi^*(\mathcal{L})^{\otimes a} \otimes \mathcal{M}^{-1}\right) \otimes_{\mathbb{Z}} \mathbb{Q},$$

there exists a nonzero section s_2 of $H^0\left(\widetilde{\mathcal{X}}, \pi^*(\mathcal{L})^{\otimes a} \otimes \mathcal{M}^{-1}\right)$. Let C_2' be the sup-norm of s_2. Set $C_2 = \max(C_2', 1)$.

(iii) For any $x \in \mathcal{X}_{\mathbb{Q}}(\mathbb{C})$ and a function ϕ defined over $\pi^{-1}(x)$, we set $\|\phi\| = \sup_{y\in\pi^{-1}(y)} |\phi(y)|$. Then, the coherent sheaf $\pi_*(\mathcal{O}_{\widetilde{\mathcal{X}}})$ over $\mathcal{X}$ is equipped with norm. Hence, the coherent sheaf $\mathcal{F} := \mathcal{H}om_{\mathcal{O}_{\mathcal{X}}}\left(\pi_*(\mathcal{O}_{\widetilde{\mathcal{X}}}), \mathcal{O}_{\mathcal{X}}\right)$ is also equipped with norm. We fix an integer $b \gg 1$ with $H^0\left(\mathcal{F}_{\mathbb{Q}} \otimes \mathcal{L}_{\mathbb{Q}}^b\right) \neq 0$. Since $H^0\left(\mathcal{X}, \mathcal{F}_{\mathbb{Q}} \otimes \mathcal{L}_{\mathbb{Q}}^b\right) = H^0\left(\mathcal{X}, \mathcal{F} \otimes \mathcal{L}^b\right) \otimes_{\mathbb{Z}} \mathbb{Q}$, there exists a nonzero section s_3 of $H^0\left(\mathcal{X}, \mathcal{F} \otimes \mathcal{L}^b\right)$. Let C_3' be the sup-norm of s_3. Set $C_3 = \max(C_3', 1)$.

(iv) Let C_4' be the constant in Lemma 3.6.3 for $\pi : \widetilde{\mathcal{X}} \to \mathcal{X}$ and $(\overline{\mathcal{L}}, \overline{\mathcal{M}})$. We set $C_4 = \max(C_4', 2)$.

First, we sketch the proof. In the rest of this subsection, we write additively the tensor product of line bundles. (For example, we write $\mathcal{L} + \mathcal{M}$ for $\mathcal{L} \otimes \mathcal{M}$.) The theorem is proven from the following: For any positive number ϵ,

$$\text{(3.6.4.1)} \qquad \frac{1}{n^{d+1}}\left|\chi_{\text{sup}}\left(\mathcal{X}, n\overline{\mathcal{L}}\right) - \frac{\widehat{\deg}(\widehat{c}_1(\overline{\mathcal{L}})^{d+1})}{(d+1)!}n^{d+1}\right| \leq \epsilon \qquad (n \to \infty).$$

In order to show (3.6.4.1), we take sufficiently large N depending on ϵ, and consider $\pi^*(\mathcal{L}) + \frac{1}{N}\mathcal{M}$. Roughly speaking, one can show (3.6.4.1) by comparing $\chi_{\text{sup}}\left(\widetilde{\mathcal{X}}, n(\pi^*(\overline{\mathcal{L}}) + \frac{1}{N}\overline{\mathcal{M}})\right)$ and $\chi_{\text{sup}}\left(\mathcal{X}, n\overline{\mathcal{L}}\right)$ for $n \to \infty$, and by using Theorem 3.3.3 to estimate $\chi_{\text{sup}}\left(\widetilde{\mathcal{X}}, n(\pi^*(\overline{\mathcal{L}}) + \frac{1}{N}\overline{\mathcal{M}})\right)$.

More precisely, since $\pi^*(\mathcal{L}) + \frac{1}{N}\mathcal{M}$ is a $\mathbb{Q}$-line bundle, we multiply it by N, i.e., we compare $\chi_{\text{sup}}\left(\mathcal{X}, (nN+i)\overline{\mathcal{L}}\right)$ and $\chi_{\text{sup}}\left(\widetilde{\mathcal{X}}, n(N\pi^*(\overline{\mathcal{L}}) + \overline{\mathcal{M}}) + \pi^*(i\overline{\mathcal{L}})\right)$ for i $(0 \leq i \leq N-1)$.

We remark that, in the following, N is fixed (according to ϵ) and only n goes to ∞.

Let N, n be positive integers, and i an integer with $0 \leq i \leq N-1$. Assume $N > a + b$.

Step 1: Multiplying by s_1^n, one has a morphism

$$\alpha : \Gamma_1 := H^0\left(\mathcal{X}, (Nn+i)\mathcal{L}\right) \longrightarrow \Gamma_2 := H^0\left(\widetilde{\mathcal{X}}, (Nn+i)\pi^*(\mathcal{L}) + n\mathcal{M}\right).$$

Put $V_1 = \Gamma_1 \otimes_{\mathbb{Z}} \mathbb{R}$ and $V_2 = \Gamma_2 \otimes_{\mathbb{Z}} \mathbb{R}$. Then, V_1 and V_2 become normed spaces with the norm induced by the sup-norm of $\overline{\mathcal{L}}$ and $\overline{\mathcal{M}}$. The norm of α is bounded from above by C_1^n.

Step 2: Multiplying by s_2^n, one has a morphism

$$H^0\left(\widetilde{\mathcal{X}}, \{(N-a-b)n+i\}\pi^*(\mathcal{L}) + n\mathcal{M}\right) \longrightarrow H^0\left(\widetilde{\mathcal{X}}, \{(N-b)n+i\}\pi^*(\mathcal{L})\right).$$

Also, multiplying by s_2^n, one has a morphism

$$H^0\left(\widetilde{\mathcal{X}}, \{(N-b)n+i\}\pi^*(\mathcal{L})\right) = H^0\left(\mathcal{X}, \{(N-b)n+i\}\mathcal{L} + \pi_*(\mathcal{O}_{\mathcal{X}})\right) \longrightarrow H^0\left(\mathcal{X}, (Nn+i)\mathcal{L}\right).$$

Let β be the composite of these two morphisms:

$$\beta : \Gamma_3 = H^0\left(\widetilde{\mathcal{X}}, \{(N-a-b)n+i\}\pi^*(\mathcal{L}) + n\mathcal{M}\right) \longrightarrow \Gamma_1 = H^0\left(\mathcal{X}, (Nn+i)\mathcal{L}\right).$$

Put $V_3 = \Gamma_3 \otimes_{\mathbb{Z}} \mathbb{R}$. Then, V_3 becomes a normed space with the norm induced by the sup-norm of $\overline{\mathcal{L}}$ and $\overline{\mathcal{M}}$. The norm of β is bounded from above by $C_2^n C_3^n$.

Step 3: We will bound $\chi_{\sup}(\Gamma_1)$ from above. We equip $\alpha(V_1) \subset V_2$ with the induced norm. Since α is injective, we have

$$\chi_{\sup}(\Gamma_1) \leq \chi_{\sup}(\alpha(\Gamma_1)) + n\,\mathrm{rk}(\Gamma_1)\log C_1.$$

Applying Lemma 3.6.1 to $(V, \Gamma) := (V_2, \Gamma_2)$ and $(V', \Gamma') := (\alpha(V_1), \alpha(\Gamma_1))$, we have

$$\chi_{\sup}(\alpha(\Gamma_1)) \leq \chi_{\sup}(\Gamma_2) + \log\left(\mathrm{rk}(\Gamma_2)!\right) + \left(\mathrm{rk}(\Gamma_2) - \mathrm{rk}(\Gamma_1)\right)(Nn+N+n)\log\left(\frac{1}{2}C_4\right).$$

Let $\widetilde{f} : \widetilde{\mathcal{X}} \to \mathrm{Spec}(\mathbb{Z})$ be the structural morphism. Since $(Nn+i)\pi^*(\mathcal{L}) + n\mathcal{M}$ is $\widetilde{f}$-ample and $c_1((Nn+i)\pi^*(\overline{\mathcal{L}}) + n\overline{\mathcal{M}})$ is everywhere positive, we can apply Theorem 3.3.3 to $(Nn+i)\pi^*(\mathcal{L}) + n\mathcal{M}$ and we get

$$\begin{aligned}\chi_{\sup}(\Gamma_2) &= \frac{n^{d+1}}{(d+1)!}\widehat{\deg}\left(\widehat{c}_1(N\pi^*\overline{\mathcal{L}} + \overline{\mathcal{M}})^{d+1}\right) + o_N(n^{d+1}) \\ &= \frac{(Nn+i)^{d+1}}{(d+1)!}\widehat{\deg}(\widehat{c}_1(\overline{\mathcal{L}})^{d+1}) + O(N^d n^{d+1}) + o_N(n^{d+1}).\end{aligned}$$

Here, $O(N^d n^{d+1})$ means that $\left|\frac{O(N^d n^{d+1})}{N^d n^{d+1}}\right|$ is bounded from above by a constant that is independent of N, n, i. And $o_N(n^{d+1})$ means that, if we fix N, then $\left|\frac{o_N(n^{d+1})}{n^{d+1}}\right|$ converges to 0 as n goes to infinity.

On the other hand, by Lemma 3.6.4, for any $n \gg 1$, $N \gg 1$, and $0 \leq i \leq N-1$, we have $H^q(Y, \pi^*(L)^{\otimes(Nn+i)} \otimes M^{\otimes n} \otimes H) = 0 \quad (\forall q > 0)$. Then, by the Riemann-Roch theorem for algebraic varieties, we find

$$\begin{aligned}\mathrm{rk}(\Gamma_2) &= \dim_{\mathbb{Q}}(V_2) \\ &= \frac{n^d}{d!}\deg\left(((Nn+i)\pi^*(\mathcal{L}_{\mathbb{C}}) + \mathcal{M}_{\mathbb{C}})^d\right) + o_N(n^{d-1}) \\ &= \frac{(Nn+i)^d}{d!}\deg(\mathcal{L}_{\mathbb{C}}^d) + O(N^{d-1}n^d) + o_N(n^{d-1}),\end{aligned}$$

and

$$\begin{aligned}\operatorname{rk}(\Gamma_1) &= \dim_{\mathbb{Q}}(V_1) \\ &= \frac{n^d}{d!}\deg((N\mathcal{L}_{\mathbb{C}})^d) + o_N(n^{d-1}) \\ &= \frac{(Nn+i)^d}{d!}\deg((\mathcal{L}_{\mathbb{C}})^d) + O(N^{d-1}n^d) + o_N(n^{d-1}).\end{aligned}$$

Therefore, we get

$$n\operatorname{rk}(\Gamma_1)\log C_1 = O(N^d n^{d+1}) + o_N(n^d)$$
$$\log(\operatorname{rk}(\Gamma_2)!) = O(N^d n^d)\log(O(N^d n^d)) = o_N(n^{d+1})$$
$$(\operatorname{rk}(\Gamma_2) - \operatorname{rk}(\Gamma_1))(Nn+N+n)\log\frac{C_4}{2} = O(N^d n^{d+1}) + o_N(n^d).$$

Combining these estimates, we obtain

$$\chi_{\sup}(\Gamma_1) \leq \frac{(Nn+i)^{d+1}}{(d+1)!}\widehat{\deg}(\widehat{c}_1(\overline{\mathcal{L}})^{d+1}) + O(N^d n^{d+1}) + o_N(n^{d+1}).$$

Step 4: We will bound $\chi_{\sup}(\Gamma_1)$ from below. We equip $\beta(V_3) \subset V_1$ with the induced norm. Since β is injective, we have

$$\chi_{\sup}(\Gamma_3) \leq \chi_{\sup}(\beta(\Gamma_3)) + n\operatorname{rk}(\Gamma_3)\log C_2C_3$$

Applying Lemma 3.6.1 to $(V,\Gamma) := (V_1,\Gamma_1)$ and $(V',\Gamma') := (\beta(V_3),\beta(\Gamma_3))$, we have

$$\chi_{\sup}(\Gamma_1) \geq \chi_{\sup}(\beta(\Gamma_3)) - \log(\operatorname{rk}(\Gamma_1)!) - (\operatorname{rk}(\Gamma_1) - \operatorname{rk}(\Gamma_3))(Nn+N+n)\log\left(\frac{1}{2}C_4\right).$$

Thus we find

$$\begin{aligned}\chi_{\sup}(\Gamma_1) \geq \chi_{\sup}(\Gamma_3) &- n\operatorname{rk}(\Gamma_3)\log C_2C_3 - \log(\operatorname{rk}(\Gamma_1)!) \\ &+ (\operatorname{rk}(\Gamma_1) - \operatorname{rk}(\Gamma_3))(Nn+N+n)\log\left(\frac{1}{2}C_4\right).\end{aligned}$$

By a similar argument as in Step 3, it follows from Theorem 3.3.3 that

$$\chi_{\sup}(\Gamma_3) = \frac{(Nn+i)^{d+1}}{(d+1)!}\widehat{\deg}(\widehat{c}_1(\overline{\mathcal{L}})^{d+1}) + O(N^d n^{d+1}) + o_N(n^{d+1}).$$

For $\operatorname{rk}(\Gamma_3)$, we have a similar estimate as for $\operatorname{rk}(\Gamma_1)$ and $\operatorname{rk}(\Gamma_2)$: For any $n \gg 1$, $N \gg 1$, and $0 \leq i \leq N-1$, we have

$$\operatorname{rk}(\Gamma_3) = \frac{(Nn+i)^d}{d!}\deg((\mathcal{L}_{\mathbb{C}})^d) + O(N^{d-1}n^d) + o_N(n^{d-1}).$$

Combining these estimates, we obtain

$$\chi_{\sup}(\Gamma_1) \geq \frac{(Nn+i)^{d+1}}{(d+1)!}\widehat{\deg}(\widehat{c}_1(\overline{\mathcal{L}})^{d+1}) + O(N^d n^{d+1}) + o_N(n^{d+1}).$$

Step 5 Recall that $\Gamma_1 = H^0\left(\widetilde{\mathcal{X}}, (Nn+i)\mathcal{L}\right)$. By Steps 3 and 4, we have

$$\chi_{\sup}\left(\mathcal{X}, (Nn+i)\overline{\mathcal{L}}\right) = \frac{(Nn+i)^{d+1}}{(d+1)!}\widehat{\deg}(\widehat{c}_1(\overline{\mathcal{L}})^{d+1}) + O(N^d n^{d+1}) + o_N(n^{d+1}).$$

Let ϵ be any positive number. First, we take a large N such that $O(N^d n^{d+1})$ is bounded from above by $\frac{\epsilon}{2}N^{d+1}n^{d+1}$. Next, we take a large n_1 such that, for any $n \geq n_1$, $o_N(n^{d+1})$ is bounded from above by $\frac{\epsilon}{2}N^{d+1}n^{d+1}$. Then, for any $n \geq Nn_1$, we have

$$\left|\chi_{\sup}\left(\mathcal{X}, n\overline{\mathcal{L}}\right) - \frac{\widehat{\deg}(\widehat{c}_1(\overline{\mathcal{L}})^{d+1})}{(d+1)!}n^{d+1}\right| \leq \epsilon n^{d+1}.$$

Hence, finally, we get Theorem 3.3.1. □

4. Adelic metric and admissible metric

4.1. **Adelic metric and intersection number.** Let X be a projective variety over an algebraically closed field F. A *metric* on a line bundle L on X is a collection of F-norms on the fibers $L(x)$ at all F-valued points x.

Let us consider the case where F is an algebraically closed non-archimedian valuation field. Let A denote its valuation ring. For a projective variety X over F and a line bundle L on X, let $(\mathcal{X}, \mathcal{L})$ be its model over A, that is, $\mathcal{X}$ is an integral projective scheme flat over A with the generic fiber X, and $\mathcal{L}$ is a line bundle on $\mathcal{X}$ with $\mathcal{L}^{\otimes n}|_X = L^{\otimes n}$ for some positive integer n. Then we can put a metric $\|\cdot\|_{\mathcal{L}}$ on L as follows. For each $x \in X(F)$, let $\tilde{x}$ denote the corresponding section of $\mathcal{X} \to \operatorname{Spec}(A)$. Note that we have $x^*L^{\otimes n} = \tilde{x}^*(\mathcal{L}^{\otimes n}) \otimes_A F$. We define *the metric* $\|\cdot\|_{\mathcal{L}}$ *associated with a model* $(\mathcal{X}, \mathcal{L})$ by putting

$$\|l\|_{\mathcal{L}} := \inf_{a \in F^\times}\{|a|^{-\frac{1}{n}} \mid al^n \in \tilde{x}^*(\mathcal{L}^{\otimes n})\}$$

for each $l \in x^*L = L(x)$. It is not difficult to see that it is independent of the choice of n. We say that an F-metric $\|\cdot\|$ on L is *continuous and bounded* if there exist a model $(\mathcal{X}, \mathcal{L})$ of (X, L) such that

$$\log\frac{\|\cdot\|}{\|\cdot\|_{\mathcal{L}}} : X(F) \to \mathbb{R}$$

is continuous and bounded in F-topology.

Let K be a number field and O_K the ring of integers of K. Let $K(\mathbb{C})$ be the set of embeddings of K into $\mathbb{C}$ and put

$$M_K := (\operatorname{Spec}(O_K) \setminus \{(0)\}) \coprod K(\mathbb{C}).$$

For each $v \in M_K$, we define a valuation $|\cdot|_v$ on K by

$$|a|_v := \begin{cases} (\#O_K/P)^{-\operatorname{ord}_P(a)} & \text{if } v = P \in \operatorname{Spec}(O_K) \setminus \{(0)\} \\ |\sigma(a)|_{\mathbb{C}} & \text{if } v = \sigma \in K(\mathbb{C}), \end{cases}$$

where $|*|_{\mathbb{C}}$ is the absolute value of $*$. Let K_v be the completion of K with respect to $|\cdot|_v$ and let $\overline{K}_v$ be its algebraic closure. A metric $\|\cdot\|$ of a line bundle L on a projective variety X over K is just a collection $\{\|\cdot\|_v\}_{v \in M_K}$ of $\overline{K}_v$-metrics $\|\cdot\|_v$ on $L \otimes_K \overline{K}_v$ for all $v \in M_K$.

Now let X be a projective variety over K and L a line bundle on X.

Definition 4.1.1. Let U be a non-empty open subscheme of $\operatorname{Spec}(O_K)$. A *model over* U of (X, L) is a pair of an integral projective scheme $\mathcal{X}_U$ flat over U and a $\mathbb{Q}$-line bundle $\bar{\mathcal{L}}$ on $\mathcal{X}_U$ with a continuous hermitian metric at the infinite places with the following properties.

(a) X is the generic fiber of $\mathcal{X}_U$.
(b) There exists a positive integer n such that $\mathcal{L}^{\otimes n}$ is a line bundle and $\mathcal{L}^{\otimes n}|_X = L^{\otimes n}$.

The following kind of metrics, called *adelic metric*, is important for our purpose.

Definition 4.1.2. A metric $\|\cdot\| = \{\|\cdot\|_v\}_{v \in M_K}$ on L is called an *adelic metric* if it has the following properties.

(a) $\|\cdot\|_P$ is continuous and bounded for any $P \in \operatorname{Spec}(O_K) \setminus \{(0)\}$ and $\|\cdot\|_\sigma$ is continuous for any $\sigma \in K(\mathbb{C})$.
(b) There exist a non-empty open subscheme $U \subset \operatorname{Spec}(O_K)$ and a model $(\mathcal{X}_U, \mathcal{L})$ of (X, L) over U such that $\mathcal{L}$ is a line bundle with $\mathcal{L}|_X = L$ and that for any $P \in U \setminus \{(0)\}$, the metric $\|\cdot\|_P$ is the one associated with the model

$$(\mathcal{X}_P, \mathcal{L}_P) = (\mathcal{X} \times \operatorname{Spec}(\overline{O}_{K,P}), \mathcal{L} \otimes \overline{O}_{K,P}),$$

where $\overline{O}_{K,P}$ is the ring of integers of $\overline{K}_P$.

If (X, L) has a global model $(\mathcal{X}, \bar{\mathcal{L}})$, i.e., a model over $\operatorname{Spec} O_K$, then the metric on L associated with this model is an adelic metric. In fact, it obviously has the property (a). To see (b), let U be a non-empty open subset of $\operatorname{Spec}(O_K)$ such that there exists a line bundle $\mathcal{M}$ on $\mathcal{X}_U$ with $\mathcal{M}|_X = L$ and that, for some n, the generic canonical isomorphism $\mathcal{M}^{\otimes n}|_X \cong \mathcal{L}^{\otimes n}|_X$ extends to an isomorphism of line bundles over $\mathcal{X}_U$. Then, we can check (b) using this U and $\mathcal{M}$.

Definition 4.1.3. Let $\|\cdot\|$ be an adelic metric on L and let $\{\|\cdot\|_n\}_{n=1,2,\ldots}$ be a sequence of adelic metrics on L. We say $\{\|\cdot\|_n\}_{n=1,2,\ldots}$ *converges to* $\|\cdot\|$ if there exists a non-empty open subscheme $U \subset \operatorname{Spec} O_K$ such that $\|\cdot\|_{n,P} = \|\cdot\|_P$ for any

$P \in U \setminus \{(0)\}$ and

$$\log \frac{\|\cdot\|_{n,v}}{\|\cdot\|_v} : X(\overline{K}_v) \to \mathbb{R}$$

converges uniformly to 0 for any $v \in M_K$.

Note that an adelic metric approximated by C^∞-models does not necessarily have a C^∞-metric at an infinite place.

Definition 4.1.4. Let $\|\cdot\|$ be an adelic metric on L. The metric $\|\cdot\|$ is said to be *vertically ample* (*resp. vertically nef*) if there exists a sequence of vertically ample (resp. vertically nef) C^∞-models $\{(\mathcal{X}_n, \bar{\mathcal{L}}_n)\}_{n=1,2,\ldots}$ over $\mathrm{Spec}(O_K)$ such that $\|\cdot\|_{\bar{\mathcal{L}}_n}$ converges to $\|\cdot\|$ (cf. §3.3.2). An adelic metric $\|\cdot\|$ is said to be *integrable* if it satisfies the following conditions

(1) $\|\cdot\|$ is approximated by a sequence of C^∞-models $\{(\mathcal{X}_n, \bar{\mathcal{L}}_n)\}_{n=1,2,\ldots}$ over $\mathrm{Spec}(O_K)$.
(2) For each n, there exist vertically nef hermitian $\mathbb{Q}$-line bundles $\bar{\mathcal{A}}_n$ and $\bar{\mathcal{B}}_n$ such that $\bar{\mathcal{L}}_n = \bar{\mathcal{A}}_n \otimes \bar{\mathcal{B}}_n^{-1}$.
(3) The set $\{(\mathcal{A}_n)_K, (\mathcal{B}_n)_K\}$ is bounded in $\mathrm{NS}(X_K) \otimes \mathbb{R}$.

A vertically nef adelic metric is integrable for example.

Let X be a projective variety over a number field K of dimension $d-1$ $(d > 0)$ and let $\bar{L}_1, \ldots, \bar{L}_d$ line bundles with a integrable adelic metric on X. Let $\{(\mathcal{X}_{i,n}, \bar{\mathcal{L}}_{i,n})\}_{n=1,2,\ldots}$, $\bar{\mathcal{A}}_{i,n}$ and $\bar{\mathcal{B}}_{i,n}$ be those in Definition 4.1.4 for $\bar{L}_1$. For an $(n_1, \ldots, n_d) \in \mathbb{N}^d$, let $\mathcal{X}_{n_1,\ldots,n_d}$ be a model of X dominating any $\mathcal{X}_{i,n_i}$ $(i = 1, 2, \ldots, d)$. We denote the pullback of $\bar{\mathcal{L}}_{i,n_i}$ to $\mathcal{X}_{n_1,\ldots,n_d}$ by the same symbol $\bar{\mathcal{L}}_{i,n_i}$.

Theorem 4.1.5. *Under the above situation, we have the following.*

(1) $c_{n_1,\ldots,n_d} := \widehat{\deg}(\widehat{c}_1(\bar{\mathcal{L}}_{1,n_1}) \cdots \widehat{c}_1(\bar{\mathcal{L}}_{d,n_d}))$ *converges as* $\min\{n_1, \ldots, n_d\} \to \infty$, *and the limit does not depend on the choice of sequences of models. We denote it by* $c_1(\bar{L}_1) \cdots c_1(\bar{L}_d)$.
(2) $c_1(\bar{L}_1) \cdots c_1(\bar{L}_d)$ *is multilinear with respect to* $\bar{L}_1, \ldots, \bar{L}_d$.

Proof: Let us take two d-tuples $(n_1, \ldots, n_d), (n'_1, \ldots, n'_d) \in \mathbb{N}^d$. Let $\mathcal{X}$ be a model dominating two models $\mathcal{X}_{n_1,\ldots,n_d}$ and $\mathcal{X}_{n'_1,\ldots,n'_d}$. Let $\bar{\mathcal{L}}_i$, $\bar{\mathcal{A}}_i$ and $\bar{\mathcal{B}}_i$ be the pull-backs of $\bar{\mathcal{L}}_{i,n_i}$ $\bar{\mathcal{A}}_{i,n_i}$ and $\bar{\mathcal{B}}_{i,n_i}$ to $\mathcal{X}$ respectively, and let $\bar{\mathcal{L}}'_i$, $\bar{\mathcal{A}}'_i$ and $\bar{\mathcal{B}}'_i$ be the pull-backs of $\bar{\mathcal{L}}_{i,n'_i}$, $\bar{\mathcal{A}}_{i,n'_i}$ and $\bar{\mathcal{B}}_{i,n'_i}$ to $\mathcal{X}$ respectively. Then $(\mathcal{X}, \bar{\mathcal{L}}_i)$ and $(\mathcal{X}, \bar{\mathcal{L}}'_i)$ are models of (X, L_i) and induce the same metrics L_i on as $\bar{\mathcal{L}}_{i,n_i}$ and $\bar{\mathcal{L}}_{i,n'_i}$ respectively. We shall prove

$$|\widehat{\deg}(\widehat{c}_1(\bar{\mathcal{L}}_1) \cdots \widehat{c}_1(\bar{\mathcal{L}}_d)) - \widehat{\deg}(\widehat{c}_1(\bar{\mathcal{L}}'_1) \cdots \widehat{c}_1(\bar{\mathcal{L}}'_d))| \to 0$$

as $\min\{n_1, \ldots, n_d, n'_1, \ldots, n'_d\} \to \infty$.

By the definition of convergence of the models, we can see the following.

(1) There exists a non-empty open subset U of $\operatorname{Spec}(O_K)$ such that $\|\cdot\|_{\mathcal{L}_{i,n},P} = \|\cdot\|_{\mathcal{L}_{i,n'},P}$ for any $i = 1, \dots, d$, for any n and n', and for any finite place $P \in U$.

(2) For any positive number ϵ, if n_k and n'_k are sufficiently large, we have

$$\left|\log \frac{\|\cdot\|_{\bar{\mathcal{L}}'_{k,v}}}{\|\cdot\|_{\bar{\mathcal{L}}_{k,v}}}\right| < \begin{cases} \epsilon \log \#(O_K/P) & \text{if } v \text{ is a finite place } P, \\ \epsilon & \text{if } v \text{ is an infinite place,} \end{cases}$$

for any $v \in M_K$.

Note that there are only finitely many places at which $\log \frac{\|\cdot\|_{\bar{\mathcal{L}}'_{k,v}}}{\|\cdot\|_{\bar{\mathcal{L}}_{k,v}}}$ survive. (They are outside of U.) Let e_k be a positive integer such that $\mathcal{L}_k^{\otimes e_k}$ and ${\mathcal{L}'_k}^{\otimes e_k}$ are line bundles and coincide with $L_k^{\otimes e_k}$ on the generic fiber X. Then $\left({\mathcal{L}'_k}^{\otimes e_k}\right)\Big|_X = \mathcal{O}_X$. Let s_k be a rational section of $\mathcal{L}_k^{\otimes e_k} \otimes \left({\mathcal{L}'_k}^{\otimes e_k}\right)^{-1}$ of which restriction to X is 1. Our goal is to estimate

$$I_k := \widehat{\deg}\Big(\widehat{c}_1(\bar{\mathcal{L}}'_1) \cdots \widehat{c}_1(\bar{\mathcal{L}}'_{k-1}) \widehat{c}_1(\bar{\mathcal{L}}_{k+1}) \cdots \widehat{c}_1(\bar{\mathcal{L}}_d) \cdot \frac{1}{e_k} \widehat{\operatorname{div}}(s_k)\Big),$$

namely,

$$\begin{aligned} &\left|\widehat{\deg}\big(\widehat{c}_1(\bar{\mathcal{L}}_1) \cdots \widehat{c}_1(\bar{\mathcal{L}}_d)\big) - \widehat{\deg}\big(\widehat{c}_1(\bar{\mathcal{L}}'_1) \cdots \widehat{c}_1(\bar{\mathcal{L}}'_d)\big)\right| \\ \le &\sum_{k=1}^{d} \left|\widehat{\deg}\big(\widehat{c}_1(\bar{\mathcal{L}}'_1) \cdots \widehat{c}_1(\bar{\mathcal{L}}'_{k-1}) \widehat{c}_1(\bar{\mathcal{L}}_k \otimes (\bar{\mathcal{L}}'_k)^{-1}) \widehat{c}_1(\bar{\mathcal{L}}_{k+1}) \cdots \widehat{c}_1(\bar{\mathcal{L}}_d)\big)\right| = \sum_{k=1}^{d} |I_k|. \end{aligned}$$

For each $\boldsymbol{a} = (a_1, \dots, a_{k-1}, a_{k+1}, \dots, a_d) \in \{0,1\}^{d-1}$, we define hermitian $\mathbb{Q}$-line bundles $\bar{\mathcal{M}}_i(\boldsymbol{a})$'s as follows: for i with $1 \le i \le k-1$,

$$\bar{\mathcal{M}}_i(\boldsymbol{a}) := \begin{cases} \bar{A}'_i & \text{if } a_i = 0, \\ \bar{B}'_i & \text{if } a_i = 1, \end{cases}$$

and for i with $k+1 \le i \le d$,

$$\bar{\mathcal{M}}_i(\boldsymbol{a}) := \begin{cases} \bar{A}_i & \text{if } a_i = 0, \\ \bar{B}_i & \text{if } a_i = 1. \end{cases}$$

Note that each $\bar{\mathcal{M}}_i(\boldsymbol{a})$ is vertically nef. Now we put

$$\bar{J}_k(\boldsymbol{a}) := \widehat{\deg}\Big(\widehat{c}_1(\bar{\mathcal{M}}_1(\boldsymbol{a})) \cdots \widehat{c}_1(\bar{\mathcal{M}}_{k-1}(\boldsymbol{a})) \widehat{c}_1(\bar{\mathcal{M}}_{k+1}(\boldsymbol{a})) \cdots \widehat{c}_1(\bar{\mathcal{M}}_d(\boldsymbol{a})) \cdot \frac{1}{e_k} \widehat{\operatorname{div}}(s_k)\Big).$$

It is straightforward that

$$\sum_{k=1}^{d} |I_k| \le \sum_{k=1}^{d} \sum_{\boldsymbol{a} \in \{0,1\}^{d-1}} |\bar{J}_k(\boldsymbol{a})|.$$

We shall examine how much each place contributes to $\bar{J}_k(\boldsymbol{a})$.

If P is a finite place, then

$$\#(O_K/P)^{-\epsilon e_k} < \|s_k(x)\|_P < \#(O_K/P)^{\epsilon e_k}$$

for any $x \in X(\overline{K})$. By the definition of the metric associated with a model, this implies both $D_{P,1} := [\mathrm{div}(s_k)]_P + \epsilon e_k[\mathcal{X}_P]$ and $D_{P,2} := -[\mathrm{div}(s_k)]_P + \epsilon e_k[\mathcal{X}_P]$ are effective divisors, where $[\mathcal{X}_P]$ is the divisor of fiber over P and $[\mathrm{div}(s_k)]_P$ is the part of $\mathrm{div}(s_k)$ supported in $\mathcal{X}_P$. Since each $\bar{\mathcal{M}}_i(\boldsymbol{a})$ is vertically nef, we have

$$c_1(\mathcal{M}_1(\boldsymbol{a})|_{D_{P,j}})\cdots c_1(\mathcal{M}_{k-1}(\boldsymbol{a})|_{D_{P,j}})c_1(\mathcal{M}_{k+1}(\boldsymbol{a})|_{D_{P,j}})\cdots c_1(\mathcal{M}_d(\boldsymbol{a})|_{D_{P,j}}) \geq 0$$

for $j = 1, 2$. This implies that the absolute value of the contribution of P to $\bar{J}_k(\boldsymbol{a})$ is not greater than

$$\epsilon\big(\log \#(O_K/P)\big)c_1(M_1(\boldsymbol{a}))\cdots c_1(M_{k-1}(\boldsymbol{a}))c_1(M_{k+1}(\boldsymbol{a}))\cdots c_1(M_d(\boldsymbol{a})),$$

where $M_j(\boldsymbol{a})$ is the restriction of $\mathcal{M}_1(\boldsymbol{a})$ to the generic fiber. By the assumption that $\{(\mathcal{A}_{i,n})_K, (\mathcal{B}_{i,n})_K\}_{i=1,\ldots,d, n\in\mathbb{N}}$ is bounded in $\mathrm{NS}(X_K)\otimes\mathbb{R}$, there exists a positive number C independent of the choice of $(n_1, \ldots, n_d)$ or $(n'_1, \ldots, n'_d)$ such that

$$c_1(M_1(\boldsymbol{a}))\cdots c_1(M_{k-1}(\boldsymbol{a}))c_1(M_{k+1}(\boldsymbol{a}))\cdots c_1(M_d(\boldsymbol{a})) \leq C.$$

Consequently, the absolute value of the contribution of a finite place P ($\notin U$) to $\bar{J}_k(\boldsymbol{a})$ is not beyond

$$\epsilon \log \#(O_K/P)C.$$

Next let us consider the contribution of infinite places. The contribution of an infinite place σ to $\bar{J}_k(\boldsymbol{a})$ is

$$\frac{1}{e_k}\int \log \|s_k\|_\sigma c_1'(\bar{\mathcal{M}}_{1,\sigma}(\boldsymbol{a}))\cdots c_1'(\bar{\mathcal{M}}_{k-1,\sigma}(\boldsymbol{a}))c_1'(\bar{\mathcal{M}}_{k+1,\sigma}(\boldsymbol{a}))\cdots c_1'(\bar{\mathcal{M}}_{d,\sigma}(\boldsymbol{a})),$$

where $c_1'(\cdot)$ denotes the curvature form. Taking account that all the curvature forms which appear above are positive semi-definite and that $|\log\|s_k\|_\sigma| < \epsilon e_k$, we can similarly observe that its absolute value is bounded with

$$c_1(M_1(\boldsymbol{a}))\cdots c_1(M_{k-1}(\boldsymbol{a}))c_1(M_{k+1}(\boldsymbol{a}))\cdots c_1(M_d(\boldsymbol{a}))$$

and we conclude that it is not greater than ϵC.

Accordingly, we have

$$\bar{J}_k(\boldsymbol{a}) \leq \epsilon C\left(\sum_{P\notin U} \log \#(O_K/P) + [K:\mathbb{Q}]\right),$$

and thus we obtain

$$\begin{aligned}&\left|\widehat{\deg}(\widehat{c}_1(\bar{\mathcal{L}}_1)\cdots\widehat{c}_1(\bar{\mathcal{L}}_d)) - \widehat{\deg}(\widehat{c}_1(\bar{\mathcal{L}}'_1)\cdots\widehat{c}_1(\bar{\mathcal{L}}'_d))\right| \\ &\qquad \leq \epsilon 2^{d-1}dC\left(\sum_{P\notin U} \log \#(O_K/P) + [K:\mathbb{Q}]\right),\end{aligned}$$

which implies the convergence.

Let c denote this value. If $\{(\mathcal{X}'_{i,n}, \bar{\mathcal{L}}'_{i,n})\}_n$ is another sequence of models, then we can obtain another c'. If we consider the third one

$$(\mathcal{X}_{i,1}, \bar{\mathcal{L}}_{i,1}), (\mathcal{X}'_{i,1}, \bar{\mathcal{L}}'_{i,1}), (\mathcal{X}_{i,2}, \bar{\mathcal{L}}_{i,2}), (\mathcal{X}'_{i,2}, \bar{\mathcal{L}}'_{i,2}), (\mathcal{X}_{i,3}, \bar{\mathcal{L}}_{i,3}), \ldots,$$

we have the third c''. We must have $c = c''$ and $c' = c''$, hence we conclude $c = c'$, which completes the proof of (1). (2) is immediate from the definition. □

4.2. **Admissible metric and cubic metric.** Let X be a projective variety over an algebraically closed field F and $\bar{L} = (L, \|\cdot\|)$ be a line bundle on X with a continuous and bounded metric. Suppose a surjective endomorphism $f : X \to X$ and an isomorphism $\phi : L^{\otimes d} \xrightarrow{\sim} f^*L$ $(d > 1)$ are given. We define a metric $\|\cdot\|_n$ on L as follows:

$$\|\cdot\|_1 = \|\cdot\|, \quad \|\cdot\|_n = \left(\phi^* f^* \|\cdot\|_{n-1}\right)^{\frac{1}{d}}.$$

Then we have the following theorem.

Theorem 4.2.1. (1) $\{\|\cdot\|_n\}_{n=1,2,\ldots}$ *converges uniformly to a metric, say* $\|\cdot\|_0$.

(2) $\|\cdot\|_0$ *is the only continuous and bounded metric with* $\|\cdot\|_0 = \left(\phi^* f^* \|\cdot\|_0\right)^{1/d}$.

(3) *If we replace* ϕ *by* $\lambda\phi$, $\|\cdot\|_0$ *is replaced by* $|\lambda|^{1/(d-1)}\|\cdot\|_0$.

Proof: (1) Put

$$h := \log \frac{\|\cdot\|_2}{\|\cdot\|_1}.$$

Then we have

$$\log \frac{\|\cdot\|_n}{\|\cdot\|_{n-1}} = \left(\frac{1}{d}\phi^* f^*\right)^{n-2} (h)$$

and hence

$$\log \frac{\|\cdot\|_n}{\|\cdot\|_1} = \sum_{k=0}^{n-2} \left(\frac{1}{d}\phi^* f^*\right)^{k} (h).$$

Since $\|\left((1/d)\phi^* f^*\right)^k (h)\|_{\sup} = (1/d^k)\|h\|_{\sup}$, a sequence $\log \frac{\|\cdot\|_n}{\|\cdot\|_1}$ converges uniformly to a function h_0 on $X(F)$. Therefore, $\|\cdot\|_n$ converges uniformly to $\|\cdot\|_0 := e^{h_0}\|\cdot\|_1$.

(2) ¿From the construction, we see $\|\cdot\|_0$ is a required continuous and bounded metric. Let $\|\cdot\|'_0$ be another continuous and bounded metric satisfying the conditions. Then $g := \log \frac{\|\cdot\|_0}{\|\cdot\|'_0}$ is a continuous and bounded function on $X(F)$ and satisfies $\left((1/d)\phi^* f^*\right)(g) = g$. Therefore $\|g\|_{\sup} = (1/d)\|g\|_{\sup}$, and since $d > 1$, we have $\|g\|_{\sup} = 0$.

(3) Let $\alpha\|\cdot\|_0$ be the metric satisfying the condition (1) with respect to $\lambda\phi$. Let $\|\cdot\|_{0,f}$ denote the metric $f^*\|\cdot\|_0$ on f^*L. Then for any $x \in X(F)$ and a local section l around x, we have

$$\begin{aligned}\alpha\|l(x)\|_0 &= \left(\alpha\|\lambda\phi(l^d)(x)\|_{0,f}\right)^{\frac{1}{d}}\\ &= \left(\alpha|\lambda|\right)^{\frac{1}{d}}\left(\|\phi(l^d)(x)\|_{0,f}\right)^{\frac{1}{d}}\\ &= \left(\alpha|\lambda|\right)^{\frac{1}{d}}\|l(x)\|_0,\end{aligned}$$

which implies $\alpha = \left(\alpha|\lambda|\right)^{1/d}$. Thus we have the conclusion. □

Let X be a projective variety over a number field K and L an ample line bundle on X. Suppose that we are given a surjective morphism $f : X \to X$ and isomorphism $\phi : L^{\otimes d} \xrightarrow{\sim} f^*L$ $(d > 1)$. Since L is ample, f is a finite morphism of degree $d^{\dim X}$. Fix a model $(\mathcal{X}, \bar{\mathcal{L}})$ over $\operatorname{Spec} O_K$ of (X, L) such that $\bar{\mathcal{L}}$ is vertically nef. Let $\|\cdot\|$ be the metric on L associated with $(\mathcal{X}, \bar{\mathcal{L}})$, and let e be a natural number such that $\mathcal{L}^{\otimes e}$ is a line bundle and $\mathcal{L}^{\otimes e}|_X = L^{\otimes e}$. Then, there exist a non-empty open subscheme $U \subset \operatorname{Spec} O_K$, a surjective morphism $f_U : \mathcal{X}_U \to \mathcal{X}_U$ and an isomorphism $\phi_U^{\otimes e} : (\mathcal{L}^{\otimes e})^{\otimes d} \to f_U^*(\mathcal{L}^{\otimes e})$. By virtue of the above theorem, we obtain for each place v a metric $\|\cdot\|_{v,0}$ with $\|\cdot\|_{v,0} = \left(\phi^* f^*\|\cdot\|_{v,0}\right)^{1/d}$. On the other hand, we can see from the construction that $\|\cdot\|_P = \left(\phi^* f^*\|\cdot\|_P\right)^{1/d}$ for each $P \in U \setminus \{(0)\}$. Therefore, the sequence of adelic metrics defined by $\|\cdot\|_n = \left(\phi^* f^*\|\cdot\|_{n-1}\right)^{1/d}$ converges to an adelic metric $\|\cdot\|_0$ with $\|\cdot\|_0 = \left(\phi^* f^*\|\cdot\|_0\right)^{1/d}$. We claim that this metric $\|\cdot\|_0$ is vertically nef. Indeed, let $\tilde{f}_n : \mathcal{X}_n \to \mathcal{X}$ be the normalization of the composite morphism

$$\mathcal{X}_U \xrightarrow{f_U^n} \mathcal{X}_U \hookrightarrow \mathcal{X}$$

and put $\bar{\mathcal{L}}_n := \tilde{f}_n^*\bar{\mathcal{L}}$. Then the metric $\|\cdot\|_n$ is nothing more than the metric associated with the model $\left(\mathcal{X}_n, \bar{\mathcal{L}}_n^{\otimes(1/d^n)}\right)$, and $\bar{\mathcal{L}}_n^{\otimes(1/d^n)}$ is vertically nef. Thus, each $\|\cdot\|_n$ is vertically nef and so is $(L, \|\cdot\|_0)$.

If X is an abelian variety, then we have the n-times morphism. Using this, let us define the cubic metric (up to constant) on any line bundle as follows.

Let A be an abelian variety and let L be a line bundle on A. We define morphisms $s_{1,2,3}$, $s_{1,2}$, $s_{2,3}$, $s_{3,1}$ and s_i $(i = 1, 2, 3)$ from $A \times A \times A$ to A by

$$\begin{aligned}s_{1,2,3}(x_1, x_2, x_3) &= x_1 + x_2 + x_3,\\ s_{i,j}(x_1, x_2, x_3) &= x_i + x_j,\\ s_i(x_1, x_2, x_3) &= x_i.\end{aligned}$$

Now we put

$$\operatorname{Cub}(L) := s_{1,2,3}^*(L)\otimes s_{1,2}^*(L^{\otimes(-1)})\otimes s_{2,3}^*(L^{\otimes(-1)})\otimes s_{3,1}^*(L^{\otimes(-1)})\otimes s_1^*(L)\otimes s_2^*(L)\otimes s_3^*(L).$$

Then, the cubic lemma gives us an isomorphism $\phi_L : \mathcal{O}_{A\times A\times A} \to \mathrm{Cub}(L)$. If L is equipped with a metric $\|\cdot\|$, we can canonically endow a metric $\mathrm{Cub}(\bar{L}) = \mathrm{Cub}((L, \|\cdot\|))$ on $\mathrm{Cub}(L)$ by pulling back the metric, and we further have a metric $\|\cdot\|_{\phi_L}$ on $\mathcal{O}_{A\times A\times A}$ via ϕ_L.

Definition 4.2.2. Let $|\cdot|_{\mathrm{can}}$ be the canonical metric on $\mathcal{O}_{A\times A\times A}$. We call a continuous and bounded metric $\|\cdot\|$ on L a *cubic metric* if there exists a positive number c such that $\|\cdot\|_{\phi_L} = c|\cdot|_{\mathrm{can}}$.

Note that it is independent of the choice of ϕ_L. This notion appears in [22] in case of archimedean valuation.

We have a basic result on cubic metrics.

Proposition 4.2.3. *Let L be a line bundle on an abelian variety A.*

(1) *There exists a cubic metric on L.*
(2) *If $\|\cdot\|_1$ and $\|\cdot\|_2$ are cubic metrics on L, then there exists a positive number c such that $\|\cdot\|_1 = c\|\cdot\|_2$.*

Proof: (1) First we suppose L is even, i.e., $[-1]^*L \cong L$. Fix an isomorphism $\psi : [2]^*L \to L^{\otimes 4}$. Let $\|\cdot\|$ be the admissible metric with respect to $[2] : A \to A$ and ψ. Then, we see

$$\begin{aligned}
[2]^*\mathrm{Cub}(\bar{L}) &= [2]^*\big(s_{1,2,3}^*(\bar{L}) \otimes s_{1,2}^*(\bar{L}^{\otimes -1}) \otimes s_{2,3}^*(\bar{L}^{\otimes -1}) \otimes s_{3,1}^*(\bar{L}^{\otimes -1}) \\
&\quad \otimes s_1^*(\bar{L}) \otimes s_2^*(\bar{L}) \otimes s_3^*(\bar{L})\big) \\
&= s_{1,2,3}^*[2]^*(\bar{L}) \otimes s_{1,2}^*[2]^*(\bar{L}^{\otimes -1}) \otimes s_{2,3}^*[2]^*(\bar{L}^{\otimes -1}) \otimes s_{3,1}^*[2]^*(\bar{L}^{\otimes -1}) \\
&\quad \otimes s_1^*[2]^*(\bar{L}) \otimes s_2^*[2]^*(\bar{L}) \otimes s_3^*[2]^*(\bar{L})\big) \\
&\cong s_{1,2,3}^*(\bar{L}^{\otimes 4}) \otimes s_{1,2}^*(\bar{L}^{-4}) \otimes s_{2,3}^*(\bar{L}^{-4}) \otimes s_{3,1}^*(\bar{L}^{-4}) \\
&\quad \otimes s_1^*(\bar{L}^{\otimes 4}) \otimes s_2^*(\bar{L}^{\otimes 4}) \otimes s_3^*(\bar{L}^{\otimes 4})\big) \\
&= \mathrm{Cub}(\bar{L})^{\otimes 4}.
\end{aligned}$$

This implies the metric of $\mathrm{Cub}(\bar{L})$ is the admissible metric with respect to

$$[2] : A \times A \times A \to A \times A \times A$$

and

$$s_{1,2,3}^*(\psi)\otimes s_{1,2}^*(\psi')\otimes s_{2,3}^*(\psi')\otimes s_{3,1}^*(\psi')\otimes s_1^*(\psi)\otimes s_2^*(\psi)\otimes s_3^*(\psi) : [2]^*\mathrm{Cub}(\bar{L}) \to \mathrm{Cub}(\bar{L})^{\otimes 4},$$

where ψ' is the canonical isomorphism $\psi' : [2]^*L^{-1} \to L^{\otimes -4}$ induced by ψ. The metric of $\mathrm{Cub}(\bar{L})$ is also regarded as an admissible metric on $\mathcal{O}_{A\times A\times A}$ via ϕ_L, which must be $c|\cdot|_{\mathrm{can}}$. Thus we see that $\|\cdot\|$ is a cubic metric on L.

If L is odd, i.e., $[-1]^*L \cong L^{-1}$, then we can construct a cubic metric in the same way using $[2]^*L \cong L^2$.

For a general M, we can find an even line bundle M_1 and an odd one M_2 such that $M = M_1 \otimes M_2$. Taking account that $\mathrm{Cub}(\bar{M_1}) \otimes \mathrm{Cub}(\bar{M_2}) = \mathrm{Cub}(\bar{M_1} \otimes \bar{M_2})$, we can easily see that the metric on M defined as a product of a cubic metric on M_1 and that on M_2 is a cubic metric.

(2) It is enough to show that if the metric of $\mathrm{Cub}(\mathcal{O}_A, \|\cdot\|)$ is the canonical metric up to constant, then so is $\|\cdot\|$. For a global section 1 of $\mathcal{O}_A$, we define a function $\lambda : A(F) \longrightarrow \mathbb{R}$ by $x \mapsto \log \|1\|(x)$. Since $\|\cdot\|$ is continuous and bounded, so is λ. By the assumption that the metric of $\mathrm{Cub}(\mathcal{O}_A, \|\cdot\|)$ is the canonical one up to constant, we have

$$\lambda(x+y+z) + \lambda(x) + \lambda(y) + \lambda(z) - \lambda(x+y) - \lambda(y+z) - \lambda(z+x) = \lambda(0),$$

and hence, the map from $A \times A$ to $\mathbb{R}$ defined by

$$(x, y) \mapsto \lambda(x+y) - \lambda(x) - \lambda(y) + \lambda(0)$$

comes out to be $\mathbb{Z}$-bilinear. Taking account that it is also continuous and bounded, we must have

$$\lambda(x+y) - \lambda(x) - \lambda(y) + \lambda(0) \equiv 1.$$

This implies that the function

$$x \mapsto \lambda(x) - \lambda(0)$$

is $\mathbb{Z}$-linear, and by continuity and boundedness again, we have $\lambda(x) \equiv \lambda(0)$ and hence $\|\cdot\| = \lambda(0)|\cdot|_{\mathrm{can}}$. □

Note that the above proof of (1) tells us that the admissible metric with respect to a given isomorphism $\psi : [n]^*L \cong L^{\otimes n}$ is a cubic metric.

Finally, let us consider the case of $F = \mathbb{C}$. Let A be an abelian variety over $\mathbb{C}$. Let $\pi : \mathbb{C}^g \longrightarrow A$ be the universal covering of A compatible with the natural group law. Let us fix the coordinates $z_1, \ldots, z_g$ of $\mathbb{C}^g$. Then the 1-forms $dz_1, \ldots, dz_g, d\bar{z}_1, \ldots, d\bar{z}_g$ are naturally regarded as 1-forms on A. If h is a C^∞-hermitian metric on a line bundle L, there exist a matrix $(c_{i,j})_{i,j=1,\ldots,g}$ with $c_{i,j} = \bar{c}_{i,j}$, and a 1-form η such that

$$c_1(L, h) = \sum_{i,j} c_{i,j}\sqrt{-1}dz_i \wedge d\bar{z}_j + d\eta.$$

By dd^c-lemma, there exists an $\mathbb{R}$-valued C^∞-function f such that

$$c_1(L, h) = \sum_{i,j} c_{i,j}\sqrt{-1}dz_i \wedge d\bar{z}_j + dd^c f.$$

Hence if we put $h_1 = e^f h$, we have

$$c_1(L, h_1) = \sum_{i,j} c_{i,j}\sqrt{-1}dz_i \wedge d\bar{z}_j.$$

It is straightforward to see that h_1 is a cubic metric. Moreover, we see any cubic metric is necessarily C^∞-class, and if L is ample, then the curvature form of its cubic metric is positive definite. In particular, if L is in addition even, the admissible metric with respect to an isomorphism $[2]^*L \cong L^{\otimes 4}$ is C^∞-class and its curvature form is positive definite.

5. Arithmetic height function

5.1. Definition of arithmetic height functions and their properties. First of all, let us recall the definition of height functions.

Let K be a number field and O_K the ring of integers of K. We denote by $K(\mathbb{C})$ the set of all embeddings K into $\mathbb{C}$. Let $\mathcal{L}$ be a line bundle on $\operatorname{Spec}(O_K)$, i.e., $\mathcal{L}$ is a projective and finitely generated O_K-module of rank 1. For $\sigma \in K(\mathbb{C})$, the tensor product $\mathcal{L} \otimes_{O_K} \mathbb{C}$ in terms of the embedding $\sigma : O_K \to \mathbb{C}$ is denoted by $\mathcal{L}_\sigma$. We give a hermitian metric $\|\cdot\|_\sigma$ of $\mathcal{L}_\sigma$ for each $\sigma \in K(\mathbb{C})$. The collection $\{\|\cdot\|_\sigma\}_{\sigma \in K(\mathbb{C})}$ is called a hermitian metric of $\mathcal{L}$ and is simply denoted by $\|\cdot\|$. Moreover, a pair $(\mathcal{L}, \|\cdot\|)$ of a line bundle $\mathcal{L}$ and a hermitian metric $\|\cdot\|$ of $\mathcal{L}$ is called a hermitian line bundle on $\operatorname{Spec}(O_K)$ and is often denoted by $\overline{\mathcal{L}}$. For $s \in \mathcal{L} \setminus \{0\}$, let us consider the following:

$$\log \#(\mathcal{L}/s\mathcal{L}) - \sum_{\sigma \in K(\mathbb{C})} \log \|s\|_\sigma \in \mathbb{R}.$$

Then, by virtue of the product formula, the above number does not depend on the choice of s, so that it is denoted by $\widehat{\deg}(\overline{\mathcal{L}})$.

Let $\mathcal{X}$ be an integral projective scheme over $\operatorname{Spec}(O_K)$. For $x \in \mathcal{X}_K(\overline{K})$, let Δ_x be the closure of the image $\operatorname{Spec}(\overline{K}) \to \mathcal{X}_K \to \mathcal{X}$. Let $\overline{\mathcal{L}}$ be a continuous hermitian line bundle on $\mathcal{X}$. Namely, for each $\sigma \in K(\mathbb{C})$, we endow a continuous hermitian metric to the line bundle $\mathcal{L}_\sigma = \mathcal{L} \otimes \mathbb{C}$ on $\mathcal{X}_\sigma = \mathcal{X} \otimes \mathbb{C}$ as before. Then, we set

$$h_{(\mathcal{X},\overline{\mathcal{L}})}(x) = \frac{\widehat{\deg}\left((\mathcal{X},\overline{\mathcal{L}})|_{\Delta_x} \right)}{[K(x):K]}.$$

More precisely, letting $\widetilde{\Delta}_x$ be the normalization of Δ_x, $\widehat{\deg}\left((\mathcal{X},\overline{\mathcal{L}})|_{\Delta_x} \right)$ is given by $\widehat{\deg}\left((\mathcal{X},\overline{\mathcal{L}})|_{\widetilde{\Delta}_x} \right)$. We call this the height of x with respect to $(\mathcal{X},\overline{\mathcal{L}})$ (cf. §1.6).

Here we consider the following proposition. For simplicity, we set $X = \mathcal{X}_K$ and $L = \mathcal{L}_K$.

Proposition 5.1.1. *Let* $\operatorname{Bs}(L)$ *be the base locus of* L*, i.e.,* $\operatorname{Bs}(L) = \operatorname{Supp}\left(\operatorname{Coker}(H^0(X,L) \otimes \mathcal{O}_X \to L)\right)$. *Then, there is a constant* C *such that* $h_{(\mathcal{X},\overline{\mathcal{L}})}(x) \geq C$ *for all* $x \in (X \setminus \operatorname{Bs}(L))(\overline{K})$.

Proof: Let $\{s_1, \ldots, s_l\}$ be a system of generators of $H^0(\mathcal{X},\mathcal{L})$ as O_K-module. We set

$$a = \max_{\substack{1 \leq i \leq l \\ \sigma \in K(\mathbb{C})}} \{\sup \|s_i\|_\sigma\}.$$

For $x \in (X \setminus \mathrm{Bs}(L))(\overline{K})$, we can find a generator s_i with $s_i(x) \neq 0$. This means that

$$\widehat{\deg}\left((\mathcal{L}, (1/a)\|\cdot\|)|_{\Delta_P} \right) \geq 0.$$

Therefore, we can see

$$h_{(\mathcal{X},\overline{\mathcal{L}})}(x) \geq -[K:\mathbb{Q}]\log(a).$$

□

Corollary 5.1.2. *If $L \simeq \mathcal{O}_X$, then there is a constant C such that $|h_{(\mathcal{X},\overline{\mathcal{L}})}(x)| \leq C$ for all $x \in X(\overline{K})$.*

Proof: Apply Proposition 5.1.1 to $\overline{\mathcal{L}}$ and $\overline{\mathcal{L}}^{\otimes -1}$. □

Recall that a model of (X, L) is a pair $(\mathcal{X}, \overline{\mathcal{L}})$ such that $\mathcal{X}$ is an integral projective scheme flat over $\mathrm{Spec}(\mathcal{O}_K)$ with the generic fiber X, and $\overline{\mathcal{L}}$ is a continuous hermitian line bundle on $\mathcal{X}$ with $\mathcal{L}^{\otimes n}|_X = L^{\otimes n}$ for some positive integer n (cf. §4.1).

Corollary 5.1.3. *Let $(\mathcal{X}, \overline{\mathcal{L}})$ and $(\mathcal{X}', \overline{\mathcal{L}}')$ be two models of (X, L). Then, there is a constant C such that*

$$|h_{(\mathcal{X},\overline{\mathcal{L}})}(x) - h_{(\mathcal{X}',\overline{\mathcal{L}}')}(x)| \leq C$$

for all $x \in X(\overline{K})$.

Proof: Note that $\mathcal{X}$ is birational to $\mathcal{X}'$. Let $\mathcal{X}''$ be the graph of the birational correspondence $\mathcal{X} \dashrightarrow \mathcal{X}'$ in $\mathcal{X} \times \mathcal{X}'$. Let $\mu : \mathcal{X}'' \to \mathcal{X}$ and $\mu' : \mathcal{X}'' \to \mathcal{X}'$ be the natural morphisms. Then, it is easy to see

$$h_{(\mathcal{X}'',\mu^*(\overline{\mathcal{L}}))} = h_{(\mathcal{X},\overline{\mathcal{L}})} \quad \text{and} \quad h_{(\mathcal{X}'',(\mu')^*(\overline{\mathcal{L}}'))} = h_{(\mathcal{X}',\overline{\mathcal{L}}')}.$$

On the other hand, $\mu^*(\overline{\mathcal{L}})$ coincides with $(\mu')^*(\overline{\mathcal{L}}')$ on the generic fiber $\mathcal{X}'' \to \mathrm{Spec}(O_K)$. Thus, Corollary 5.1.2, there is a constant C such that

$$|h_{(\mathcal{X}'',\mu^*(\overline{\mathcal{L}}))} - h_{(\mathcal{X}'',(\mu')^*(\overline{\mathcal{L}}'))}| \leq C.$$

Hence we get our corollary. □

Let us denote by $Fun(X(\overline{K}), \mathbb{R})$ (resp. $BFun(X(\overline{K}), \mathbb{R})$) the set of all (resp. bounded) real valued functions on $X(\overline{K})$. By the above corollary, the class of $h_{(\mathcal{X},\overline{\mathcal{L}})}$ in

$$Fun(X(\overline{K}), \mathbb{R})/BFun(X(\overline{K}), \mathbb{R})$$

does not depend on the choice of the model $(\mathcal{X}, \overline{\mathcal{L}})$ of (X, L), so that the class of $h_{(\mathcal{X},\overline{\mathcal{L}})}$ is denoted by $h_{(X,L)}$ or h_L. The following Northcott's theorem is very important (for details, see Lang [21] or Serre [31]).

Theorem 5.1.4. *If L is ample, then the set*

$$\{x \in X(K) \mid [K(x) : K] \leq d, \quad h_L(x) \leq M\}$$

is finite for all M and d. Here by abuse of notation, h_L is a representative of the class h_L.

5.2. **Height functions on abelian varieties.** Let K be a number field, A an abelian variety over K, and L a line bundle on A. Let h_L be a height function associated to (X, L). Let us consider a problem to find a good representative of the class of h_L. By virtue of the cubic theorem and Corollary 5.1.2, we can see that

$$h_L(x+y+z) - h_L(x+y) - h_L(y+z) - h_L(z+x) + h_L(x) + h_L(y) + h_L(z)$$

is a bounded function on $A(\overline{K}) \times A(\overline{K}) \times A(\overline{K})$. The following lemma is well known (for example, see [21]).

Lemma 5.2.1. *Let G be an abelian group, and $f : G \to \mathbb{R}$ a function on G. If*

$$f(x+y+z) - f(x+y) - f(y+z) - f(z+x) + f(x) + f(y) + f(z)$$

is a bounded function on G, then there exists a quadric form q and a linear form l uniquely such that

$$f(x) = q(x) + l(x) + O(1).$$

By the above lemma, there are a quadric form q_L and a linear form l_L on $A(\overline{K})$ such that

$$h_L \equiv q_L + l_L \qquad (\text{mod } BFun(A(\overline{K}), \mathbb{R})).$$

$q_L + l_L$ is denoted by $\hat{h}_L$, and is called the *canonical height function* (or *Nérol Tate height function*). If L is symmetric (i.e., $[-1]^*(L) = L$), then $l_L = 0$. Thus, in this case, $\hat{h}_L$ is a quadric form. Finally, let us consider the following proposition.

Proposition 5.2.2. *We assume that L is symmetric and ample. Then we have the following.*

(1) *For all $x \in A(\overline{K})$, $\hat{h}_L(x) \geq 0$.*
(2) *For $x \in A(\overline{K})$, $\hat{h}_L(x) = 0$ if and only if $x \in A(\overline{K})_{tor}$.*

Proof: (1) By the definition of height functions, $\hat{h}_{L^{\otimes n}} = n\hat{h}_L$. Thus, we may assume that L is generated by global sections. Therefore, by Proposition 5.1.1, there is a constant C such that $\hat{h}_L(x) \geq C$ for all $x \in A(\overline{K})$. Here, since $\hat{h}_L$ is a quadric form, for all $n > 0$,

$$n^2 \hat{h}_L(x) = \hat{h}_L(nx) \geq C.$$

Therefore, taking $n \to \infty$, we obtain $\hat{h}_L(x) \geq 0$.

(2) Since $\hat{h}_L$ is a quadric form, it is obvious that if $x \in A(\overline{K})_{tor}$, then $\hat{h}_L(x) = 0$. Conversely, we assume that $\hat{h}_L(x) = 0$. Let us consider a subgroup $\langle x \rangle$ generated

by x. Note that if $x \in A(K')$ for a finite extension field K', then $\langle x \rangle \subseteq A(K')$. Moreover, $\hat{h}_L(nx) = n^2 \hat{h}_L(x) = 0$ for all n. Thus, by Northcott's theorem (cf. Theorem 5.1.4), $\langle x \rangle$ is a finite group. Therefore, $x \in A(\overline{K})_{tor}$. □

5.3. **Adelic metric and height function.** Let X be a projective variety over K, and L a line bundle. Let $\| \cdot \|$ be an adelic metric of L. We assume that the adelic metric $(X, \overline{L})$ is given by the limit of a sequence of continuous models $\{(\mathcal{X}_n, \overline{\mathcal{L}}_n)\}$. Here $\mathcal{X}_n$ is a projective arithmetic variety over $\mathrm{Spec}(O_K)$ whose generic fiber is X. Moreover, $\overline{\mathcal{L}}_n$ is a continuous hermitian $\mathbb{Q}$-line bundle on $\mathcal{X}_n$ which gives rise to L on the generic fiber. Then, we have the following proposition.

Proposition 5.3.1. *For any positive ϵ, there is a positive number N such that, for all $n, m \geq N$ and all $x \in X(\overline{K})$,*

$$|h_{(\mathcal{X}_n, \overline{\mathcal{L}}_n)}(x) - h_{(\mathcal{X}_m, \overline{\mathcal{L}}_m)}(x)| \leq \epsilon.$$

In particular, If we set $h_{(X, \overline{L})}(x) = \lim_{n \to \infty} h_{(\mathcal{X}_n, \overline{\mathcal{L}}_n)}(x)$, then

$$\lim_{n \to \infty} \sup_{x \in X(\overline{K})} |h_{(\mathcal{X}_n, \overline{\mathcal{L}}_n)}(x) - h_{(X, \overline{L})}(x)| = 0.$$

Proof: Its proof is essentially the same as that of Theorem 4.1.5. For reader's convenience, we dare to give it.

First of all, there is a non-empty Zariski open set U of $\mathrm{Spec}(O_K)$ such that all $(\mathcal{X}_n, \mathcal{L}_n)$ are same over U. We set $S = \mathrm{Spec}(O_K) \setminus U$. For $P \in S$, the metric over P induced by the model $(\mathcal{X}_n, \mathcal{L}_n)$ is denoted by $\| \cdot \|_{P,n}$. Since $\#(S) < \infty$, for any positive number ϵ, there is N such that, for all $n, m \geq N$ and all $P \in S$,

$$\sup \left| \log \left(\frac{\| \cdot \|_{P,n}}{\| \cdot \|_{P,m}} \right) \right| \leq \epsilon \log \#(O_K/P).$$

Moreover, taking a larger N if necessarily, we may assume that for all $n, m \geq N$ and all $\sigma \in K(\mathbb{C})$,

$$\sup \left| \log \left(\frac{\| \cdot \|_{\sigma,n}}{\| \cdot \|_{\sigma,m}} \right) \right| \leq \epsilon.$$

For n, m with $n, m \geq N$, let $\mathcal{Y}$ be the graph of the birational correspondence $\mathcal{X}_n \dashrightarrow \mathcal{X}_m$ in $\mathcal{X}_n \times \mathcal{X}_m$. Moreover, let $\pi_n : \mathcal{Y} \to \mathcal{X}_n$ and $\pi_m : \mathcal{Y} \to \mathcal{X}_m$ be the natural morphisms. Let s be a rational section of $\pi_n^*(\mathcal{L}_n) \otimes \pi_m^*(\mathcal{L}_m^{-1})$ such that s gives rise to 1 over U. Then, for each $P \in S$ and each $\sigma \in K(\mathbb{C})$, we have

$$\sup |\log \|s\|_P| \leq \epsilon \log \#(O_K/P) \quad \text{and} \quad \sup |\log \|s\|_\sigma| \leq \epsilon.$$

¿From the above observation, we can easily see that the contribution to evaluate

$$\widehat{\deg} \left(\pi_n^*(\overline{\mathcal{L}}_n) \otimes \pi_m^*(\overline{\mathcal{L}}_m^{-1}) \Big|_{\Delta_x} \right)$$

over P is bounded by $\epsilon[K(x):K]\log\#(O_K/P)$, which is the intersection number of Δ_x and ϵ(the fiber of $\mathcal{Y}$ over P). Moreover, the contribution over σ is bounded by $\epsilon[K(x):K]$. Therefore,

$$\left|\widehat{\deg}\left(\pi_n^*(\overline{\mathcal{L}}_n)\otimes\pi_m^*(\overline{\mathcal{L}}_m^{-1})\Big|_{\Delta_x}\right)\right| \leq \epsilon[K(x):K]\left([K:\mathbb{Q}]+\sum_{P\in S}\log\#(O_K/P)\right).$$

Thus, if we set $B=[K:\mathbb{Q}]+\sum_{P\in S}\log\#(O_K/P)$, then

$$|h_{(\mathcal{X}_n,\overline{\mathcal{L}}_n)}(x)-h_{(\mathcal{X}_m,\overline{\mathcal{L}}_m)}(x)|\leq\epsilon B.$$

Therefore, we get our proposition. □

5.4. **Intersection number of nef C^∞-hermitian line bundles.** Let $\mathcal{X}$ be a projective arithmetic variety with $d=\dim\mathcal{X}_{\mathbb{Q}}$. Let $\overline{\mathcal{L}}$ be a C^∞-hermitian line bundle on $\mathcal{X}$. As in §3.3, we say $\overline{\mathcal{L}}$ is vertically nef if $\mathcal{L}$ is relatively nef with respect to $X\to\operatorname{Spec}(\mathbb{Z})$ and the Chern form $c_1(\overline{\mathcal{L}})$ is semi-positive. Moreover, $\overline{\mathcal{L}}$ is said to be horizontally nef if for any 1-dimensional subscheme C flat over $\operatorname{Spec}(\mathbb{Z})$, $\widehat{\deg}\left(\overline{\mathcal{L}}|_C\right)\geq 0$. Further, we say $\overline{\mathcal{L}}$ is nef if $\overline{\mathcal{L}}$ is horizontally and vertically nef. Then, we have the following theorem.

Theorem 5.4.1. *Let $\overline{\mathcal{L}}_1,\ldots,\overline{\mathcal{L}}_{d+1}$ be nef C^∞-hermitian line bundles on $\mathcal{X}$. Then*

$$\widehat{\deg}\left(\widehat{c}_1(\overline{\mathcal{L}}_1)\cdots\widehat{c}_1(\overline{\mathcal{L}}_{d+1})\right)\geq 0.$$

Proof: Let us begin with the following lemma.

Lemma 5.4.2. *Let $\overline{\mathcal{L}}$ be a nef C^∞-hermitian line bundle on $\mathcal{X}$, and $\overline{\mathcal{M}}$ a vertically nef C^∞-hermitian line bundle such that for all integral subscheme Γ flat over $\mathbb{Z}$,*

$$\widehat{\deg}\left(\widehat{c}_1(\overline{\mathcal{M}}|_\Gamma)^{\dim(\Gamma_{\mathbb{Q}})+1}\right)>0.$$

Then, for all $0\leq i\leq d+1$,

$$\widehat{\deg}\left(\widehat{c}_1(\overline{\mathcal{L}})^i\cdot\widehat{c}_1(\overline{\mathcal{M}})^{d+1-i}\right)\geq 0$$

Proof: We prove it by induction of d. If $d=0$, the our assertion is obvious. Here we assume that $d>0$. First, let us consider the case where $0\leq i<d+1$. Since $\widehat{\deg}\left(\widehat{c}_1(\overline{\mathcal{M}})^{d+1}\right)>0$, replacing $\overline{\mathcal{M}}$ by $\overline{\mathcal{M}}^{\otimes n}$ $(n>0)$, we may assume that there is a small section s of $\mathcal{M}$, namely, $s\in H^0(\mathcal{X},\mathcal{M})\setminus\{0\}$ with $\|s\|_{\sup}<1$. Let $\operatorname{div}(s)=a_1\Gamma_1+\cdots+a_e\Gamma_e$ be a decomposition as a cycle. Here $a_j>0$. Then,

$$\widehat{\deg}\left(\widehat{c}_1(\overline{\mathcal{L}})^i\cdot\widehat{c}_1(\overline{\mathcal{M}})^{d+1-i}\right)=\sum_j a_j\widehat{\deg}\left(\widehat{c}_1(\overline{\mathcal{L}}|_{\Gamma_j})^i\cdot\widehat{c}_1(\overline{\mathcal{M}}|_{\Gamma_j})^{d-i}\right)$$
$$-\int_{\mathcal{X}(\mathbb{C})}\log(\|s\|)c_1(\overline{\mathcal{L}})^i\cdot\widehat{c}_1(\overline{\mathcal{M}})^{d-i}.$$

Thus it is positive by the hypothesis of induction.

Next let us consider the case where $i = d+1$. For a rational number t, we set $\overline{\mathcal{L}}_t = \overline{\mathcal{L}} + t\overline{\mathcal{M}}$. Let us consider a polynomial $P(t)$ given by $P(t) = \widehat{\deg}\left(\widehat{c}_1(\overline{\mathcal{L}}_t)^{d+1}\right)$. Here we claim the following.

Claim 5.4.2.1. *If $t > 0$ and $P(t) > 0$, then $P(t) \geq t^{d+1}\widehat{\deg}\left(\widehat{c}_1(\overline{\mathcal{M}})^{d+1}\right)$.*

First of all, by the hypothesis of induction and the assumption $P(t) > 0$, we get

$$\widehat{\deg}\left(\widehat{c}_1(\overline{\mathcal{L}}_t|_{\Gamma})^{\dim(\Gamma_{\mathbb{Q}})+1}\right) > 0$$

for all integral subscheme Γ flat over $\mathbb{Z}$. Therefore, in the same as before, we can see that $\widehat{\deg}\left(\widehat{c}_1(\overline{\mathcal{L}}) \cdot \widehat{c}_1(\overline{\mathcal{L}}_t)^d\right) \geq 0$. Then

$$\begin{aligned} P(t) &= \widehat{\deg}\left((\widehat{c}_1(\overline{\mathcal{L}}) + t\widehat{c}_1(\overline{\mathcal{M}})) \cdot \widehat{c}_1(\overline{\mathcal{L}}_t)^d\right) \\ &\geq t\widehat{\deg}\left(\widehat{c}_1(\overline{\mathcal{M}}) \cdot \widehat{c}_1(\overline{\mathcal{L}}_t)^d\right) \\ &\geq t^{d+1}\widehat{\deg}\left(\widehat{c}_1(\overline{\mathcal{M}})^{d+1}\right). \end{aligned}$$

Thus we get our claim.

Let us go back to the proof of the lemma. We set $t_0 = \max\{t \in \mathbb{R} \mid P(t) = 0\}$. We assume $t_0 > 0$. Then, by the above claim, for all $t > t_0$,

$$P(t) \geq t^{d+1}\widehat{\deg}\left(\widehat{c}_1(\overline{\mathcal{M}})^{e+1}\right).$$

Therefore, taking $t \to t_0$,

$$0 = P(t_0) \geq t_0^{e+1}\widehat{\deg}\left(\widehat{c}_1(\overline{\mathcal{M}})^{d+1}\right) > 0.$$

This is a contradiction, that is, $t_0 \leq 0$. In particular, $P(0) \geq 0$. Hence, we get the lemma. □

Let us go back to the proof of the theorem. We also prove this by induction of d. If $d = 0$, then our assertion is obvious, so that we assume that $d > 0$. Let us fix a vertically nef C^∞-hermitian line bundle $\overline{\mathcal{M}}$ such that, for all integral subscheme Γ flat over $\mathbb{Z}$,

$$\widehat{\deg}\left(\widehat{c}_1(\overline{\mathcal{M}}|_{\Gamma})^{\dim(\Gamma_{\mathbb{Q}})+1}\right) > 0.$$

For a positive rational number t, by using the above lemma,

$$\widehat{\deg}\left((\widehat{c}_1(\overline{\mathcal{L}}_{d+1}) + t\widehat{c}_1(\overline{\mathcal{M}}))^{d+1}\right) > 0.$$

Therefore, if we take a large number n, $n(\mathcal{L} + t\mathcal{M})$ has a small section s. Let $\operatorname{div}(s) = a_1\Gamma_1 + \cdots + a_e\Gamma_e$ be a decomposition as cycle. Here $a_j > 0$. Then,

$$n\widehat{\deg}\left(\widehat{c}_1(\overline{\mathcal{L}}_1)\cdots\widehat{c}_1(\overline{\mathcal{L}}_d)\cdot\widehat{c}_1(\overline{\mathcal{L}}_{d+1} + t\widehat{c}_1(\overline{\mathcal{M}}))\right) = \sum_j a_i\widehat{\deg}\left(\widehat{c}_1(\overline{\mathcal{L}}_1\big|_{\Gamma_j})\cdots\widehat{c}_1(\overline{\mathcal{L}}_d\big|_{\Gamma_j})^{d-i}\right)$$
$$- \int_{\mathcal{X}(\mathbb{C})} \log(\|s\|)c_1(\overline{\mathcal{L}}_1)\cdots\widehat{c}_1(\overline{\mathcal{L}}_d).$$

Thus, by the hypothesis of induction, we can see

$$\widehat{\deg}\left(\widehat{c}_1(\overline{\mathcal{L}}_1)\cdots\widehat{c}_1(\overline{\mathcal{L}}_d)\cdot\widehat{c}_1(\overline{\mathcal{L}}_{d+1} + t\widehat{c}_1(\overline{\mathcal{M}}))\right) \geq 0.$$

Thus we get our theorem because t is an arbitrary positive rational number. □

5.5. **Relation of height function and intersection number.** Here let $\mathcal{X}$ be a projective arithmetic variety with $d = \dim \mathcal{X}_{\mathbb{Q}}$.

Theorem 5.5.1. *Let $\overline{\mathcal{L}}$ be a C^∞-hermitian line bundle on $\mathcal{X}$ such that $\overline{\mathcal{L}}$ is vertically nef and $\mathcal{L}_{\mathbb{Q}}$ is ample. Then, we have the following inequalities:*

$$\sup_{Y\subsetneq\mathcal{X}_{\mathbb{Q}}}\left\{\inf_{x\in(\mathcal{X}_{\mathbb{Q}}\setminus Y)(\overline{\mathbb{Q}})} h_{(\mathcal{X},\overline{\mathcal{L}})}(x)\right\} \geq \frac{\widehat{\deg}(\widehat{c}_1(\overline{\mathcal{L}})^{d+1})}{(d+1)\deg(\mathcal{L}_{\mathbb{Q}}^d)} \geq \inf_{x\in\mathcal{X}(\overline{\mathbb{Q}})} h_{(\mathcal{X},\overline{\mathcal{L}})}(x)$$

Proof: Let us begin with the following lemma.

Lemma 5.5.2. *If $\widehat{\deg}(\widehat{c}_1(\overline{\mathcal{L}})^{d+1}) > 0$, then*

$$\sup_{Y\subsetneq\mathcal{X}_{\mathbb{Q}}}\left\{\inf_{x\in(\mathcal{X}_{\mathbb{Q}}\setminus Y)(\overline{\mathbb{Q}})} h_{(\mathcal{X},\overline{\mathcal{L}})}(x)\right\} \geq 0.$$

Proof: By our assumption, replacing $\overline{\mathcal{L}}$ by $\overline{\mathcal{L}}^{\otimes n}$ $(n > 0)$, by Corollary 3.3.2 we may assume $\mathcal{L}$ has a small section s. We set $Y = \operatorname{div}(s)$. Then, for any $x \in (\mathcal{X}_{\mathbb{Q}} \setminus Y)(\overline{\mathbb{Q}})$,

$$h_{(\mathcal{X},\overline{\mathcal{L}})}(x) = \frac{1}{\deg(\Delta_x \to B)}\widehat{\deg}\left(\widehat{\operatorname{div}(s)}\Big|_{\Delta_x}\right).$$

Thus, $h_{(\mathcal{X},\overline{\mathcal{L}})}(x) \geq 0$. □

Let us start the proof of the theorem. Let c be a real number with $0 < c < 1$. Let $\overline{A}$ be a hermitian line bundle on $\operatorname{Spec}(\mathbb{Z})$ given by $\overline{A} = (\mathcal{O}_{\operatorname{Spec}(\mathbb{Z})}, c|\cdot|)$. First, let us consider the left inequality. Let λ be a rational number with

$$\lambda < \frac{\widehat{\deg}(\widehat{c}_1(\overline{\mathcal{L}})^{d+1})}{(d+1)\deg(\mathcal{L}_{\mathbb{Q}}^d)\widehat{\deg}(\overline{A})}.$$

Then, by an easy calculation,

$$\widehat{\deg}\left((\widehat{c}_1(\overline{\mathcal{L}}) - \lambda \widehat{c}_1(\pi^*(\overline{A})))^{d+1}\right) > 0,$$

where $\pi : \mathcal{X} \to \mathrm{Spec}(\mathbb{Z})$ is the natural morphism. Therefore, noting that $\overline{\mathcal{L}} - \lambda\pi^*(\overline{A})$ is vertically nef and ample on $\mathcal{X}_{\mathbb{Q}}$, by the above lemma, we can see

$$\sup_{Y \subsetneq \mathcal{X}_{\mathbb{Q}}} \left\{ \inf_{x \in (\mathcal{X}_{\mathbb{Q}} \setminus Y)(\overline{\mathbb{Q}})} h_{(\mathcal{X}, \overline{\mathcal{L}} - \lambda\pi^*(\overline{A}))}(x) \right\} \geq 0.$$

Thus,

$$\sup_{Y \subsetneq \mathcal{X}_{\mathbb{Q}}} \left\{ \inf_{x \in (\mathcal{X}_{\mathbb{Q}} \setminus Y)(\overline{\mathbb{Q}})} h_{(\mathcal{X}, \overline{\mathcal{L}})}(x) \right\} \geq \lambda \widehat{\deg}(\overline{A}).$$

Hence, we get the left inequality.

Next let us consider the right inequality. Let μ be a rational number with

$$\mu \leq \frac{\inf_{x \in \mathcal{X}(\overline{\mathbb{Q}})} h_{(\mathcal{X}, \overline{\mathcal{L}})}(x)}{\widehat{\deg}(\overline{A})}.$$

Then,

$$\inf_{x \in \mathcal{X}(\overline{\mathbb{Q}})} h_{(\mathcal{X}, \overline{\mathcal{L}} - \mu\pi^*(\overline{A}))}(x) \geq 0.$$

This means that $\overline{\mathcal{L}} - \mu\pi^*(\overline{A})$ is horizontally nef, so that it is nef. Thus, by Theorem 5.4.1,

$$\widehat{\deg}\left((\widehat{c}_1(\overline{\mathcal{L}}) - \mu \widehat{c}_1(\pi^*(\overline{A})))^{d+1}\right) \geq 0.$$

Hence

$$\widehat{\deg}\left(\widehat{c}_1(\overline{\mathcal{L}})^{d+1}\right) \geq \mu(d+1)\deg(\mathcal{L}_{\mathbb{Q}}^d)\widehat{\deg}(\overline{A}).$$

Thus, we get the right inequality. □

Corollary 5.5.3. *Let X be a projective variety over $\mathbb{Q}$ and $\overline{L}$ an ample adelic line bundle on X. Then*

$$\sup_{Y \subsetneq X} \left\{ \inf_{x \in (X \setminus Y)(\overline{\mathbb{Q}})} h_{(X, \overline{L})}(x) \right\} \geq \frac{\widehat{\deg}(\widehat{c}_1(\overline{L})^{d+1})}{(d+1)\deg(L^d)} \geq \inf_{x \in X(\overline{\mathbb{Q}})} h_{(X, \overline{L})}(x).$$

Proof: Let $(\mathcal{X}_n, \overline{\mathcal{L}}_n)$ be a sequence of C^∞-models which gives rise to the adelic line bundle $\overline{L}$. Here $\overline{L}$ is vertically nef, so that we may assume that $\overline{\mathcal{L}}_n$ is vertically nef on $\mathcal{X}_n$. Thus, by the previous theorem,

$$\sup_{Y \subsetneq X} \left\{ \inf_{x \in (X \setminus Y)(\overline{\mathbb{Q}})} h_{(\mathcal{X}_n, \overline{\mathcal{L}}_n)}(x) \right\} \geq \frac{\widehat{\deg}(\widehat{c}_1(\overline{\mathcal{L}}_n)^{d+1})}{(d+1)\deg(L^d)} \geq \inf_{x \in X(\overline{\mathbb{Q}})} h_{(\mathcal{X}_n, \overline{\mathcal{L}}_n)}(x).$$

Here

$$\lim_{n\to\infty} \widehat{\deg}(\widehat{c}_1(\overline{\mathcal{L}}_n)^{d+1}) = \widehat{\deg}(\overline{L}^{d+1}) \quad \text{and} \quad \lim_{n\to\infty} h_{(\mathcal{X}_n,\overline{\mathcal{L}}_n)}(x) = h_{(X,\overline{L})}(x).$$

Note that the left hand is a uniform convergence, namely, for any positive ϵ, there is a number n such that

$$\left| h_{(\mathcal{X}_n,\overline{\mathcal{L}}_n)}(x) - h_{(X,\overline{L})}(x) \right| \leq \epsilon$$

for all $x \in X(\overline{\mathbb{Q}})$. By using this, we can see

$$\lim_{n\to\infty} \inf_{x\in X(\overline{\mathbb{Q}})} h_{(\mathcal{X}_n,\overline{\mathcal{L}}_n)}(x) = \inf_{x\in X(\overline{\mathbb{Q}})} h_{(X,\overline{L})}(x)$$

and

$$\lim_{n\to\infty} \left(\sup_{Y\subsetneq X} \left\{ \inf_{x\in(X\setminus Y)(\overline{\mathbb{Q}})} h_{(\mathcal{X}_n,\overline{\mathcal{L}}_n)}(x) \right\} \right) = \sup_{Y\subsetneq X} \left\{ \inf_{x\in(X\setminus Y)(\overline{\mathbb{Q}})} h_{(X,\overline{L})}(x) \right\}.$$

Thus, we get our corollary. □

6. Bogomolov's conjecture

6.1. **Equidistribution theorem.** Let K be a number field, and X a geometrically irreducible projective variety over K. For an embedding $\sigma : K \to \mathbb{C}$, let $X_\sigma = X \otimes_K \mathbb{C}$ in terms of the embedding σ. Let $\bar{\sigma} : \overline{K} \to \mathbb{C}$ be an extension of $\sigma : K \to \mathbb{C}$. Using $\bar{\sigma}$, we have a natural map $X(\overline{K}) \to X_\sigma(\mathbb{C})$. For $x \in X(\overline{K})$, let $O(x)$ be the orbit of x by the action $\operatorname{Gal}(\overline{K}/K)$. We denote by $O_\sigma(x)$ the image of $O(x)$ by the map $\bar{\sigma} : X(\overline{K}) \to X_\sigma(\mathbb{C})$. Note that $O_\sigma(x)$ does not depend on the choice of the extension $\bar{\sigma}$.

Let $\{x_m\}_{m=1}^\infty$ be a sequence in $X(\overline{K})$. We say $\{x_m\}_{m=1}^\infty$ is *generic* if every subsequence of $\{x_m\}_{m=1}^\infty$ is Zariski dense in X.

Let L be an ample line bundle on X and $\|\cdot\|$ an adelic metric of L such that $\overline{L} = (L, \|\cdot\|)$ is vertically nef, that is, $\|\cdot\|$ is given by a sequence of vertically nef models $\{(\mathcal{X}_n, \overline{\mathcal{L}}_n)\}_{n=1}^\infty$. Then, we have the following theorem.

Theorem 6.1.1 (cf. [35], [36], [39]). *Let $\{x_m\}_{m=1}^\infty$ be a generic sequence in $X(\overline{K})$. We assume the following.*

(1) *$c_1(\overline{L}_\sigma)$ is positive on X_σ. (For positivity, see* §3.3.*)*
(2) *$h_{(X,\overline{L})}(x) \geq 0$ for all $x \in X(\overline{K})$.*
(3) *$\lim_{m\to\infty} h_{(X,\overline{L})}(x_m) = 0$.*

Then, as currents on X_σ, we have a weak convergence

$$\lim_{m\to\infty} \left(\frac{1}{\#O_\sigma(x_m)} \sum_{z\in O_\sigma(x_m)} \delta_z \right) = \frac{c_1(\overline{L}_\sigma)^{\dim X}}{\deg(L^{\dim X})}.$$

Namely, for any C^∞-function f on X_σ,

$$\lim_{m\to\infty}\left(\frac{1}{\#O_\sigma(x_m)}\sum_{z\in O_\sigma(x_m)} f(z)\right) = \int_{X_\sigma}\frac{fc_1(\overline{L}_\sigma)^{\dim X}}{\deg(L^{\dim X})}.$$

Proof: Let f be a C^∞-function on X_σ. Clearly, we may assume that f is real valued. For a real number λ, we change the metric of $\overline{L}_\lambda$ as follows. On X_σ, we change it by $\exp(-\lambda f)\|\cdot\|$ and on the other component of $X(\mathbb{C})$, we keep the metric. Then, since $c_1(\overline{L}_\sigma)$ is positive and X_σ is compact, there is a positive number λ_0 such that, for any λ with $|\lambda|\le\lambda_0$, $\overline{L}_\lambda$ is vertically nef. Thus, by Corollary 5.5.3,

$$\sup_{Y\subsetneq X}\left\{\inf_{x\in(X\setminus Y)(\overline{K})} h_{(X,\overline{L}_\lambda)}(x)\right\}\ge\frac{\widehat{\deg}(\widehat{c}_1(\overline{L}_\lambda)^{d+1})}{(d+1)\deg(L^d)}\ge\inf_{x\in X(\overline{K})} h_{(X,\overline{L}_\lambda)}(x).$$

Here $d=\dim X$. Applying the above inequality for the generic sequence $\{x_m\}$ in the case where $\lambda=0$, we can see $\widehat{\deg}(\widehat{c}_1(\overline{L})^{d+1})=0$. Moreover, by easy calculations, we can see

$$h_{(X,\overline{L}_\lambda)}(x)=h_{(X,\overline{L})}(x)+\frac{\lambda}{\#O_\sigma(x)}\sum_{z\in O_\sigma(x)} f(z)$$

and

$$\frac{\widehat{\deg}(\widehat{c}_1(\overline{L}_\lambda)^{d+1})}{(d+1)\deg(L^d)}=\frac{\lambda}{\deg(L^d)}\int_{X_\sigma} fc_1(\overline{L}_\sigma)^d+O(\lambda^2).$$

Therefore, noting $\{x_m\}$ is generic and $\lim_{m\to\infty} h_{(X,\overline{L})}(x_m)=0$, we get

$$\lambda\liminf_m\left(\frac{1}{\#O_\sigma(x_m)}\sum_{z\in O_\sigma(x_m)} f(z)\right)\ge\frac{\lambda}{\deg(L^d)}\int_{X_\sigma} fc_1(\overline{L}_\sigma)^d+O(\lambda^2).$$

Thus, taking $\lambda\to 0$,

$$\liminf_m\left(\frac{1}{\#O_\sigma(x_m)}\sum_{z\in O_\sigma(x_m)} f(z)\right)\ge\frac{1}{\deg(L^d)}\int_{X_\sigma} fc_1(\overline{L}_\sigma)^d.$$

This inequality holds even if we replace f by $-f$, namely,

$$\limsup_m\left(\frac{1}{\#O_\sigma(x_m)}\sum_{z\in O_\sigma(x_m)} f(z)\right)\le\frac{1}{\deg(L^d)}\int_{X_\sigma} fc_1(\overline{L}_\sigma)^d.$$

Thus we get our theorem. □

6.2. **The proof of Bogomolov's conjecture.** Let K be a number field, A an abelian variety over K and L a symmetric ample line bundle on A. Moreover, let X be a subvariety of $A_{\overline{K}}$. Then the following theorem is called Bogomolov's conjecture.

Theorem 6.2.1. *Let $\hat{h}_L$ be the canonical height associated with L. If for any positive number ϵ, the set $\{x \in X(\overline{K}) \mid \hat{h}_L(x) \leq \epsilon\}$ is Zariski dense in X, then there are an abelian subvariety B of $A_{\overline{K}}$ and a torsion point b with $X = B + b$.*

We set a subgroup $G(X) = \{a \in A(\overline{K}) \mid a + X = X\}$. Let us begin with the following lemma.

Lemma 6.2.2. *For an integer m with $m \geq 2$, We give a morphism $\alpha_m : A_K^m \to A_K^{m-1}$ by*

$$\alpha_m(x_1, \ldots, x_m) = (x_1 - x_2, \ldots, x_{m-1} - x_m).$$

If $G(X) = \{0\}$ and m is sufficiently large, then $\alpha_m|_{X^m} : X^m \to \alpha_m(X^m)$ is birational.

Proof: For $x_1, \ldots, x_m \in X$, let $G(x_1, \ldots, x_m)$ be a subgroup given by

$$G(x_1, \ldots, x_m) = \{a \in A_K \mid a + x_1, \ldots, a + x_m \in X\}.$$

Note that $G(X) = \{a \in A_K \mid a + x \in X \text{ for all } x \in X\}$. Thus, there are a positive integer m_0 and $x_1, \ldots, x_{m_0} \in X$ such that $G(x_1, \ldots, x_{m_0}) = \{0\}$. Hence, it is sufficient to show the following claim.

Claim 6.2.2.1. *For $x_1, \ldots, x_m \in X$,*

$$(\alpha_m|_{X^m})^{-1}(\alpha_m(x_1, \ldots, x_m)) = \{(x_1 + a, \ldots, x_m + a) \mid a \in G(x_1, \ldots, x_m)\}.$$

We assume $(y_1, \ldots, y_m) \in (\alpha_m|_{X^m})^{-1}(\alpha_m(x_1, \ldots, x_m))$. Then, $x_i - x_{i+1} = y_i - y_{i+1}$ for all $1 \leq i < m$. Thus, $y_1 - x_1 = \cdots = y_m - x_m$. Therefore, $y_i = x_i + a$ for some $a \in G(x_1, \ldots, x_m)$. The other inclusion is obvious. □

Let us start the proof of the theorem. First we consider the case where $G(X) = \{0\}$. In this case, it is sufficient to show that $\dim X = 0$. For, if we set $X = \{x\}$, then $0 \leq \hat{h}_L(x) \leq \epsilon$ for all $\epsilon > 0$, that is, $\hat{h}_L(x) = 0$. Thus, $x \in A(\overline{K})_{tor}$. This indicates the theorem in this case.

For this purpose, we assume that $\dim X > 0$. Since $G(X) = \{0\}$, by the previous lemma, if we take a large integer m, then $\alpha_m|_{X^m} : X^m \to \alpha_m(X^m)$ is birational. By abuse of notation, we denote $\alpha_m|_{X^m}$ by α_m. In order to prove our assertion, we may replace K by a finite extension field of K, so that we may assume that X is defined over K and $X^m \to \alpha_m(X^m)$ is birational over K. Let $\|\cdot\|$ be the admissible metric induced by $[2] : A \to A$. Then, $\overline{L} = (L, \|\cdot\|)$ is vertically nef and $c_1(\overline{L})$ is a positive C^∞-form. Moreover, $h_{(A,\overline{L})} = \hat{h}_L$. Let $p_i : A^m \to A$ be the projection to the i-th factor and $\overline{L}_m = p_1^*(\overline{L}) \otimes \cdots \otimes p_m^*(\overline{L})$. The metric of $\overline{L}_m$ is an admissible metric with respect to $[2] : A^m \to A^m$. Moreover, note that

(1) $c_1(\overline{L}_m) = p_1^*(c_1(\overline{L})) + \cdots + p_m^*(c_1(\overline{L}))$.
(2) $h_{(A^m, \overline{L}_m)}(a_1, \ldots, a_m) = \hat{h}_L(a_1) + \cdots + \hat{h}_L(a_m)$.

In the same way as before, we define $\overline{L}_{m-1}$ on A^{m-1}.

Since the set of all proper subvarieties of X is countable, we denote it by $\{Y_n\}_{n=1}^{\infty}$. Then, by our assumption, for each n, we can find $x_n \in X(\overline{K})$ with $x_n \notin \bigcup_{i=1}^n Y_i$ and $\hat{h}_L(x_n) \leq 1/n$. Then, the sequence $\{x_n\}$ is generic and $\lim_{n\to\infty} \hat{h}_L(x_n) = 0$. Let $\mathbb{N}$ be the set of all natural numbers and let us fix a bijection $\beta : \mathbb{N} \to \mathbb{N}^m$. Denoting the i-th entry of $\beta(n)$ by $\beta_i(n)$, we set $x(n) = (x_{\beta_1(n)}, \ldots, x_{\beta_m(n)})$. Thus, we get a sequence $\{x(n)\}$ in X^m. Then, $\lim_{n\to\infty} h_{(X^m, \overline{L}_m)}(x(n)) = 0$. Moreover, $\{x(n)\}$ is Zariski dense in X^m. Indeed, note that

$$\{x(n) \mid n \in \mathbb{N}\} = \{(x_{e_1}, \ldots, x_{e_m}) \mid (e_1, \ldots, e_m) \in \mathbb{N}^m\}.$$

Thus, we can see by induction of m that $\{(x_{e_1}, \ldots, x_{e_m}) \mid (e_1, \ldots, e_m) \in \mathbb{N}^m\}$ is Zariski dense in X^m. Here let $\{Z_i\}_{i=1}^{\infty}$ be the set of all proper subvarieties of X^m. We choose $x(n_i)$ with $x(n_i) \notin \bigcup_{j=1}^i Z_j$. Then, $\{x(n_i)\}$ is a generic sequence and $\lim_{i\to\infty} h_{(X^m, \overline{L}_m)}(x(n_i)) = 0$. Here since $\alpha_m : X^m \to \alpha_m(X^m)$ is birational over K, there is a Zariski open set U of $\alpha_m(X^m)$ over K such that $\alpha_m : \alpha_m^{-1}(U) \xrightarrow{\sim} U$. Considering a subsequence of $\{x(n_i)\}$, we may assume that $x(n_i) \in \alpha_m^{-1}(U)$. Let us consider $\{\alpha_m(x(n_i))\}$. This is generic on $\alpha_m(X^m)$ and

$$\lim_{i\to\infty} h_{(A^{m-1}, \overline{L}_{m-1})}(\alpha_m(x(n_i))) = 0.$$

Therefore, if we fix an embedding $\sigma : K \to \mathbb{C}$ and we set $d = \dim X^m = \dim \alpha_m(X^m)$, then, by Theorem 6.1.1, we have

$$\lim_{i\to\infty} \left(\frac{1}{\#O_\sigma(x(n_i))} \sum_{z \in O_\sigma(x(n_i))} \delta_z \right) = \frac{\left(c_1(\overline{L}_{m\sigma})\big|_{X_\sigma^m} \right)^{\wedge d}}{\deg\left((L_m|_{X^m})^d \right)}$$

on X_σ^m, and

$$\lim_{i\to\infty} \left(\frac{1}{\#O_\sigma(\alpha_m(x(n_i)))} \sum_{z \in O_\sigma(\alpha_m(x(n_i)))} \delta_z \right) = \frac{\left(c_1(\overline{L}_{m-1\sigma})\big|_{\alpha_m(X^m)_\sigma} \right)^{\wedge d}}{\deg\left(\left(L_{m-1}|_{\alpha_m(X^m)} \right)^d \right)}$$

on $\alpha_m(X^m)_\sigma$. Thus, each convergence holds on $\alpha_m^{-1}(U)_\sigma$ and U_σ respectively. Here since $\alpha_m^{-1}(U)_\sigma \simeq U_\sigma$, we have

$$\frac{\left(c_1(\overline{L}_{m\sigma})\big|_{X_\sigma^m} \right)^{\wedge d}}{\deg\left((L_m|_{X^m})^d \right)} = \alpha_m^* \left(\frac{\left(c_1(\overline{L}_{m-1\sigma})\big|_{\alpha_m(X^m)_\sigma} \right)^{\wedge d}}{\deg\left(\left(L_{m-1}|_{\alpha_m(X^m)} \right)^d \right)} \right)$$

on $\alpha_m^{-1}(U)_\sigma$. Moreover, each side is a C^∞-form. Thus, the equality holds on X_σ^m. On the other hand, $\alpha_m : X^m \to \alpha_m(X^m)$ is not an isomorphism on the diagram $\{(x, \ldots, x) \mid x \in X\}$. Thus, the right hand is not a positive form on the diagram. This is a contradiction because the left hand side is positive.

Finally, let us consider a general case. Let $A' = A/G(X)$ and $\pi : A \to A/G(X)$ be the natural morphism. Moreover, we set $X' = \pi(X)$. Then, $G(X') = \{0\}$ and $\pi^{-1}(X') = X$. Let L' be a symmetric ample line bundle on A'. Choose a positive integer a such that $L^{\otimes a} \otimes \pi^*(L)^{\otimes -1}$ is ample, which implies $\pi^*(\hat{h}_{L'}) \leq a\hat{h}_L$. Then, for every positive ϵ, $\{x' \in X'(\overline{K}) \mid \hat{h}_{L'}(x') \leq \epsilon\}$ is Zariski dense. Indeed,

$$\{x \in X(\overline{K}) \mid \hat{h}_L(x) \leq \epsilon/a\} \subseteq \pi^{-1}\left(\{x' \in X'(\overline{K}) \mid \hat{h}_{L'}(x') \leq \epsilon\}\right).$$

Therefore, by the previous observation, there is a $x' \in A'(\overline{K})_{tor}$ with $X' = \{x'\}$. Thus, $X = \pi^{-1}(x')$. Hence we can easily get our theorem.

7. A generalization of conjectures of Bogomolov and Lang

In this section, we discuss a generalization of conjectures of Bogomolov and Lang, which Poonen [26] and Zhang [40] independently proved and which Moriwaki [25] extended for finitely generated fields.

7.1. **Statement of Bogomolov plus Lang.** Let K be a number field. Let A be an abelian variety over K and L a symmetric ample line bundle on A.

As in §5.2, let $\widehat{h}_L : A(\overline{K}) \to \mathbb{R}$ be the Néron-Tate height function associated with L. For $x, y \in A(\overline{K})$, put

$$\langle x, y\rangle_L = \frac{1}{2}\left(\widehat{h}_L(x+y) - \widehat{h}_L(x) - \widehat{h}_L(y)\right).$$

Since $\widehat{h}_L$ is quadric, $\langle\ ,\ \rangle_L : A(\overline{K}) \times A(\overline{K}) \to \mathbb{R}$ is a symmetric bilinear form, which is called the *Néron-Tate height pairing* associated with L. By Proposition 5.2.2, $\langle\ ,\ \rangle_L$ gives a hermitian inner product on $A(\overline{K})/A(\overline{K})_{tor}$. For $x_1, \cdots, x_l \in A(\overline{K})$, we set

$$\delta_L(x_1, \cdots, x_l) = \det\left(\langle x_i, x_j\rangle_L\right).$$

Let Γ be a subgroup of finite rank in $A(\overline{K})$, i.e., Γ is a subgroup of $A(\overline{K})$ with $\dim_{\mathbb{Q}}(\Gamma \otimes \mathbb{Q}) < \infty$. Note that

$$\Gamma_{\text{div}} = \{x \in A(\overline{K}) \mid nx \in \Gamma \text{ for some positive integer } n\}$$

is also a subgroup of finite rank in $A(\overline{K})$.

The following theorem is a generalization of the conjectures of Bogomolov and Lang.

Theorem 7.1.1 ([26], [40]). *Let the notation be as above. Let X be a subvariety of $A_{\overline{K}}$. Fix $\gamma_1, \cdots, \gamma_n \in \Gamma$ such that they form a basis of $\Gamma \otimes \mathbb{Q}$. Assume that, for any positive number $\epsilon > 0$,*

$$\{x \in X(\overline{K}) \mid \delta_L(\gamma_1, \cdots, \gamma_l, x) \leq \epsilon\}$$

is Zariski dense in X. Then, X is a translation of an abelian subvariety of $A_{\overline{K}}$ by an element of Γ_{div}.

If we take $\Gamma = 0$ in the above theorem, then we have Bogomolov's conjecture (Theorem 6.2.1). As we will see below, Theorem 7.1.1 is also a generalization of the following Lang conjecture, which Falting proved. (For its proof, we refer to [7], [8].)

Theorem 7.1.2 (Lang's conjecture over a number field). *Let K be a number field, and A an abelian variety over K. Let X be a subvariety of $A_{\overline{K}}$. Then, there exist abelian subvarieties $C_1, \cdots, C_n$ of $A_{\overline{K}}$ and $\gamma_1, \cdots, \gamma_n \in X(\overline{K})$ such that*

$$\overline{X(K)} = \bigcup_{i=1}^{n} (C_i + \gamma_i),$$

where $\overline{X(K)}$ is the Zariski closure of $X(K)$.

In the following subsections, following [25] (which underlines Poonen's idea and uses a geometric trick), we will prove Theorem 7.1.1. Its proof is based on Bogomolov's conjecture and Lang's conjecture.

In the meanwhile, let us check that Theorem 7.1.1 is indeed a generalization of Lang's conjecture. Namely, assuming Theorem 7.1.1, let us show Theorem 7.1.2. If $\dim(\overline{X(K)}) = 0$, we are done. Let Y be a positive dimensional irreducible component of $\overline{X(K)}$. Then $Y(K)$ is Zariski dense in Y. On the other hand, Mordell-Weil's theorem says that $A(K)$ is a finitely generated group. Then, applying Theorem 7.1.1 to Y and $\Gamma = A(K)$, we find that Y is a translation of an abelian subvariety. Thus, we get Theorem 7.1.2.

7.2. **Small points with respect to a group of finite rank.** As in §7.1, let K be a number field, A an abelian variety over K, L a symmetric ample line bundle on A, and Γ a subgroup of finite rank in $A(\overline{K})$.

In this subsection, we assume that there exists a subgroup Γ_0 of Γ with $\Gamma_0 \subseteq A(K)$ and $\Gamma_0 \otimes \mathbb{Q} = \Gamma \otimes \mathbb{Q}$. Note that, if we replace K by a suitable finite extension field, then there always exists such a Γ_0.

A nonempty subset $S \subseteq A(\overline{K})$ is said to be *small with respect to* Γ if there exists a decomposition $s = \gamma(s) + z(s)$ for each $s \in S$ with the following properties:

(i) $\gamma(s) \in \Gamma$ for every $s \in S$;

(ii) For any $\epsilon > 0$, there is a finite proper subset M of S such that $\langle z(s), z(s)\rangle_L \leq \epsilon$ for all $s \in S \setminus M$. (Warning: M is proper, i.e., $S \setminus M \neq \emptyset$.)

A nonempty subset $S \subseteq A(\overline{K})$ is said to be *small* if S is small with respect to $\{0\}$. For example, if S is a finite set, then S is small if and only if S contains a torsion point.

Let us state Bogomolov's conjecture (Theorem 6.2.1) in terms of small sets:

Theorem 7.2.1. *Let $S \subseteq A(\overline{K})$ be small. Then, there are abelian subvarieties $C_1, \cdots, C_r$, torsion points $c_1, \cdots, c_r$, and finite nontorsion points $b_1, \cdots, b_m$ such that*

$$\overline{S} = \bigcup_{i=1}^{r} (C_i + c_i) \cup \{b_1, \cdots, b_m\},$$

where $\overline{S}$ is the Zariski closure of S in A.

Proof: Let X be the positive dimensional irreducible component of $\overline{S}$. Put $S' = S \cap X$. Then, S' is small. By Theorem 6.2.1, X is a translation of an abelian subvariety by a torsion point. □

Let T be a subset of $A(\overline{K})$, and F a finite extension field of K in $\overline{K}$. For $x \in A(\overline{K})$, set $O_F(x) = \{\sigma(x) \in A(\overline{K}) \mid \sigma \in \mathrm{Gal}(\overline{K}/F)\}$. For $n \geq 2$, let $\beta_n : A^n \to A^{n-1}$ be the homomorphism defined by $\beta_n(x_1, \cdots, x_n) = (x_2 - x_1, x_3 - x_1, \cdots, x_n - x_1)$. Then, we define the subset $\mathcal{D}_n(T, F)$ of $A(\overline{K})^{n-1}$ by

$$\mathcal{D}_n(T, F) = \bigcup_{s \in T} \beta_n\left(O_F(s)^n\right).$$

Let $\overline{\mathcal{D}}_n(T, F)$ denote the Zariski closure of $\mathcal{D}_n(T, F)$ in A^{n-1}.

Notice that, if T' is a subset of T and F' is a finite extension field of F, then $\mathcal{D}_n(T', F') \subseteq \mathcal{D}_n(T, F)$ and thus $\overline{\mathcal{D}}_n(T', F') \subseteq \overline{\mathcal{D}}_n(T, F)$. Let $[N] : A \to A$ be the homomorphism given by $x \to Nx$. Then, since $[N](\mathcal{D}_n(T, F)) = \mathcal{D}_n([N](T), F)$, we have $\overline{\mathcal{D}}_n([N](T), F) = [N](\overline{\mathcal{D}}_n(T, F))$.

We say that a pair (T, F) is *n-minimized* if the following conditions are satisfied:

(0) T is an infinite set.

(1) For any infinite subset T' of T and finite extension field F' of F, $\overline{\mathcal{D}}_n(T', F') = \overline{\mathcal{D}}_n(T, F)$;

(2) For any positive integer N, $[N](\overline{\mathcal{D}}_n(T, F)) = \overline{\mathcal{D}}_n(T, F)$. (This condition is equivalent to $\overline{\mathcal{D}}_n([N](T), F) = \overline{\mathcal{D}}_n(T, F)$.)

Before we state Lemma 7.2.2, let us make a few remarks. Let $f : A \to A'$ be a homomorphism of abelian varieties over $\overline{K}$. Let L' be a symmetric ample line bundle on A. Since L is ample, we can take a positive integer a such that $L^{\otimes a} \otimes f^*(L')^{-1}$ is base-point free. Then, using Proposition 5.1.1, we get

$$\langle f(x), f(x) \rangle_{L'} = \langle x, x \rangle_{f^*(L')} \leq a \langle x, x \rangle_L.$$

In particular, the property that S is small with respect to Γ does not depend on the choice of symmetric ample line bundles L on A.

On A^n, we can give the Néron-Tate height pairing associated with $\bigotimes_{i=1}^n p_i^*(L)$, where $p_i : A^n \to A$ is the projection to the i-th factor. For $x = (x_1, \cdots, x_n) \in A(\overline{K})^n$, $\langle x, x\rangle_{\bigotimes_{i=1}^n p_i^*(L)}$ is given by

$$\langle x, x\rangle_{\bigotimes_{i=1}^n p_i^*(L)} = \langle x_1, x_1\rangle_L + \cdots + \langle x_n, x_n\rangle_L.$$

By abuse of notation, we denote $\langle\ ,\ \rangle_{\bigotimes_{i=1}^n p_i^*(L)}$ by $\langle\ ,\ \rangle_L$.

Lemma 7.2.2. *Let $S \subset A(\overline{K})$ be a subset that is small with respect to Γ, and F a finite extension field of K. Let $f : A \to A'$ be a homomorphism of abelian varieties over $\overline{K}$ and $f^{n-1} : A^{n-1} \to A^{n-1}$ the homomorphism given by $(x_1, \cdots, x_{n-1}) \mapsto (f(x_1), \cdots, f(x_{n-1}))$. Then, we have the following:*

(1) *The set $f^{n-1}(\overline{\mathcal{D}}_n(S, F))$ is small;*
(2) *Let $b_1, \cdots, b_l$ be nontorsion points in $f^{n-1}(\overline{\mathcal{D}}_n(S, F))$; Then there exists a finite proper subset M such that $b_i \notin f^{n-1}(\overline{\mathcal{D}}_n(S \setminus M, F))$ for any i.*

Proof: Recall that we have assumed the existence of a finitely generated subgroup Γ_0 of Γ with Γ_0 in $A(K)$ and $\Gamma_0 \otimes \mathbb{Q} = \Gamma \otimes \mathbb{Q}$. For simplicity, we put $\|\cdot\|_L = \sqrt{\langle\cdot,\cdot\rangle_L}$.

Let $s = \gamma(s) + z(s) \in S$. Since $m\gamma(s) \in \Gamma_0 \subseteq A(K)$ for some positive integer m, $\sigma(\gamma(s)) - \tau(\gamma(s))$ is a torsion point for any $\sigma, \tau \in \mathrm{Gal}(\overline{K}/F)$. Then, $\|\sigma(s) - \tau(s)\|_L = \|\sigma(z(s)) - \tau(z(s))\|_L \le 2\|z(s)\|_L$. Therefore, we have, for any $x \in O_F(s)^n$, $\|\beta_n(x)\|_L \le 2\sqrt{n-1}\|z(s)\|_L$. Let L' be a symmetric ample line bundle on A'. Then, there is a positive constant a with $\langle f(x), f(x)\rangle_{L'} \le a\langle x, x\rangle_L$. Thus,

$$\|f^{n-1}(\beta_n(x))\|_{L'} < 2\sqrt{a(n-1)}\|z(s)\|_L \qquad (\forall x \in O_F(s)^n). \tag{7.2.2.1}$$

First, let us see (2). Put $\mu = \min_{i=1,\cdots,l}\{\|b_i\|_{L'}\}$. Then, there is a finite proper subset M of S with

$$\|z(s)\|_L < \frac{\mu}{2\sqrt{a(n-1)}} \qquad (\forall s \in S \setminus M).$$

Thus, by (7.2.2.1), we have

$$\|f^{n-1}(\beta_n(x))\|_{L'} < \mu \qquad (\forall x \in \textstyle\bigcup_{S\setminus M} O_F(s)^n).$$

Hence, $b_i \notin f^{n-1}(\overline{\mathcal{D}}_n(S \setminus M, F))$ for any i.

Next, we consider (1). If $f^{n-1}(\overline{\mathcal{D}}_n(S, F))$ is infinite, then the assertion of (1) is obvious by (7.2.2.1). Otherwise, let $\{b_1, \cdots, b_l\}$ be the set of all nontorsion points in $f^{n-1}(\overline{\mathcal{D}}_n(S, F))$. Then, by (2), we can find a proper subset M of S with

$$\emptyset \ne f^{n-1}(\overline{\mathcal{D}}_n(S \setminus M, F)) \subseteq f^{n-1}(\overline{\mathcal{D}}_n(S, F)) \setminus \{b_1, \cdots, b_l\}.$$

Hence, $f^{n-1}(\overline{\mathcal{D}}_n(S, F))$ contains a torsion point. Thus, $f^{n-1}(\overline{\mathcal{D}}_n(S, F))$ is small. □

Proposition 7.2.3. *Let S be an infinite subset of $A(\overline{K})$ that is small with respect to Γ. Then, there exist an infinite subset T of S, a finite extension field F of K, and a positive integer N such that $([N](T), F)$ is n-minimized.*

Proof: Since $A^{n-1}_{\overline{K}}$ is a noetherian space,

$$\{\overline{\mathcal{D}}_n(T, F) \mid T \text{ is an infinite subset of } S \text{ and } F \text{ is a finite extension field of } K\}$$

has a minimal element. We fix a pair (T, F) for which $\overline{\mathcal{D}}_n(T, F)$ is minimal.

By Theorem 7.2.1, there are abelian subvarieties $C_1, \cdots, C_r$, torsion points $c_1, \cdots, c_r$, and finite nontorsion points $b_1, \cdots, b_m$ such that $\overline{\mathcal{D}}_n(T, F) = \bigcup_{i=1}^r (C_i + c_i) \cup \{b_1, \cdots, b_m\}$. By Lemma 7.2.2, we can find a finite subset M of T with $\overline{\mathcal{D}}_n(T \setminus M, F) \subseteq \bigcup_{i=1}^r (C_i + c_i)$. Then by the minimality of (T, F), we have

$$\overline{\mathcal{D}}_n(T, F) = \bigcup_{i=1}^{r} (C_i + c_i). \tag{7.2.3.1}$$

Take a positive integer N with $Nc_i = 0$ for all i. Here, we claim that $([N](T), F)$ is n-minimal. Indeed, since T is an infinite set and $[N]$ is a finite morphism, $[N](T)$ is an infinite set. Let $T_1 \subseteq [N](T)$ be an infinite subset, and F' be a finite extension field of F. Take $T' \subseteq T$ with $[N](T') = T_1$. Then, we find

$$\begin{aligned} \overline{\mathcal{D}}_n(T_1, F') = \overline{\mathcal{D}}_n([N](T'), F') &= [N](\overline{\mathcal{D}}_n(T', F')) \\ &= [N](\overline{\mathcal{D}}_n(T, F)) = \overline{\mathcal{D}}_n([N](T), F). \end{aligned}$$

Moreover, for any positive integer N', we get

$$\begin{aligned} [N'](\overline{\mathcal{D}}_n([N](T), F)) &= [N']\left(\bigcup_{i=1}^{r} C_i\right) \\ &= \bigcup_{i=1}^{r} C_i = \overline{\mathcal{D}}_n([N](T), F)). \end{aligned}$$

Therefore, $([N](T), F)$ is n-minimal. □

The following theorem plays a key role in the proof of Theorem 7.1.1 in the next subsection.

Theorem 7.2.4 ([26], [25])**.** *Let S be an infinite subset of $A(\overline{K})$ that is small with respect to Γ. Let F be a finite extension field of K. Assume that (S, F) is 2-minimized. Then, there exists an abelian subvariety C of $A_{\overline{K}}$ such that $\overline{\mathcal{D}}_n(S, F) = C^{n-1}$ for any $n \geq 2$.*

Proof: Let us begin with the following lemma.

Lemma 7.2.5. *Let the notation and the assumption be as in the theorem. Let B an abelian subscheme of A_F over F. We assume that there exists a positive integer e with the following properties: For each $s \in S$, there exists a subset $T(s)$ of $O_F(s) \times O_F(s)$ such that $\beta_2(T(s)) \subseteq B(\overline{K})$ and ${}^\#T(s) \geq {}^\#(O_F(s) \times O_F(s))/e$. Then, there exists a finite subset M of S and a positive integer N with $\overline{\mathcal{D}}_2([N](S \setminus M, F)) \subseteq B(\overline{K})$.*

Proof: Let $\pi : A \to A/B$ be the natural homomorphism. Fix $s \in S$. Take a finite extension field F' of $F(s)$ such that F'/F is Galois. Let $\phi : G := \mathrm{Gal}(F'/F) \to O_F(s)$ be the map given by $\sigma \mapsto \sigma(s)$.

Let $G_{\pi(s)}$ be the stabilizer of $\pi(s)$ by the action of G. We set

$$R = \{(\sigma, \tau) \in G \times G \mid \sigma(\pi(s)) = \tau(\pi(s))\}.$$

Then, we have ${}^\#R = {}^\#G_{\pi(s)} \cdot {}^\#G$. Moreover, since $(\phi \times \phi)^{-1}(T(s)) \subseteq R$, we get ${}^\#R \geq {}^\#(G \times G)/e$. Thus, $[G : G_{\pi(s)}] \leq e$, which means that $[F(\pi(s)) : F] \leq e$.

By Lemma 7.2.2, $\pi(\overline{\mathcal{D}}_2(S, F))$ is small. Then, by Northcott's theorem (Theorem 5.1.4), $\pi(\overline{\mathcal{D}}_2([N](S, F)))$ is a finite set. Thus, by Lemma 7.2.2 again, there exists a finite subset $M \subseteq S$ such that $\pi(\overline{\mathcal{D}}_2(S \setminus M, F))$ consists of torsion points. Hence, if we take a positive integer N with $[N](\pi(\overline{\mathcal{D}}_2(S \setminus M, F))) = \{0\}$, then we have $\overline{\mathcal{D}}_2([N](S \setminus M, F)) \subseteq B(\overline{K})$. □

Let us go back to the proof of Theorem 7.2.4. First, let us consider the case $n = 2$. By the same reason as in (7.2.3.1), we find abelian subvarieties $C_1, \cdots, C_e$ with

$$\overline{\mathcal{D}}_2(S, F) = \bigcup_{i=1}^{e} C_i,$$

because $\overline{\mathcal{D}}_2(S, F)$ is stable under any $[N]$. In order to see $e = 1$, it is sufficient to find C_i, a positive integer N_1, an infinite subset S_1 of S, and a finite extension field F_1 of F such that

$$\overline{\mathcal{D}}_2([N_1](S_1), F_1) \subseteq C_i(\overline{K}).$$

Let F_1 be a finite extension field of F such that every C_i is defined over F_1. For each $s \in S$, put $T_i(s) = \{x \in O_{F_1}(s) \times O_{F_1}(s) \mid \beta_2(x) \subseteq C_i(\overline{K})\}$. Then, by the pigeonhole principle, there exist i and an infinite subset S' of S such that, for any $s \in S'$, ${}^\#T_i(s) \geq {}^\#(O_{F_1}(s) \times O_{F_1}(s))/e$. Then, by Lemma 7.2.5, there exist an infinite subset S_1 of S and a positive integer N_i with $\overline{\mathcal{D}}_2([N_1](S_1), F_1) \subseteq C_i(\overline{K})$.

In what follows, we denote C_i by C. Then, $\overline{\mathcal{D}}_2(S, F) = C$. Let us show that $\overline{\mathcal{D}}_n(S, F) = C^{n-1}$ for any $n \geq 2$. Clearly, $\overline{\mathcal{D}}_n(S, F) \subseteq C^{n-1}$. Thus, it is sufficient to find a positive integer N_2, an infinite subset S_2 of S, and a finite extension field F_2 of F such that

$$\overline{\mathcal{D}}_n([N_2](S_2), F_2) = C^{n-1}.$$

By Proposition 7.2.3, there exist an infinite subset S_2 of S, a finite extension field F_2 of F, and a positive integer N_2 such that $([N_2](S_2), F_2)$ is n-minimized. Then, as before, there are abelian subvarieties $B_1, \cdots, B_l$ with $\overline{\mathcal{D}}_2([N_2](S_2), F_2) = \bigcup_{j=1}^{l} B_j$. Moreover, since $([N_2](S_2), F_2)$ is n-minimized, by replacing F_2 by a suitable finite extension field, we may assume that B_i's and C are defined over F_2. Now, let us show that $\overline{\mathcal{D}}_n([N_2](S_2), F_2) = C^{n-1}$.

In the same way as before, we can find B_j and an infinite subset S' of $[N_2](S_2)$ such that, for any $s \in S'$, there exists a subset $U(s)$ of $O_{F_2}(s)^n$ with $\#U(s) \geq \#(O_{F_2}(s)^n)/l$ and $\beta_n(U(s)) \subseteq B_j(\overline{K})$. In what follows, we denote B_j by B. Let $C^{(q)} = 0 \times \cdots \times C \times \cdots \times 0$ be the q-th factor of C^{n-1}, and put $B^{(q)} = B \cap C^{(q)}$. Since $B \subseteq C^{n-1}$ and B is an abelian subcheme, it suffices to see the following claim to conclude the proof.

Claim 7.2.5.1. *For each q $(1 \leq q \leq n-1)$, $B^{(q)} = C^{(q)}$.*

Without loss of generality, we may assume $q = n-1$. Fix $s \in S'$. For $t = (t_1, \cdots, t_{n-1}) \in O_{F_2}(s)^{n-1}$, we set

$$L_t(s) = \{x \in O_F(s) \mid (t_1, \cdots, t_{n-1}, x) \in U(s)\}.$$

We choose $t' = (t'_1, \cdots, t'_{n-1}) \in O_{F_2}(s)^{n-1}$ such that $\#L_{t'}(s)$ attains the maximum among $\{\#L_t(s)\}_{t \in O_{F_2}(s)^{n-1}}$. Then,

$$\#L_{t'}(s) \cdot \#O_{F_2}(s)^{n-1} \geq \#U(s) \geq \#(O_{F_2}(s)^n)/l,$$

so that $\#L_{t'}(s) \geq \#O_F(s)/l$. Set $T(s) = L_{t'}(s) \times L_{t'}(s) \subseteq O_F(s) \times O_F(s)$. Then, for $(x, y) \in T(s)$, we have

$$\beta_n(0, \cdots, 0, y - x) = \beta_n(t'_1, \cdots, t'_{n-1}, y) - \beta_n(t'_1, \cdots, t'_{n-1}, x) \in B(\overline{K}).$$

This means that $\beta_2(x, y) \in B^{(n-1)}(\overline{K})$ if we view $B^{(n-1)}$ as a subscheme of A. Since $\beta_2(T(s)) \subseteq B^{(n-1)}(\overline{K})$ and $\#T(s) \geq \#(O_F(s) \times O_F(s))/l^2$, we get by Lemma 7.2.5 that there are an infinite subset S'' of S' and a positive integer N'' with $\overline{\mathcal{D}}_2([N''](S'', F_2)) \subseteq B^{(n-1)}(\overline{K})$. Then, since $\overline{\mathcal{D}}_2([N''](S'', F_2)) = C$, we have $B^{(n-1)} = C^{(n-1)}$. □

7.3. **Proof of Bogomolov plus Lang.** We note that, in order to prove Theorem 7.1.1, K can be replaced by any finite extension field of K. Thus, enlarging K, we may assume that X is defined over K and that there exists a subgroup Γ_0 of Γ with $\Gamma_0 \subseteq A(K)$ and $\Gamma_0 \otimes \mathbb{Q} = \Gamma \otimes \mathbb{Q}$.

As in §6.2.1, let $G(X) = \{a \in A(\overline{K}) \mid a + X = X\}$ be the stabilizer of X. By comparing A with the quotient $A/G(X)$, we see that it suffices to prove Theorem 7.1.1 when $\dim G(X) = 0$. (For a detailed argument on this comparison, we refer to [25, §3].) And when $\dim G(X) = 0$, what we need to prove is the following.

Claim 7.3.1. *If $\dim G(X) = 0$, then X is a finite subset of Γ_{div}.*

Proof: If $\dim X = 0$, then $\delta_L(\gamma_1, \cdots, \gamma_n, x) = 0$ for each $x \in X$, so that $x \in \Gamma_{\text{div}}$. Thus, it suffices to show $\dim(X) = 0$. In order to lead a contradiction, let us assume the contrary, i.e., $\dim(X) > 0$. Then we can find a sequence $\{s_l\}_{l=1}^{\infty}$ in $X(\overline{K})$ with the following properties:

(a) $s_l \neq s_{l'}$ for $l \neq l'$;
(b) Any infinite subset of $\{s_l \mid l = 1, 2, \ldots\}$ is Zariski dense in X;
(c) $\delta_L(\gamma_1, \cdots, \gamma_l, s_l) \leq 1/l$ for all l.

Indeed, by the assumption of Theorem 7.1.1, there exists a sequence $\{s'_l\}_{l=1}^{\infty}$ in $X(\overline{K})$ with the properties (a) and (c) such that $\{s'_l \mid l = 1, 2, \ldots\}$ is Zariski dense in X. Since the number of all subvarieties of $X_{\overline{K}}$ is countable, a suitable subsequence of $\{s'_l\}_{l=1}^{\infty}$ satisfies the property (b).

By virtue of (c), $\{s_l \mid l = 1, 2, \ldots\}$ is small with respect to Γ_{div}. Note that for any positive integer N, $\dim G(X) = 0$ if and only if $\dim G([N](X)) = 0$. Then, by Proposition 7.2.3, replacing K by a finite extension field, X by $[N](X)$, and $\{s_l \mid l = 1, 2, \ldots\}$ by an infinite subset of $[N](\{s_l \mid l = 1, 2, \ldots\})$, we get an infinite subset S of $X(\overline{K})$ with the following properties:

(1) S is small with respect to Γ_{div};
(2) S is Zariski dense in X;
(3) (S, K) is 2-minimized.

Then, by Theorem 7.2.4, there exists an abelian subvariety C of $A_{\overline{K}}$ such that $\overline{\mathcal{D}}_n(S, K) = C^{n-1}$ for every $n \geq 2$.

If $\dim C = 0$, then $S \subseteq A(K)$. Here we use Lang's conjecture over a number field:

> If $X(K)$ is Zariski dense in X, then X is a translation of an abelian subvariety of A.

(For its proof, we refer to [7], [8]. See also Theorem 7.1.2.) Then, X is a translation of an abelian subvariety B of A. Then $G(X) = B$. This contradicts with $\dim X > 0$ and $\dim G(X) = 0$.

Next we assume $\dim(C) > 0$. We fix an integer n with $n > 2\dim(A)$. Let $\pi : A \to A/C$ be the natural homomorphism, and put $T = \pi(X)$. Let X_T^n be the fiber product of X over T in X^n. Then, we have the morphism $\beta_n : X_T^n \to A^{n-1}$ given by $\beta_n(x_1, \cdots, x_n) = (x_2 - x_1, \cdots, x_n - x_1)$. Noting that $O_K(s)^n \subseteq X_T^n$, let Y be the Zariski closure of $\bigcup_{s \in S} O_K(s)^n$ in X_T^n. Then,

$$\beta_n(Y) = \overline{\beta_n(Y)} \supseteq \overline{\left(\bigcup_{s \in S} O_K(s)^n\right)} = C^{n-1}.$$

Therefore, we get

$$\dim(X_T^n) \geq \dim(Y) \geq \dim(C^{n-1}).$$

On the other hand, by the assumption that $\dim G(X) = 0$, we have $\dim(X/T) \leq \dim(C) - 1$. Thus,

$$\begin{aligned}\dim(X_T^n) - \dim(C^{n-1}) &= (n\dim(X/T) + \dim(T)) - (n-1)\dim(C)\\ &\leq (n(\dim(C) - 1) + \dim(T)) - (n-1)\dim(C)\\ &= \dim(C) + \dim(T) - n\\ &\leq 2\dim(A) - n < 0.\end{aligned}$$

This is a contradiction. □

Appendix

We show the equality used in the proof of the Poincaré-Lelong formula (Theorem 1.2.8):

Lemma A.1. *Let $z_1, \cdots, z_d$ be the standard coordinates of $\mathbb{C}^d$, and U an open subset of $\mathbb{C}^d$. Then, for any $\eta \in A^{d-1,d-1}(U)$ with compact support, we have*

$$\int_U \log|z_1|^2 \, dd^c\eta = \int_{z_1=0} \eta.$$

First we prove the following lemmas.

Lemma A.2. *Let X be a complex manifold of dimension d. Then, for any $\alpha \in A^{p,q}(X)$ and $\beta \in A^{d-(p+1),q-(p+1)}(X)$, we have*

$$d\alpha \wedge d^c\beta = -d^c\alpha \wedge d\beta.$$

Proof: Since $\dim X = d$, we have $\partial\alpha \wedge \partial\beta = 0$, $\overline{\partial}\alpha \wedge \overline{\partial}\beta = 0$. Then, we get the assertion. □

Lemma A.3. *Let $z_1, \cdots, z_d$ be the standard coordinates of $\mathbb{C}^d$. Then, for any $\eta \in A^{d-1,d-1}(\mathbb{C}^d)$ with compact support, we have*

(i) $\lim_{\epsilon\downarrow 0} \int_{|z_1|=\epsilon} \log|z_1|^2 \, d^c\eta = 0$,

(ii) $\lim_{\epsilon\downarrow 0} \int_{|z_1|=\epsilon} d^c \log|z_1|^2 \wedge \eta = \int_{z_1=0} \eta$.

Proof: (i) Since $d^c\eta$ is a $(2d-1)$ form, each term contains dz_1 or $d\bar{z}_1$. Set $z_1 = \epsilon e^{\sqrt{-1}\theta}$. Then, we have

$$\log|z_1|^2 \, d^c\eta = \epsilon \log \epsilon^2 \, \omega.$$

The coefficient of ω is bounded when $\epsilon \downarrow 0$. This shows (i).

(ii) We have

$$d^c \log|z_1|^2 = \frac{1}{4\pi\sqrt{-1}}(\partial - \overline{\partial})\log|z_1|^2 = \frac{1}{4\pi\sqrt{-1}}\left(\frac{dz_1}{z_1} - \frac{d\bar{z}_1}{\bar{z}_1}\right).$$

Set $z_1 = \epsilon e^{\sqrt{-1}\theta}$. Then,

$$\int_{|z_1|=\epsilon} \frac{dz_1}{z_1} \wedge \eta = \int_0^{2\pi} \int_{\mathbb{C}^{d-1}} \sqrt{-1} d\theta \wedge (\eta|_{z_1=\epsilon e^{\sqrt{-1}\theta}}) \xrightarrow{\epsilon \downarrow 0} 2\pi\sqrt{-1} \int_{z_1=0} \eta$$

$$\int_{|z_1|=\epsilon} \frac{d\bar{z}_1}{\bar{z}_1} \wedge \eta = \int_0^{2\pi} \int_{\mathbb{C}^{d-1}} -\sqrt{-1} d\theta \wedge (\eta|_{z_1=\epsilon e^{\sqrt{-1}\theta}}) \xrightarrow{\epsilon \downarrow 0} -2\pi\sqrt{-1} \int_{z_1=0} \eta.$$

Thus, we have

$$\lim_{\epsilon \downarrow 0} \int_{|z_1|=\epsilon} d^c \log |z_1|^2 \wedge \eta = \int_{z_1=0} \eta.$$

□

Proof of Lemma A.1: We have

$$\int_U \log |z_1|^2 \, dd^c \eta = \lim_{\epsilon \downarrow 0} \int_{|z_1| \leq \epsilon} \log |z_1|^2 \, dd^c \eta.$$

Moreover, using Stokes's theorem and Lemma A.2, we get

$$\begin{aligned}
\int_{|z_1| \leq \epsilon} \log |z_1|^2 \, dd^c \eta &= \int_{|z_1| \leq \epsilon} d\left(\log |z_1|^2 \, d^c \eta\right) - \int_{|z_1| \leq \epsilon} d \log |z_1|^2 \wedge d^c \eta \\
&= - \int_{|z_1|=\epsilon} \log |z_1|^2 \, d^c \eta - \int_{|z_1| \leq \epsilon} d \log |z_1|^2 \wedge d^c \eta \qquad \text{(Stokes)} \\
&= - \int_{|z_1|=\epsilon} \log |z_1|^2 \, d^c \eta + \int_{|z_1| \leq \epsilon} d^c \log |z_1|^2 \wedge d\eta \qquad \text{(Lemma A.2)} \\
&= - \int_{|z_1|=\epsilon} \log |z_1|^2 \, d^c \eta - \int_{|z_1| \leq \epsilon} d\left(d^c \log |z_1|^2 \wedge \eta\right) \\
&\qquad + \int_{|z_1| \leq \epsilon} dd^c \log |z_1|^2 \wedge \eta \\
&= - \int_{|z_1|=\epsilon} \log |z_1|^2 \, d^c \eta + \int_{|z_1|=\epsilon} d^c \log |z_1|^2 \wedge \eta \\
&\qquad + \int_{|z_1| \leq \epsilon} dd^c \log |z_1|^2 \wedge \eta \qquad \text{(Stokes).}
\end{aligned}$$

If we let $\epsilon \downarrow 0$, then the first term converges to 0 by Lemma A.3(i) and the second term converges to $\int_{z_1=0} \eta$ by Lemma A.3(ii). Moreover, since $dd^c \log |z_1|^2 = 0$, the third term equals to 0. Thus, we have

$$\int_U \log |z_1|^2 \, dd^c \eta = \int_{z_1=0} \eta.$$

□

REFERENCES

[1] A. Abbes and T. Bouche, Théorème de Hilbert-Samuel "arithmétique", Ann. Inst. Fourier(Grenoble) **45** (1995), 375–401

[2] S. Ju. Arakelov, An intersection theory for divisors on an arithmetic surface, Izv. Akad. Nauk SSSR Ser. Mat. **38** (1974), 1179–1192 AMS Translation 8, (1974), 1167–1180

[3] J.-M. Bismut, H. Gillet and C. Soulé, Analytic torsion and holomorphic determinant bundles. I, II, III. Comm. Math. Phys. **115** (1988), 49–78, 79–126, 301–351

[4] J.-M. Bismut and É. Vasserot, The asymptotics of the Ray-Singer analytic torsion associated with high powers of a positive line bundle, Comm. Math. Phys. **125** (1989), 355–367

[5] G. Faltings, Calculus on arithmetic surfaces, Ann. of Math. (2) **119** (1984), 387–424

[6] G. Faltings, *Lectures on the arithmetic Riemann-Roch theorem*, Notes taken by Shouwu Zhang. Princeton Univ. Press, Princeton, NJ, 1992

[7] G. Faltings, Diophantine approximation on abelian varieties, Ann. of Math. (2) **133** (1991), 549–576

[8] G. Faltings, The general case of S. Lang's conjecture in *Barsotti Symposium in Algebraic Geometry (Abano Terme, 1991)*, 175–182, Academic Press, 1994

[9] H. Gillet and C. Soulé, Intersection theory using Adams operations, Invent. Math. **90** (1987), 243–277

[10] H. Gillet and C. Soulé, Un théorème de Riemann-Roch-Grothendieck arithmétique, C. R. Acad. Sci. Paris Sér. I Math. **309** (1989), 929–932

[11] H. Gillet and C. Soulé, Arithmetic intersection theory, Inst. Hautes Études Sci. Publ. Math. No.72 (1990), 93–174 (1991)

[12] H. Gillet and C. Soulé, Characteristic classes for algebraic vector bundles with Hermitian metric. I, II, Ann. of Math. (2) 131 (1990), 163–203; ibid (2) **131** (1990), 205–238

[13] H. Gillet and C. Soulé, An arithmetic Riemann-Roch theorem, Invent. Math. **110** (1992), 473–543

[14] P. Griffiths and J. Harris, *Principles of algebraic geometry*, Pure and Applied Mathematics. Wiley-Intersci., New York, 1978

[15] P. M. Gruber and C. G. Lekkerkerker, *Geometry of numbers*, Second edition, North-Holland, Amsterdam, 1987

[16] R. Hartshorne, *Algebraic geometry*, Graduate Texts in Mathematics, No. 52. Springer, New York, 1977

[17] H. Hironaka, Resolution of singularities of an algebraic variety over a field of characteristic zero. I, II, Ann. of Math. (2) 79 (1964), 109–203; ibid. (2) **79** (1964), 205–326

[18] S. Kawaguchi and A. Moriwaki, Inequalities for semistable families for arithmetic varieties, J. Math. Kyoto Univ **41** (2001), 97–182.

[19] S. L. Kleiman, Toward a numerical theory of ampleness, Ann. of Math. (2) **84** (1966), 293–344

[20] F. F. Knudsen and D. Mumford, The projectivity of the moduli space of stable curves. I. Preliminaries on "det" and "Div", Math. Scand. **39** (1976), 19–55.

[21] S. Lang, *Fundamentals of Diophantine geometry*, Springer, New York, 1983

[22] L. Moret-Bailly, Métriques permises, Séminaire sur les pinceaux arithmétiques: La Conjecture de Mordell, Astérisque 127 (1985), 29-87.

[23] A. Moriwaki, Inequality of Bogomolov-Gieseker type on arithmetic surfaces, Duke Math. J. **74** (1994), 713–761

[24] A. Moriwaki, Arithmetic height functions over finitely generated fields, Invent. Math. **140** (2000), 101–142

[25] A. Moriwaki, A generalization of Bogomolov and Lang over finitely generated fields, Duke Math. J. **107** (2001), 85–102

[26] B. Poonen, Mordell-Lang plus Bogomolov, Invent. Math. **137** (1999), 413–425

[27] D. Quillen, Determinants of Cauchy-Riemann operators over a Riemann surface, Funct. Anal. Appl. (1985), 31–34

[28] D. B. Ray and I. M. Singer, Analytic torsion for complex manifolds, Ann. of Math. (2) **98** (1973), 154–177

[29] M. Raynaud, Courbes sur une variété abélienne et points de torsion, Invent. Math. **71** (1983), 207–233

[30] M. Raynaud, Sous-variétés d'une variété abélienne et points de torsion, in *Arithmetic and geometry, Vol. I*, 327–352, Progr. Math., 35, Birkhäuser, Boston, Boston, Mass., 1983

[31] J.-P. Serre, Lectures on the Mordell-Weil Theorem, (2nd edition) Vieweg (1990).

[32] C. Soulé, *Lectures on Arakelov geometry*, With the collaboration of D. Abramovich, J.-F. Burnol and J. Kramer. Cambridge Univ. Press, Cambridge, 1992

[33] C. Soulé, Hermitian vector bundles on arithmetic varieties, Algebraic geometry—Santa Cruz 1995, Proc. Sympos. Pure Math. **62** (1997), 383–419

[34] L. Szpiro, Algebraic geometry over $\overline{\mathbf{Q}}$, in *Representation theory and algebraic geometry (Waltham, MA, 1995)*, 117–123, Cambridge Univ. Press, Cambridge, 1997

[35] L. Szpiro, E. Ullmo and S. Zhang, Équirépartition des petits points, Invent. Math. **127** (1997), 337–347

[36] E. Ullmo, Positivité et discrétion des points algébriques des courbes, Ann. of Math. (2) **147** (1998), 167–179

[37] S. Zhang, Positive line bundles on arithmetic varieties, J. Amer. Math. Soc. **8** (1995), 187–221

[38] S. Zhang, Small points and adelic metrics, J. Algebraic Geom. **4** (1995), 281–300

[39] S. Zhang, Equidistribution of small points on abelian varieties, Ann. of Math. (2) **147** (1998), 159–165

[40] S. Zhang, Distribution of almost division points, Duke Math. J. **103** (2000), 39–46

DEPARTMENT OF MATHEMATICS, FACULTY OF SCIENCE, KYOTO UNIVERSITY, KYOTO, 606-8502, JAPAN

E-mail address, Shu Kawaguchi: `kawaguch@kusm.kyoto-u.ac.jp`

E-mail address, Atsushi Moriwaki: `moriwaki@kusm.kyoto-u.ac.jp`

E-mail address, Kazuhiko Yamaki: `yamaki@kusm.kyoto-u.ac.jp`

Double coverings of smooth algebraic curves

Changho Keem *

0 Introduction.

Let C be a smooth projective irreducible algebraic curve over the field of complex numbers $\mathbb{C}$ or a compact Riemann surface of genus g. Let $\mathcal{M}(C)$ be the field of rational(or meromorphic) functions on C. We consider the Lüroth's semigroup of C which is defined by

$$\begin{aligned} S(C) : &= \{d \in \mathbb{N} : \exists \quad f \in \mathcal{M}(C) \quad \text{s.t.} \quad \text{ord}(f) = d\} \\ &= \{d \in \mathbb{N} : \exists \quad \text{base-point-free} \quad g^1_d \quad on \quad C\}. \end{aligned}$$

The minimal element in $S(C)$ - which is called the gonality of C and denoted by gon(C) - is a measurement of the speciality of the given curve. For example one has gon$(C) = 1$ if and only if $C \cong \mathbb{P}^1$ and for $g \geq 1$, $2 \leq \text{gon}(C) \leq [\frac{g+3}{2}]$. Furthermore, denoting by $\mathcal{M}_{g,k}$ the irreducible subvariety of the moduli space $\mathcal{M}_g$ of smooth curves of genus g consisting of classes of curves with gonality at most k, the gonality of curves gives a stratification of the moduli space $\mathcal{M}_g$

$$\mathcal{M}_{g.2} \subset \mathcal{M}_{g.3} \subset \cdots \subset \mathcal{M}_{g.k} \subset \cdots \subset \mathcal{M}_{g.[\frac{g+3}{2}]} = \mathcal{M}_g$$

with the most special curve at one end and the the curves of general moduli at the other end.

In the Lüroth's semigroup $S(C)$, there is no maximal member. However, given an algebraic curve C one may consider the integer

$$d_0(C) := \min\{d \in S(C) : m \geq d \Rightarrow m \in S(C)\},$$

*Partially supported by KRF Grant # DS0003 (2001). The author is grateful to the Organizing Committee of the "Algebraic Geometry in East Asia" for the kind invitation. Currently the author is affiliated with the RIM-SNU to which he is grateful for providing an indirect support.

i.e. the least integer in $S(C)$ so that all the numbers from the integer is contained in $S(C)$. Similarly, given a subvariety $\mathcal{M} \subset \mathcal{M}_g$ with some geometric meaning, it seems also reasonable to consider the number

$$d_0(\mathcal{M}) := \max\{d_0(C) : C \in \mathcal{M}\}.$$

In this fairly expository note, we would like to focus on (special) linear systems of relatively high degrees in order to understand some geometry of curves in connection with the invariants such as gonality and the number $d_0(C)$ which we have newly introduced. Also, we will dig out the existence of a certain special linear systems of some particular degrees suggested by the so-called Castelnuovo-Severi inequality.

The organization of this paper is as follows. In the Section 1, we will illustrate the invariants $d_0(C)$ as well as $d_0(\mathcal{M})$ for several different types of curves of genus g. We will also investigate the existence of base-point-free pencils of degree near to g on a given curve. In Section 2, we will concentrate on bi-elliptic curves and present two proofs (which are totally different from each other in nature) for the reducibility of the $W^1_{g-1}(C)$. In Section 3, we will proceed one step further concerning the existence of a base-point-free pencil of degree $g-2$ on a double covering of a curve of genus two. In Section 4, we will treat double coverings of a curve of higher genus and try to combine those results obtained in earlier sections. In Section 5, we will determine the range of the degree d for the irreducibility of $W^1_d(C)$ for double coverings C over a general curve. In the final section we will deal with general k-gonal curves and determine the invariant $d_0(C)$ and $d_0(\mathcal{M}_{g,k})$.

We use standard notation for divisors, linear series, invertible sheaves and line bundles on algebraic curves following [ACGH]. $J(C)$ is the Jacobian variety of the curve C, which is a g-dimensional abelian variety parameterizing all the line bundles of given degree d on C. We denote by $W^r_d(C)$ the subvariety of the Jacobian variety $J(C)$ consisting of line bundles of degree d with $r+1$ or more independent global sections; in other words, $W^r_d(C)$ is the subvariety of $J(C)$ consisting of all linear equivalence classes of divisors of degree d which move in linear systems of projective dimension at least r. As usual, g^r_d is an r-dimensional linear system of degree d on C, which may be possibly incomplete. If D is a divisor on C, we write $|D|$ for the associated complete linear series on C. Also, $\mathcal{O}_C(D)$ or $\mathcal{O}(D)$ is the associated invertible sheaf on C. By K_C or K we denote a canonical divisor on C. If L is a line bundle (or an invertible sheaf) we sometimes abbreviate the notation $H^i(C, L)$ (resp. $\dim H^i(C, L)$) by $H^i(L)$ (resp. $h^i(L)$) for simplicity when no confusion is likely to occur.

A base-point-free g^r_d on C defines a morphism $f : C \to \mathbb{P}^r$ onto a non-degenerate irreducible (possibly singular) curve in $\mathbb{P}^r$. If f is birational onto its image $f(C)$ the given g^r_d is called simple or birationally very ample. In case the given g^r_d is not simple, let C' be the normalization of $f(C)$. Then there is a morphism (a non-trivial covering map) $C \to C'$ and we use the same notation f for this covering map of some degree k induced by the original morphism $f : C \to \mathbb{P}^r$. The gonality of C which we introduced earlier is indeed the minimal sheet number of a covering over $\mathbb{P}^1$.

In several places, we shall make the use of various (and standard) techniques in the theory of linear series on smooth algebraic curves such as the Castelnuovo-Severi inequality and excess linear series results.

1 Existence of base-point-free pencils of degree $g - 1$.

We begin with estimating the number $d_0(C)$ for a given curve C introduced in the previous section.

Example 1.1 *For any curve C of genus g, $d_0(C) \leq g + 1$.*

Proof. One can see this by using a well-known classical theorem of Halphen which says that a curve C of genus $g \geq 2$ has a nonspecial very ample divisor if and only if $d \geq g+3$. Therefore we may take a very ample nonspecial linear system g_d^{d-g} for any $d \geq g+3$. By taking general pencil of the very ample g_d^{d-g}, we find that $d \in S(C)$ for any $d \geq g+3$. Furthermore, by considering $g^3_{g+3}(-p)$, for general $p \in C$, one has $g+2 \in S(C)$. Since this $g^2_{g+2} = g^3_{g+3}(-p)$ is still birationally very ample, we may take off one further general point $q \in C$ to get a base-point-free $g^1_{g+1} = g^2_{g+2}(-q)$. Therefore, $d \in S(C)$ for any $d \geq g+1$ whence $d_0(C) \leq g+1$. ■

Example 1.2 *If C is not hyperelliptic, $d_0(C) \leq g$.*

Proof. Note that $|K_C| = g^{g-1}_{2g-2}$ is very ample for a non-hyperelliptic curve C and that a general hyperplane $H_0 \subset \mathbb{P}^{g-1}$ meets the canonical curve $C \subset \mathbb{P}^{g-1}$ in $2g-2$ points any $g-1$ of which are linearly independent by the uniform position property; cf. [ACGH, page 113]. Let $p_1 + \cdots + p_{g-2} + r_1 + \cdots + r_g$ be the hyperplane divisor determined by the general hyperplane H_0. We choose another hyperplane $H \subset \mathbb{P}^{g-1}$ different from H_0 such that $p_i \in H$ for all $i = 1, \cdots g-2$. Note that $r_i \notin H$ for any $i = 1, \cdots g$: Suppose otherwise, i.e. $p_1, \cdots, p_{g-2}, r_i \in H \cap H_0 = \mathbb{P}^{g-3}$ for some i. Then these $g-1$ points of the general hyperplane section of the canonical curve would violate the uniform position property.

Therefore by taking off any $g-2$ points $p_1, p_2, \cdots, p_{g-2}$ of the general hyperplane section from the canonical series, we see that $|K_C - p_1 - p_2 - \cdots - p_{g-2}| = g^1_g$ is base-point-free. ■

If C is a hyperelliptic curve, one can determine the number $d_0(C)$ fairly easily as follows.

Example 1.3 *Let C be a hyperelliptic curve of genus $g \geq 4$.*
(i) *For $d \leq g$, any complete linear system g^r_d on C for $d \leq g$ is of the form $|rg^1_2 + p_1 + \cdots + p_{d-2r}|$ where no two of the p_i's are conjugate under the hyperelliptic involution.*
(ii) *If the genus of C is even, there does not exist a base-point-free pencil of degree $g-1$. In case g is odd, there exists a base-point-free pencil of degree $g-1$ and such a base-point-free g^1_{g-1} is incomplete, which is a subseries of the complete $|\frac{g-1}{2}g^1_2|$.*
(iii) *$d_0(\mathcal{M}_{g,2}) = d_0(C) = g$ if g is even and $d_0(\mathcal{M}_{g,2}) = d_0(C) = g+1$ if g is odd.*

Proof. (i) is an immediate consequence of the Geometric Version of the Riemann-Roch Theorem; cf. [ACGH, page 12–13]. Alternatively, one may verify (i) in the following way. Suppose there exists a base-point-free pencil (either complete or incomplete) g^1_d which is not composed with the hyperelliptic involution. In other words, we assume that there exists $f \in \mathcal{M}(C)$, $\mathrm{ord}(f) = d$ with $d \leq g$ on a hyperelliptic curve $\pi : C \longrightarrow \mathbb{P}^1$ such that $f \neq \pi^* e$ for any $e \in \mathcal{M}(\mathbb{P}^1)$. Then one has a morphism

$$\phi = \pi \times f : C \longrightarrow \mathbb{P}^1 \times \mathbb{P}^1$$

which is birational onto its image. Since $\phi(C) \in |2L + dM|$, where L and M are two different rulings of the smooth quadric surface $\mathbb{P}^1 \times \mathbb{P}^1$, one has $g \leq p_a(\phi(C)) = d-1$, contrary to the assumption $d \leq g$. Statement (ii) and (iii) readily follows from (i). ■

Having seen that the invariant $d_0(C)$ or $d_0(\mathcal{M}_{g,2})$ is equal to g or $g+1$ depending on the parity of g for any hyperelliptic curve, we would like to turn our attention to the case of non-hyperelliptic curves. Specifically one may ask if there exists a base-point-free pencil of degree $g-1$ on every non-hyperelliptic curve of genus g. Unfortunately, the existence of a base-point-free g^1_{g-1} on a non-hyperelliptic curve does not seem to be so obvious as in the hyperelliptic case. However the following theorem of J. Harris provides an affirmative answer to our question.

Theorem 1.4 (Harris, [ACGH]) *Let C be a non-hyperelliptic curve of genus $g \geq 4$. Then there exists a base-point-free complete pencil g^1_{g-1} on C.*

Before presenting the proof of Theorem 1.4, one needs to invoke the following theorems of H. Martens as well as its extension by Mumford regarding the variety of special linear systems on a fixed smooth projective algebraic curve; cf. [ACGH, page 191–195] and [MH].

Theorem 1.5 (H. Martens) *Let C be a smooth curve of genus $g \geq 3$. Let d and r be integers such that $d \leq g + r - 2$, $r \geq 1$. Then*

$$\dim W_d^r(C) \leq d - 2r.$$

Furthermore, equality holds if and only if C is hyperelliptic.

Theorem 1.6 (Mumford) *Let d and r be integers such that $d \leq g + r - 3$, $r \geq 1$. Suppose that $\dim W_d^r(C) = d - 2r - 1$. Then C is either trigonal, bi-elliptic or a smooth plane quintic.*

Proof of Theorem 1.4. Given a non-hyperelliptic curve C of genus g, we assume a component $\Sigma \subset W_{g-1}^1(C)$, such that for a general element $L \in \Sigma$, the complete linear system associated to L has a base point. Therefore we have

$$\Sigma = \Lambda + W_1(C),$$

where Λ is an irreducible sublocus of $W_{g-2}^1(C)$. Recall that the singular locus of the theta divisor of the Jacobian variety $J(C)$ which is isomorphic to $W_{g-1}^1(C)$ is of pure dimension $g - 4$ for a non-hyperelliptic curve C. Therefore it follows that

$$\dim W_{g-2}^1(C) = \dim\Lambda = g - 5.$$

Hence by the Theorem 1.6, we deduce that one of the following holds:

(i) C is a smooth plane quintic and

$$\Sigma = W_5^1(C) = \{\mathcal{O}_C(1) \otimes \mathcal{O}_C(p - q) : p, q \in C.\}$$

(ii) C is a trigonal curve and

$$\Sigma = \{\mathcal{O}(g_3^1 + p_1 \cdots + p_{g-4}) : p_1, \cdots, p_{g-4} \in C\}.$$

(iii) C is a double covering of an elliptic curve E with the two sheeted morphism $\pi : C \longrightarrow E$ and

$$\Sigma = \{\pi^* g_2^1(E) \otimes \mathcal{O}(p_1 + \cdots + p_{g-5}) : p_1, \cdots, p_{g-5} \in C,\ g_2^1(E) \in W_2^1(E)\}.$$

Note that a smooth plane quintic has a (incomplete) base-point-free pencil of degree 5, which is a subsystem of the hyperplane system g_5^2.

If C is a trigonal curve with the unique trigonal pencil g^1_3 (uniqueness follows from the assumption $g \geq 4$), one consider a locus

$$\Sigma' = \{\mathcal{O}_C(K_C - g^1_3 - p_1 - \cdots - p_{g-4}) : p_1, \cdots, p_{g-4} \in C\},$$

which is a component of $W^1_{g-1}(C)$. We now claim that $\Sigma' \subset W^1_{g-1}(C)$ is different from Σ. Suppose

$$\Sigma' = \mathcal{O}_C(K - g^1_3) - W_{g-4}(C) = \Sigma = \mathcal{O}_C(g^1_3) + W_{g-4}(C),$$

then we must have $\mathcal{O}_C(K - 2g^1_3) - W_{g-4}(C) = W_{g-4}(C)$. But this is an absurdity by noting that $h^0(C, \mathcal{O}_C(K_C - 2g^1_3) = g - 4$ and hence $h^0(C, \mathcal{O}_C(K_C - 2g^1_3 - p_1 - \cdots - p_{g-4}) = 0$ for a general choice of the points $p_1, \cdots, p_{g-4} \in C$. Since Σ is the only component of $W^1_{g-1}(C)$ whose general element has a base point, we see that a general member of the locus Σ' is a base-point-free complete pencil of degree $g - 1$.

If C is a bi-elliptic curve, one may consider the component

$$\Sigma' = \{\mathcal{O}_C(K_C - \pi^*(g^1_2) - q_1 - \cdots - q_{g-5}) : g^1_2 \in W^1_2(E), q_i \in C\} \subset W^1_{g-1}(C).$$

As is easy to see, for any g^1_2 on E the linear series $|K_C - \pi^*(g^1_2)|$ is not birationally very ample. Instead it induces the morphism π onto E if $g \geq 6$. Therefore

$$|K_C - \pi^*(g^1_2) - q_1 - \cdots - q_{g-5}| = \pi^* g^1_2(E) \otimes \mathcal{O}(p_1 + \cdots + p_{g-5})$$

for some $g^1_2(E) \in W^1_2(E)$. where p_i and q_i are conjugate to the other under the bi-elliptic involution, and it follows that $\Sigma = \Sigma'$. Therefore we cannot assure the existence of a component which is different from Σ as in an elementary way as in the case (ii).

We would like exhibit two completely different proofs for the existence of a base-point-free pencil of degree $g - 1$ on a bi-elliptic curve. The first one is due to J. Harris using enumerative method and the second one is due to S. Park using only elementary methods. Since these proofs are somewhat lengthy and need several preparatory results, we will return to the full treatment of this bi-elliptic case in the next section. ∎

We now close this section by recalling the following Castelnuovo-Severi inequality which one can prove easily by using Hodge index theorem and the adjunction formula.

Lemma 1.7 (Castelnuovo-Severi inequality [A, Theorem 3.5]) *Assume that there exist two curves C_1 and C_2 of genus h_1 and h_2 respectively, so that C is a k_i-sheeted covering of C_i $(i = 1, 2)$ with the non-trivial morphisms $\pi_i : C \longrightarrow C_i$.*

Assume further that the morphism $\pi = \pi_1 \times \pi_2 : C \longrightarrow C_1 \times C_2$ is birational onto its image. Then

$$g \leq (k_1 - 1)(k_2 - 1) + k_1 h_1 + k_2 h_2.$$

2 Irreducibility of $W^1_{g-1}(C)$ for a bi-elliptic curve.

Before proceeding, it is worthwhile to remark that some of the other proofs which appeared in the literature for the existence of base-point-free pencil of degree $g-1$ on a bi-elliptic curve do not seem to be so kind to the reader; for example in the proof of [Ho1, Theorem 5], the author obtained a plane model of a bi-elliptic curve of degree $g+1$ with a singular point s of certain high multiplicity. He then proceeded to exhibit the existence of another singular point by using a well-known formula for the geometric genus of a singular plane curve. Unfortunately, the singular point different from s could be infinitely singular points lying over s. Therefore the projection method used in [Ho1] to obtain a complete and base-point-free pencil of degree $g-1$ which is cut out by lines through the other singular point does not work well if the singular point s of high multiplicity is not an ordinary singular point. Incidentally, the same objection applies to the proof of Shokurov [Sh]. A proof due to J. Harris, which was sketched in [ACGH, Chapter VIII; Exercise D and F], seems to be the only complete proof without a gap which appeared in the literature as far as the author knows. On the other hand, the proof due to Harris uses the so-called enumerative method as well as several advanced results in Brill-Noether theory and hence one needs a quite a bit of heavy duty machinery for a proof of this seemingly simple fact; indeed the proof in [ACGH] shows the reducibility of $W^1_{g-1}(C)$, which is a much harder problem and the existence of the base-point-free pencil follows as a corollary. On the other hand, the proof due to S. Park is much simpler using only easy geometric arguments following an idea from [CKM1, Appendix].

In order to reproduce the proof of J. Harris, we first recall some of the notations used in [ACGH]. Let C be a smooth algebraic curve, not necessarily bi-elliptic. Let $u : C_d \to J(C)$ be the abelian sum map and let θ be the class of the theta divisor in $J(C)$. Let

$$u^* : H^*(J(C), \mathbb{Q}) \to H^*(C_d, \mathbb{Q}) \text{ and } u_* : H^{2(d-g+k)}(C_d, \mathbb{Q}) \to H^{2k}(J(C), \mathbb{Q})$$

be the homomorphisms induced by u. By abusing notation, we use the same letter θ for the class $u^*\theta$. By fixing a point p on C, one has the map $\iota : C_{d-1} \to C_d$ defined by $\iota(D) = D + p$. We denote the class of $\iota(C_{d-1})$ by x.

We now list up several formulae about cycles in a symmetric product of a curve. These are given without proof, and the readers are referred to the last chapter of [ACGH].

Formula 2.1 (Push-pull formula I) *For every* $\alpha \in H^{2(d-g+k)}(C_d, \mathbb{Q})$ *and* $\beta \in H^{2(g-k)}(J(C), \mathbb{Q})$,

$$u_*(\alpha \cdot u^*\beta) = u_*\alpha \cdot \beta.$$

Formula 2.2 (Diagonal formula) *The diagonal mappings for* C_d *are defined to be the mappings*

$$\phi_a : C_{n_1} \times \cdots \times C_{n_k} \longrightarrow C_d$$

given by

$$\phi_a(D_1, \cdots, D_k) = a_1 D_1 + \cdots + a_k D_k$$

where $a = (a_1, \cdots, a_k)$ *and* $\sum_{i=1}^{k} a_i n_i = d$.

The image, via the diagonal mapping ϕ_a, *of the fundamental class of* $C_{n_1} \times \cdots \times C_{n_k}$ *is given by*

$$(\phi_a)_*[C_{n_1} \times \cdots \times C_{n_k}] = \left[\sum_{\alpha \geq \beta} \frac{(-1)^{\alpha+\beta}}{\beta!(\alpha-\beta)!} P_a(t)^{e-g+\beta} R_a(t)^{g-\beta} x^{d-e-\alpha} \theta^{\alpha} \right]_{t_1^{n_1} \cdots t_k^{n_k}}$$

where $e = n_1 \cdots + n_k$, $R_a(t) = 1 + \sum_{i=1}^{k} a_i^2 t_i$, $P_a(t) = 1 + \sum_{i=1}^{k} a_i t_i$ *and* $[F(t)]_{t_1^{n_1} \cdots t_k^{n_k}}$ *is the coefficient of* $t_1^{n_1} \cdots t_k^{n_k}$ *in the formal Laurent series of* $F(t) = F(t_1 \cdots t_k)$.

Formula 2.3 (Poincaré formula) $w_d = \frac{\theta^{g-d}}{(g-d)!}$, *where* $w_d \in H^{2(g-d)}(J(C), \mathbb{Q})$ *is the class of* $W_d(C)$.

Let's also recall that given a cycle Z in C_d, the assignments

$$Z \mapsto A_k(Z) := \{E \in C_{d+k} : E - D \geq 0 \quad \text{for some} \quad D \in Z\},$$

$$Z \mapsto B_k(Z) := \{E \in C_{d-k} : D - E \geq 0 \quad \text{for some} \quad D \in Z\}$$

induce maps

$$A_k : H^{2m}(C_d, \mathbb{Q}) \to H^{2m}(C_{d+k}, \mathbb{Q}), \quad B_k : H^{2m}(C_d, \mathbb{Q}) \to H^{2m-2k}(C_{d-k}, \mathbb{Q})$$

and the following formulas for symmetric products hold.

Formula 2.4 (Push-pull formula II) *Given* $z \in H^{2d-2m}(C_d, \mathbb{Q})$ *and* $w \in H^{2m+2}(C_{d+1}, \mathbb{Q})$

$$A(z) \cdot w = a \cdot B(w)$$

where $A = A_1$ *and* $B = B_1$.

Also the following relations hold.

Formula 2.5 (i) $A_k = \frac{1}{k!}A^k, B_k = \frac{1}{k!}B^k$
(ii) $B_k(x^\alpha\theta^\beta) = \sum_{j=0}^{k}\binom{\alpha}{k-j}\binom{\beta}{j}\binom{g-\beta+j}{j}j!x^{\alpha-k+j}\theta^{\beta-j}$
(iii) $A_k(x^\alpha\theta^\beta) = \sum_{j=0}^{k}\binom{\beta}{i}\binom{g-\beta+i}{i}\binom{d+k-\alpha-2\beta}{k-i}i!x^{\alpha+i}\theta^{\beta-i}$

We now let $C \xrightarrow{\pi} E$ be a k-sheeted covering over a curve E of genus e. We define
(a) $\tilde{\Gamma}_0 = \{\pi^*q : q \in E\} \subset C_k$
(b) $\tilde{\Gamma}_d = \{\pi^*q + D : q \in E \text{ and } D \in C_d\} = A_d(\tilde{\Gamma}_0) \subset C_{d+k}$
(c) $\tilde{\Gamma}_0 = u(\tilde{\Gamma}_0)$
(d) $\Gamma_d = u(\tilde{\Gamma}_d) = u(\tilde{\Gamma}_0) + W_d(C) = \Gamma_d + W_d(C)$
(e) $\tilde{\Sigma}_0 = \{\pi^*(p+q) : p, q \in E\} \subset C_{2k}$
(f) $\tilde{\Sigma}_d = \{\pi^*(p+q) + D : p, q \in E, D \in C_d\} = A_d(\tilde{\Sigma}_0) \subset C_{d+2k}$
(g) $\Delta = \{2p + p_1 + \cdots + p_{k-2}, p \in C\} \subset C_k$ is the main diagonal
(h) $\tilde{\gamma}_0, \tilde{\gamma}_d, \gamma_d, \tilde{\sigma}_0, \tilde{\sigma}_d, \delta$ are the fundamental classes of $\tilde{\Gamma}_0, \tilde{\Gamma}_d, \Gamma_d, \tilde{\Sigma}_0, \tilde{\Sigma}_d, \Delta$ respectively.

Now it is easy to see that $(\tilde{\gamma}_0 \cdot x)_{C_k} = 1$ and $(\tilde{\gamma}_0 \cdot \delta)_{C_k} = 2g-2-k(2e-2)$, which is the number of branch points of π. On the other hand, by the Diagonal formula 2.2, $\delta = (2g-2+2k) - 2\theta$. Therefore we have $\gamma \cdot \theta = \tilde{\gamma}_0 \cdot \theta = ke$; the first equality holds by the Push-Pull formula I (Formula 2.1). Also, by using the Push-Pull formula II (Formula 2.4) and Formula 2.5(i), we get

$$(\tilde{\gamma}_d \cdot \theta^{d+1})_{C_{d+k}} = (A_d(\tilde{\gamma}_0) \cdot \theta^{d+1})_{C_{d+k}} = (\tilde{\gamma}_0 \cdot B_d(\theta^{d+1}))_{C_k}.$$

Using Formula 2.5(ii), we finally get

$$\tilde{\gamma}_d \cdot \theta^{d+1} = \gamma_d \cdot \theta^{d+1} = (\tilde{\gamma}_0 \cdot B_d(\theta^{d+1}))_{C_k} = \frac{(g-1)!}{(g-d-1)!}(d+1)ke. \tag{2.5.1}$$

The following important results which are due to Kempf, Kleiman-Laksov, and Griffiths-Harris will be used in proving the reducibility of $W^1_{g-1}(C)$ for a bi-elliptic curve C; cf. [GH] and [KL].

Proposition 2.6 *Let C be a smooth algebraic curve of genus g. Let d, r be integers such that $d \geq 1, r \geq 0$ and $g - d + r \geq 0$. Then*

(i) *In case $\rho = \rho(g, r, d) = g - (r+1)(g - d + r) \geq 0$, $W_d^r(C)$ is not empty and every one of its components has dimension at least equal to ρ.*

(ii) *Let $X_1, \cdots, X_s$ be the irreducible components of $C_d^r := \{D \in C_d : r(D) \geq r\} \subset C_d$ and $k_1, \cdots, k_s$ their respective multiplicities. Suppose that $W_d^r(C)$ has pure dimension ρ, and denote by $w_d^r \in H^{2(g-\rho)}(JC), \mathbb{Z})$ the cohomology class of the cycle $\sum k_i u(X_i)$; u is the abelian sum map. Then*

$$w_d^r = \frac{r! \cdots 0!}{(g-d+2r)! \cdots (g-d+r)!}(g-\rho)! \ w_\rho.$$

Proof of Theorem 1.4 for bi-elliptic case due to J. Harris. Since C is a bi-elliptic curve we have $k = 2, e = 1$. For $d = g - 5$, we also have

$$u(\tilde{\Sigma}_0) = \pi^* W_2^1(E) = \pi^* J(E)$$

$$\Gamma_0 = u(\tilde{\Gamma}_0) = \pi^* W_1(E) = \pi^* J(E)$$

$$\Sigma = u(\tilde{\Sigma}_{g-5}) = u(\tilde{\Sigma}_0) + W_{g-5}(C)$$

$$\Gamma_{g-5} = \Gamma_0 + W_{g-5}(C).$$

Thus, denoting by σ the fundamental class of Σ, it follows from (2.5.1) that

$$\sigma \cdot \theta^{g-4} = \gamma_{g-5} \cdot \theta^{g-4} = \frac{(g-1)!(g-4)}{12}.$$

On the other hand,

$$w_{g-1}^1 \cdot \theta^{g-4} = \frac{g!}{12}$$

by the Poincaré formula (2.3) and Proposition 2.6. This shows that $\Sigma \neq W_{g-1}^1(C)$ and hence $W_{g-1}^1(C)$ is reducible.

It may be checked that a trigonal curve of genus at least 5 has a base-point-free g_{g-1}^1 by using this method. In fact,

$$\sigma \cdot \theta^{g-4} = w_{g-4} \cdot \theta^{g-4} = \frac{g!}{24}$$

while

$$w_{g-1}^1 \cdot \theta^{g-4} = \frac{g!}{12}.$$

Hence $\Sigma \neq W_{g-1}^1(C)$ for a trigonal curve C. ∎

For the other proof due to S. Park, we need the following proposition.

Proposition 2.7 ([ACGH, Chapter III-Exercise F]) *Let L be a line bundle of degree $d \geq 2g+2$ on a smooth curve C of genus g. Let*

$$\varphi_L : C \to \mathbb{P}^{d-g}$$

be the embedding induced by L. Then $\varphi_L(C)$ is the intersection of quadrics.

Proof of Theorem 1.4 for bi-elliptic case due to S. Park. This proof, which we will present in details for the convenience of the reader, is almost same as one can find in [P]. We break up the proof in several steps.

Step 1 : The canonical image of C lies on a cone of degree $g-1$.

Since C is non-hyperelliptic, we may identify C with its canonical image $\varphi_K(C)$ in $\mathbb{P}^{g-1}$. For $r_i \in E$, let $\pi^*(r_i) = p_i + \bar{p}_i$; $i = 1, 2$. Then for the effective divisor

$$D = \pi^*(r_1 + r_2) = p_1 + \bar{p}_1 + p_2 + \bar{p}_2 \in g^1_4 = \pi^*(g^1_2) = \pi^*(|r_1 + r_2|),$$

$\dim\overline{D} = 2$ by the geometric version of Riemann-Roch theorem; i.e. D spans a 2−plane in $\mathbb{P}^{g-1}$. Therefore for any $r, r' \in E$, the two lines spanned by $\pi^*(r)$ and $\pi^*(r')$ must intersect. Since C is non-degenerate in $\mathbb{P}^{g-1}$, all the lines spanned by $\pi^*(r), r \in E$ pass through a point $v \in \mathbb{P}^{g-1}$. Let

$$S_{g-1} = \bigcup_{r \in E} \overline{\pi^*(r)},$$

which is a cone with vertex v containing the canonical image of C. Furthermore, one sees easily that $v \notin \varphi_K(C)$; if $v \in \varphi_K(C)$, then the divisor $v + \pi^*(r_1)$ with $\pi(v) \neq r_1$ is a trisecant line hence $v + \pi^*(r_1)$ moves in a pencil which is contradictory to the Castelnuovo-Severi inequality.

Let $H \cong \mathbb{P}^{g-2}$ be a hyperplane in $\mathbb{P}^{g-1}$ not passing through v and φ be the projection away from v to H. By our construction, E is isomorphic to the hyperplane section $H \cap S_{g-1}$, which we use the same symbol E for simplicity. A hyperplane section $H_E = E \cap \mathbb{P}^{g-3} \subset H = \mathbb{P}^{g-2}$ of E is the image under φ of the intersection $\varphi_K(C) \cap < H_E, v >$, where $< H_E, v >$ is the hyperplane in $\mathbb{P}^{g-1}$ spanned by H_E and v. Since the projection φ is indeed the degree two morphism $\pi : C \longrightarrow E$,

$$\deg E = \deg(H_E) = \frac{1}{2}\deg(\varphi_K(C) \cap < H_E, v >) = \frac{1}{2}(2g-2) = g-1,$$

and hence $\deg S_{g-1} = g-1$.

Step 2 : There is a sequence of birational maps $\{\varphi_i\}_{0\le i\le g-4}$ *with the following properties.*
(1) $\varphi_0 = \varphi_K : C \longrightarrow \mathbb{P}^{g-1}$ *is the canonical map of* C.
(2) *For* $1 \le i \le g-4$, $\varphi_i : \mathbb{P}^{g-i} \dashrightarrow \mathbb{P}^{g-1-i}$ *is a projection away from a point and restricted to* $(\varphi_{i-1}\circ \ldots \circ \varphi_0)(C)$, $\varphi_i : (\varphi_{i-1}\circ\ldots\circ\varphi_0)(C) \dashrightarrow \mathbb{P}^{g-1-i}$ *is still birational onto its image.*
(3) $(\varphi_i \circ \cdots \circ \varphi_0)(C)$ *has only one singular point for every* $2 \le i \le g-4$.
(4) $(\varphi_{g-4}\circ\cdots\circ\varphi_0)(C)$ *lies on a cubic cone in* $\mathbb{P}^3$.

Choose a point $p_1 \in C_{g-1} := \varphi_0(C)$ and let φ_1 be the projection away from p_1 onto $\mathbb{P}^{g-2}$. Let $q_{g-1} := \bar{p}_1$ be the conjugate point of p_1 with respect to π and take $C_{g-2} := \varphi_1(C_{g-1})$. The image of S_{g-1}under the projection φ_1 is also a cone of degree $g-2$ with vertex $q_{g-2} := \varphi_1(q_{g-1}) = \varphi_1(v)$, which is denoted by S_{g-2}. Now we take a general point p_2 in C_{g-2} and let φ_2 be the projection away from p_2 onto $\mathbb{P}^{g-3}$. Applying this process repeatedly, we can obtain $\{(\varphi_i, S_{g-1-i}, C_{g-1-i}, p_i, q_{g-1-i}) : i = 1,\cdots,g-4\}$ as follows;

$$\begin{array}{ccccccc}
 & \varphi_1 & & \varphi_2 & & \varphi_{g-4} & \\
\mathbb{P}^{g-1} & \dashrightarrow & \mathbb{P}^{g-2} & \dashrightarrow & \cdots & \dashrightarrow & \mathbb{P}^3 \\
\cup & & \cup & & & & \cup \\
S_{g-1} & \dashrightarrow & S_{g-2} & \dashrightarrow & \cdots & \dashrightarrow & S_3 \\
\cup & & \cup & & & & \cup \\
C_{g-1} & \dashrightarrow & C_{g-2} & \dashrightarrow & \cdots & \dashrightarrow & C_3
\end{array}$$

where φ_i is the projection away from a general point $p_i \in C_{g-i}$ onto a hyperplane, $q_{g-2-i} = \varphi_i(q_{g-1-i}), S_{g-2-i} = \varphi_i(S_{g-1-i})$ and $C_{g-2-i} = \varphi_i(C_{g-1-i})$. Note that
(i) S_{g-1-i} is a cone with vertex q_{g-1-i} of degree $g-1-i$,
(ii) C_{g-1-i} is a curve of degree $2g-2-i$ with $\mathrm{mult}_{q_{g-1-i}}C_{g-1-i} = i$ where C_{g-1-i} is the image of the morphism induced by $|K - p_1 - \cdots - p_i|$,
(iii) $\dim|K - p_1 - \cdots - p_i - \bar{p}_1 - \cdots - \bar{p}_i| = \dim|K - p_1 - \cdots - p_i| - 1$ and
(iv) $|K - p_1 - \cdots - p_i - \bar{p}_1 - \cdots - \bar{p}_i|$ is base-point-free.
(v) In particular the image of C_4 in $\mathbb{P}^3$ under φ_{g-4} lies on a cubic cone S_3.

Let $E_k := S_k \cap H$ where $H \cong \mathbb{P}^{k-1}$ is a hyperplane not passing through the vertex q_k of $S_k \subset \mathbb{P}^k$. Since S_{g-1} is a cone over the elliptic curve $E \subset \mathbb{P}^{g-2}$ of degree $g-1$ and S_k is obtained by successive projections, we easily see that $\deg E_k = \deg S_k = k$ and $E_k \cong E$, i.e. $g(E_k) = 1$. Applying Proposition 2.7 to the hyperplane bundle on $E_k \subset \mathbb{P}^{k-1}$, E_k is cut out by quadrics in H and hence S_k is also cut out by quadrics in $\mathbb{P}^k$ for $k \ge 4$.

Note that, for $k \ge 3$, any singular point of C_k different from q_k may only arise from a trisecant line of $C_{k+1} \subset S_{k+1} \subset \mathbb{P}^{k+1}$ other than rulings of the cone S_{k+1}.

Since S_{k+1} is cut out by quadrics for $k \geq 3$, we see that there is no trisecant line of C_{k+1} other than rulings of S_{k+1}. Therefore C_k has no singular point other than q_k for $k = 3, \cdots, g-3$.

Step 3 : C is birational to a plane curve $C_2 \subset \mathbb{P}^2$ of degree $g+1$ with ordinary singular point of multiplicity $g-3$.

The projection away from a general point $p_{g-3} \in C_3$, denoted by φ_{g-3}, gives a birational map from C_3 onto $C_2 := \varphi_{g-3}(C_3)$ in $\mathbb{P}^2$. Note that

$$\deg \varphi_{g-3}(C_3) = \deg C_3 - 1 = g+1$$

and the point $q_2 := \varphi_{g-3}(q_3)$ is singular point with multiplicity $g-3$. We observe that q_2 being an ordinary singular point is equivalent to

$$\text{(A)} \qquad |K - p_1 - \cdots - p_{g-3} - \bar{p}_1 - \cdots - \bar{p}_{g-3} - \bar{p_i} - \bar{p_j}| = \emptyset$$

for all distinct $i, j \in \{1, 2, \cdots g-3\}$. Therefore in order to show that q_2 is an ordinary singular point, we need to choose the points $p_1, \cdots, p_{g-4} \in C$ properly in step 2 as well as p_{g-3} which satisfy the condition (A). We now set

$$T_{ij} := \{(p_1, \cdots, p_{g-3}) \in C^{g-3} : \dim|K - p_1 - \cdots - p_{g-3} - \bar{p_1} - \cdots - \bar{p}_{g-3} - \bar{p}_i - \bar{p}_j| \geq 0\}$$

for distinct $i, j \in \{1, 2, \cdots g-3\}$ and $T := \cup T_{ij}$. Since T_{ij} is closed in the $(g-3)$-fold product C^{g-3}, so is T. Therefore it is sufficient to show that each T_{ij} is a proper closed subset in C^{g-3}; then any $(p_1, \cdots, p_{g-3}) \in C^{g-3} \backslash T$ satisfies the condition (A). Accordingly, without loss of generality, we assume $(i, j) = (1, 2)$ and proceed as follows.

Claim: For distinct $p_1, p_2 \in C$ such that $\pi(p_1) \neq \pi(p_2)$, $|p_1 + 2\bar{p}_1 + p_2 + 2\bar{p}_2| = g^1_6$. For this we consider $|2p_1 + 2p_2 + 2\bar{p}_1 + 2\bar{p}_2|$ and let $\pi(p_i) = r_i \in E, i = 1, 2$. Since C cannot be hyperelliptic

$$\dim|2p_1 + 2p_2 + 2\bar{p}_1 + 2\bar{p}_2| = \dim|\pi^*(2r_1 + 2r_2)| = 3$$

by Clifford's theorem. Note that the linear series $|2p_1 + 2p_2 + 2\bar{p}_1 + 2\bar{p}_2|$ induces the double covering $\pi : C \longrightarrow E$. Therefore,

$$\dim|p_1 + 2\bar{p}_1 + p_2 + 2\bar{p}_2| = \dim|2p_1 + 2p_2 + 2\bar{p}_1 + 2\bar{p}_2| - 2 = 1$$

since $\pi(p_1) \neq \pi(p_2)$ and this finishes the proof of the claim.

By the Claim, $\dim|K - p_1 - p_2 - 2\bar{p_1} - 2\bar{p_2}| = g - 6$ and therefore we may choose general points $p_3, \cdots, p_{g-4} \in C$ so that

$$\dim|K - p_1 - p_2 - 2\bar{p_1} - 2\bar{p_2} - p_3 - \cdots - p_{g-4}| = 0.$$

Finally we take a point $p_{g-3} \in C$ such that

$$p_{g-3} \notin |K - p_1 - p_2 - 2\bar{p_1} - 2\bar{p_2} - p_3 - \cdots - p_{g-4}|$$

and p_{g-3} is not a conjugate point of p_i for any $i = 1, \cdots, g-4$. Hence

$$(p_1, \cdots, p_{g-3}) \notin T_{12}$$

and this shows that T_{12} is a proper closed subset of C^{g-3}.

Step 4 : The plane curve C_2 constructed in Step 3 has another singular point with multiplicity 2.

Since q_2 is a singular point of multiplicity $g-3$, we have

$$g \leq p_a(C_2) - \frac{(g-3)(g-4)}{2} = \frac{g(g-1)}{2} - \frac{(g-3)(g-4)}{2} = 3g - 6.$$

Note that $g < 3g - 6$ for $g \geq 6$. Since q_2 is an ordinary singular point, it follows that there exist another singular point, say $q_0 \in C_2$ besides q_2. Suppose that $\text{mult}_{q_0} C_2 \geq 3$. Recall that C_k has only one singular point q_k for every $k = 3, \cdots, g-3$. Therefore the singular point $q_0 \in C_2$ with $\text{mult}_{q_0} C_2 \geq 3$ arises from at least a 4-secant line passing through p_{g-3} other than ruling of the cone S_3. Since S_3 is a cubic cone, this is impossible. Therefore we have $\text{mult}_{q_0} C_2 = 2$ and the pencil of lines through q_0 cuts out base-point-free and complete g^1_{g-1} on C. ■

We now state the result which we have demonstrated in this section as follows.

Theorem 2.8 *Let C be a bi-elliptic curve of genus $g \geq 6$. Then $W^1_{g-1}(C)$ is reducible and there exists a base-point-free complete pencil of degree $g-1$ which is not composed with the bi-elliptic involution.*

So far what we have seen that for a non-hyperelliptic curve C, $d_0(C) \leq g - 1$, which easily yields the following corollary.

Corollary 2.9 *Let C be an algebraic curve of genus g.*

$$d_0(C) \geq g \textit{ if and only if } C \textit{ is hyperelliptic.}$$

On the other hand we may ask:

Question 2.10 *Let C be a non-hyperelliptic curve of genus g. Is $d_0(C) \leq g-2$?*

Remark 2.11 *The answer to the question shall be provided in the next section. At this point, it is worthy of noting that every base-point-free pencil on a bi-elliptic curve $C \xrightarrow{\pi} E$ of degree $d \leq g-2$ is composed with the bi-elliptic involution by Castelnuovo-Severi inequality. In other words, for $d \leq g-2$ any base-point-free g^1_d on C is of the form $\pi^*(g^1_{\frac{d}{2}})$ for some $g^1_{\frac{d}{2}}$ on E. Therefore Theorem 2.8 is quite optimal in this sense.*

3 Existence of base-point-free pencils of degree $g-2$.

In this section, we turn to the problem concerning the existence of base-point-free pencil of degree $g-2$ on a given algebraic curve of genus g. Our goal is the following:

Theorem 3.1 *Let C be a non-hyperelliptic curve of genus $g \geq 11$, and assume that C is not bi-elliptic of odd genus. Then there exists a base-point-free pencil of degree $g-2$ on C.*

Unfortunately, the things become involved much more than the case of degree $g-1$. Therefore we still need several auxiliary results and we begin with stating the following theorem which is the generalization of the H. Martens and Mumford type theorems regarding the variety of special linear systems on a fixed algebraic curve; cf. [K] and [CKO1].

Theorem 3.2 (Coppens, Cho, Keem, Ohbuchi and Mukai) *Let d and r be integers such that $d-2r-2 > \rho(d,g,r) = g-(r+1)(g-d+r)$, $r \geq 1$. Assume that*

$$\dim W^r_d(C) \geq d-2r-2 \geq 0.$$

Then C is either hyperelliptic, trigonal, bi-elliptic, tetragonal, a possibly singular plane curve of degree 6 or a double covering of a curve of genus 2.

Lemma 3.3 *If C is trigonal with $g \geq 7$, then* $\dim W^2_{g-1}(C) = g-7$.

Proof. For every $p_1, \cdots, p_{g-7} \in C$, note that $\mathcal{O}(2g^1_3+p_1+\cdots+p_{g-7}) \in W^2_{g-7}(C)$ and hence $\dim W^2_{g-1}(C) \geq g-7$. To prove the reverse inequality, consider the natural map

$$\nu : H^0(C, \mathcal{O}(K-E)) \otimes H^0(C, \mathcal{O}(K-E)) \longrightarrow H^0(C, \mathcal{O}(K-E+F))$$

where $\mathcal{O}(F) = g_3^1$ and $\mathcal{O}(E) \in W_{g-7}^2(C)$. Certainly, for any $\mathcal{O}(E) \in W_{g-7}^2(C)$, either $h^0(C, \mathcal{O}(E-F)) \leq 1$ or $h^0(C, \mathcal{O}(E-F)) \geq 2$. By the base-point-free pencil trick, it follows that either $h^0(C, \mathcal{O}(K-E-F)) = \dim(\ker \nu) \geq 2$ or $h^0(C, \mathcal{O}(E-F)) \geq 2$. Thus for every $\mathcal{O}(E) \in W_{g-7}^2(C)$, either $\mathcal{O}(K-E-F)) \in W_{g-4}^1(C)$ or $\mathcal{O}(E-F) \in W_{g-4}^1(C)$. Therefore

$$\dim W_{g-1}^2(C) \leq \dim W_{g-4}^1(C) \leq g-7$$

by H. Martens' theorem; Theorem 1.5. ■

Lemma 3.4 *If C is trigonal with $g \geq 8$ then there exists a base-point-free g_{g-2}^1 on C.*

Proof. Let Σ be a component of $W_{g-2}^1(C)$ whose general element $\mathcal{O}(D) \in \Sigma$ has a base point. In other words, $\mathcal{O}(D)$ is contained in $\Sigma_{g-3}^1 + W_1(C)$, where Σ_{g-3}^1 is a sublocus of $W_{g-3}^1(C)$. Note that Σ_{g-3}^1 is at least a $(g-7)$-dimensional family since $\dim\Sigma \geq \rho(g,1,g-2) = g-6$.

We now claim that $h^0(C, \mathcal{O}(E+F)) \leq 3$ for a general Σ_{g-3}^1. If it were not, then $h^0(C, \mathcal{O}(E+F)) = 1 + h^0(C, \mathcal{O}(K-E-F)) \geq 4$, and hence $h^0(C, \mathcal{O}(K-E-F)) \geq 3$. But then

$$g-7 \leq \dim[\mathcal{O}(K-F) - \Sigma_{g-3}^1] \leq \dim W_{g-2}^2(C) < \dim W_{g-1}^2(C) = g-7$$

by Lemma 3.3.

Now consider the natural map

$$\nu : H^0(C, \mathcal{O}(E)) \otimes H^0(C, \mathcal{O}(E)) \longrightarrow H^0(C, O(E+F)).$$

By the base-point-free pencil trick and the above claim,

$$h^0(C, \mathcal{O}(E-F)) = \dim(\ker \nu) \geq 1$$

for a general $\mathcal{O}(E) \in \Sigma_{g-3}^1$. In other words, $\mathcal{O}(E) \in (F) + W_{g-6}(C)$. Thus $\mathcal{O}(D) = \mathcal{O}(E) + \mathcal{O}(p) \in \mathcal{O}(F) + W_{g-5}(C)$ and $\Sigma = \mathcal{O}(F) + W_{g-5}(C)$.

Finally, suppose that there does not exist a base-point-free g_{g-2}^1 on C. Then $\Sigma = W_{g-2}^1(C)$ and thus

$$\mathcal{O}(K-2F) - W_{g-6}(C) \subset W_{g-2}^1(C) = \Sigma = \mathcal{O}(F) + W_{g-5}(C).$$

But this is a contradiction since $h^0(C, \mathcal{O}(3F)) = 4$ if $g \geq 8$, hence $h^0(C, \mathcal{O}(K - 3F)) = g-6$. ■

When we proved the existence of a base-point-free pencil of degree $g-1$, it was necessary to check for the bi-elliptic curve. In the same vein, we need the following theorem regarding the double covering of curve of genus 2 for the existence of a base-point-free pencil of degree $g-2$.

Theorem 3.5 *Let C be a double covering of a genus-2-curve C of genus $g \geq 11$. Then C has a complete and base-point-free pencil g^1_{g-2} of degree $g-2$.*

As we did in the previous section, we will provide two proofs which are completely different from each other. One is a proof which is totally parallel to the one in [BK2] by using a enumerative method. The other one is a proof due to Coppens-Keem-G.Martens [CKM1, Appendix] and S. Park [P]. In [P], there is an improvement of the genus bound; $g \geq 11$ compared with the bound $g \geq 13$ in [CKM1].

Proof of Theorem 3.5. We first present a proof which does not use enumerative technique. In fact, the proof given in [CKM1, Appendix] consists of two parts. In the first part it is shown that there exist a plane model of degree g with a singular point s of multiplicity $g-6$, where everything works well even with the assumption $g \geq 11$. In the second part it is shown that s is an ordinary singularity and the restricted assumption $g \geq 13$ is required when monodromy argument is used. Accordingly, we only need to argue that s is still an ordinary singularity under a slightly wider range $g \geq 11$. Let $\pi : C \to E$ be the double covering over a curve E of genus 2. We note that such a covering is unique by the Castelnuovo-Severi inequality and the assumption $g \geq 11$. We briefly recall several facts which were already shown in the first part of the proof in [CKM1, Appendix].

The series $|K - g^1_4|$ is very ample for the unique $g^1_4 = \pi^*(|K_E|) = \pi^*(g^1_2)$. For a general choice of $p_1, \cdots, p_{g-6} \in C$, the series $|K - g^1_4 - p_1 - \cdots - p_{g-6}|$ induces a singular plane model Γ of C of degree g. Denoting the conjugate points of $p_1, \cdots, p_{g-6}$ by $\bar{p}_1, \cdots, \bar{p}_{g-6}$, the series $|K - g^1_4 - p_1 - \cdots - p_{g-6} - \bar{p}_1 - \cdots - \bar{p}_{g-6}|$ is a base-point-free g^1_6 and hence there is a singularity $s \in \Gamma$ with multiplicity $g-6$. To show that s is an ordinary singularity, it is enough to prove that

$$|K - g^1_4 - p_1 - \cdots - p_{g-6} - \bar{p}_1 - \cdots - \bar{p}_{g-6} - \bar{p}_i - \bar{p}_j| = \emptyset \qquad (3.5.1)$$

for $1 \leq i < j \leq g-6$. Keeping these in mind, we now proceed as follows.
We let $T_{ij} := \{(p_1, \cdots, p_{g-6}) \in C^{g-6} : \dim|K - g^1_4 - p_1 - \cdots - p_{g-6} - \bar{p}_1 - \cdots - \bar{p}_{g-6} - \bar{p}_i - \bar{p}_j| \geq 0\} \subset C^{g-6}$ for distinct i, j and $T := \cup T_{ij}$. Since T_{ij} is closed in the $(g-6)$-fold product C^{g-6}, so is T. Therefore it is enough to show that each T_{ij} is a proper closed subset in C^{g-6}; then any $(p_1, \cdots, p_{g-6}) \in C^{g-6} \backslash T$ satisfies the condition (3.5.1). Accordingly, without loss of generality, we assume $(i, j) = (1, 2)$.

Claim. For any p_1 and $p_2 \in C$ with $f(p_1) \neq f(p_2)$, $|g^1_4 + p_1 + 2\bar{p}_1 + p_2 + 2\bar{p}_2| = g^2_{10}$.

To demonstrate the validity of the claim, we recall the well-known Riemann-Hurwitz relation for double coverings. Let E be a curve of genus h and let $\pi : C \to E$ be a double covering. Let $R \subset E$ be a branch locus of π. Then we have

$$\pi_*(\mathcal{O}_C) \cong \mathcal{O}_E \oplus \mathcal{S} \quad \text{and} \quad \mathcal{S}^{\otimes 2} \cong \mathcal{O}_E(-R). \tag{3.5.2}$$

In our case, $h = 2$ and $\deg \mathcal{S} = 3 - g \leq -8$. Let $\pi(p_i) = \pi(\bar{p}_i) = r_i \in E, i = 1, 2$ and we consider $\mathcal{O}_C(g_4^1 + 2p_1 + 2p_2 + 2\bar{p}_1 + 2\bar{p}_2)$. By (3.5.2) and the projection formula, we have

$$\begin{aligned} h^0(C, \mathcal{O}_C(g_4^1 + 2p_1 + 2p_2 + 2\bar{p}_1 + 2\bar{p}_2) &= h^0(C, \mathcal{O}_C(\pi^*(g_2^1 + 2r_1 + 2r_2)) \\ &= h^0(E, \pi_*\mathcal{O}_C(\pi^*(g_2^1 + 2r_1 + 2r_2)) \\ &= h^0(E, \mathcal{O}_E(g_2^1 + 2r_1 + 2r_2)) + h^0(E, \mathcal{O}_E(g_2^1 + 2r_1 + 2r_2) \otimes \mathcal{S})) = 5. \end{aligned}$$

Note that the linear series $|g_4^1 + 2p_1 + 2p_2 + 2\bar{p}_1 + 2\bar{p}_2|$ induces the double covering $\pi : C \longrightarrow E$. Therefore,

$$\dim|g_4^1 + p_1 + 2\bar{p}_1 + p_2 + 2\bar{p}_2| = \dim|g_4^1 + 2p_1 + 2p_2 + 2\bar{p}_1 + 2\bar{p}_2| - 2 = 3$$

since $\pi(p_1) \neq \pi(p_2)$ and this finishes the proof of the claim.

By the claim, $|K - g_4^1 - p_1 - p_2 - 2\bar{p}_1 - 2\bar{p}_2| = g_{2g-12}^{g-9}$ and hence we can choose $p_3, \cdots, p_{g-7} \in C$ such that $\dim|K - g_4^1 - p_1 - 2\bar{p}_1 - p_2 - 2\bar{p}_2 - p_3 - \cdots - p_{g-7}| = 0$. Finally, we take a point $p_{g-6} \in C$ such that $p_{g-6} \notin |K - g_4^1 - p_1 - 2\bar{p}_1 - p_2 - 2\bar{p}_2 - p_3 - \cdots - p_{g-7}|$ and p_{g-6} is not conjugate to p_i, for any $i = 1, \cdots, g-7$. Therefore $(p_1, \cdots, p_{g-6}) \notin T_{12}$ and it follows that T_{12} is proper and closed in C^{g-6}.

Enumerative proof of Theorem 3.5. We use the same notations and conventions introduced in the previous section. Let $\pi : C \to E$ be the 2-sheeted covering, genus$(E) = 2$. By H. Martens and Mumford type dimension theorems on the subvariety of $J(C)$ - e.g. Theorem 1.5, Theorem 1.6 and Theorem 3.2 - it is easy to show that $W_{g-2}^1(C)$ is of pure dimension $g - 6 = \rho(g, 1, g-2)$, hence the subvariety C_{g-2}^1 of C_{g-2} is of pure dimension $g - 5$. Also it is easy to show that the only components of $W_{g-2}^1(C)$ whose general element has a base point are $\pi^*(W_2^1(E)) + W_{g-6}(C)$ and $\pi^*(W_3^1(E)) + W_{g-8}(C)$ and hence the only components of C_{g-2}^1 consisting of divisors whose complete linear series have base points are $\pi^*(E_2^1) + C_{g-6}$ and $\pi^*(E_3^1) + C_{g-8}$ whose class in C_{g-2}^1 we denote by γ and η respectively. Because C_{g-2}^1 is of pure (and expected) dimension $\rho(g-2, g, 1) + 1$, the class c_{g-2}^1 of C_{g-2}^1 is known [ACGH, p326];

$$c_{g-2}^1 = \frac{\theta^3}{6} - \frac{x\theta^2}{2}.$$

Note that γ and η occur with multiplicity 1 in C^1_{g-2}, i.e. C^1_{g-2} is reduced at general points of $\pi^*(E^1_2) + C_{g-6}$ and $\pi^*(E^1_3) + C_{g-8}$; this follows from the description of the tangent space of the scheme C^r_d (cf. [ACGH, Lemma (1.5), page 162] and the fact that $h^0(C, K - 2D - \Delta) = 0$ where $D \in \pi^*(E^1_2)$ and $\Delta \in C_{g-6}$ general (or $D \in \pi^*(E^1_3)$ and $\Delta \in C_{g-6}$ general), which can be computed easily.

Let's also recall that given a cycle Z in C_d, the assignments

$$Z \mapsto A_k(Z) := \{E \in C_{d+k} : E - D \geq 0 \quad \text{for some} \quad D \in Z\},$$
$$Z \mapsto B_k(Z) := \{E \in C_{d-k} : D - E \geq 0 \quad \text{for some} \quad D \in Z\}$$

induce maps

$$A_k : H^{2m}(C_d, \mathbb{Q}) \to H^{2m}(C_{d+k}, \mathbb{Q}), \quad B_k : H^{2m}(C_d, \mathbb{Q}) \to H^{2m-2k}(C_{d-k}, \mathbb{Q})$$

and the so called push-pull formulas for symmetric products hold; cf. Formula 2.1, Formula 2.4 or [ACGH, page 367-369]. Thus by the push-pull formulae,

$$B_{g-6}(x^{g-5}) = (g-5)x \quad \text{and} \quad B_{g-8}(x^{g-5}) = \frac{(g-5)(g-6)(g-7)}{6}x^3.$$

Denoting $\tilde{\gamma}$ and $\tilde{\eta}$ by the classes of $\pi^*(E^1_2)$ in C_4 and of $\pi^*(E^1_3)$ in C_6 respectively, we will now check that $(\tilde{\gamma} \cdot x)_{C_4} = 1$ and $(\tilde{\eta} \cdot x^3)_{C_6} = 1$, i.e. $\tilde{\gamma}$ and x (resp. $\tilde{\eta}$ and x^3) intersects transversally in C_4 (resp. C_6). Let $D \in \tilde{\gamma} \cap x$ be general. Under the natural identification between $T_D(C_4)$ and $H^0(D, \mathcal{O}_D(D))$, the tangent space $T_D(x)$ is the kernel of

$$H^0(D, \mathcal{O}_D(D)) \longrightarrow H^0(p, \mathcal{O}_p(D))$$

with p the point defining x. One also has

$$T_D(\tilde{\gamma}) = \{s \in H^0(D, \mathcal{O}_D(D)); Z(s) \in \pi^*(E^1_2)\}.$$

Since $\tilde{\gamma} = \pi^*(E^1_2) = g^1_4$ is a base-point-free pencil, one finds that

$$T_D(x) \cap T_D(\tilde{\gamma}) = \{0\}.$$

For $\tilde{\eta}$ and x^3, define x^3 using p_1, p_2, p_3 with different images on C and fix $D' \in \tilde{\eta} \cap x^3$. Again by noting that the tangent space $T_{D'}(x^3)$ is the kernel of

$$H^0(D', \mathcal{O}_{D'}(D')) \longrightarrow H^0(p_1 + p_2 + p_3, \mathcal{O}_{p_1+p_2+p_3}(D'))$$

and

$$T_{D'}(\tilde{\eta}) = \{s \in H^0(D', \mathcal{O}_{D'}(D')); Z(s) \in \pi^*(E^1_3)\},$$

one also finds that $T_{D'}(x^3) \cap T_{D'}(\tilde{\eta}) = \{0\}$.

Since $(\tilde{\eta} \cdot x^3)_{C_6} = 1$ and $(\tilde{\gamma} \cdot x)_{C_4} = 1$, we have

$$\begin{aligned}(\gamma \cdot x^{g-5})_{C_{g-2}} &= \left(A_{g-6}(\tilde{\gamma}) \cdot x^{g-5}\right)_{C_{g-2}} = \left(\tilde{\gamma} \cdot B_{g-6}(x^{g-5})\right)_{C_4} \\ &= (\tilde{\gamma} \cdot (g-5)x)_{C_4} = g-5\end{aligned}$$

and

$$\begin{aligned}(\eta \cdot x^{g-5})_{C_{g-2}} &= \left(A_{g-8}(\tilde{\eta}) \cdot x^{g-5}\right)_{C_{g-2}} = \left(\tilde{\eta} \cdot B_{g-8}(x^{g-5})\right)_{C_6} \\ &= \left(\tilde{\eta} \cdot \frac{(g-5)(g-6)(g-7)}{6} x^3\right)_{C_6} = \frac{(g-5)(g-6)(g-7)}{6} (\tilde{\eta} \cdot x^3)_{C_6} \\ &= \frac{(g-5)(g-6)(g-7)}{6}\end{aligned}$$

On the other hand

$$(c^1_{g-2} \cdot x^{g-5})_{C_{g-2}} = \left((\frac{\theta^3}{6} - \frac{x\theta^2}{2}) \cdot x^{g-5} \right)_{C_{g-2}} = \frac{g!}{6(g-3)!} - \frac{g!}{2(g-2)!}$$

by the Poincaré formula; Formula 2.3.

Comparing the above intersection numbers we have

$$(\gamma \cdot x^{g-5})_{C_{g-2}} + (\eta \cdot x^{g-5})_{C_{g-2}} < (c^1_{g-2} \cdot x^{g-5})_{C_{g-2}}$$

and this shows that there exists a component other than $\pi^*(E^1_2) + C_{g-6}$ and $\pi^*(E^1_3) + C_{g-8}$ in C^1_{g-2} which in turn proves the existence of a divisor of degree $g-2$ which moves in a complete base-point-free pencil and whose complete linear system is not composed with the given involution. ■

Proof of Theorem 3.1. Since C is non-hyperelliptic, $\dim W^1_{g-2}(C) = g-5$ or $g-6$ by Theorem 1.5 and Proposition 2.6. If $\dim W^1_{g-2}(C) = g-5$, then C is either trigonal or bi-elliptic by Theorem 1.6. If C is trigonal then we are done by Lemma 3.4. If C is bi-elliptic and g is even, then there exists a base-point-free pencil of degree $g-2$ which is a subseries of $\pi^*(g^{s-1}_s(E))$ where $2s = g-2$ and $\pi : C \longrightarrow E$ is the bi-elliptic covering.

We now assume that C is not trigonal or bi-elliptic. In particular $W^1_{g-2}(C)$ is of pure dimension $\rho(g,1,g-2) = g-6$. If there does not exist a component of $W^1_{g-2}(C)$ whose general element has a base pint, we are done.

Suppose that Σ is a component of $W^1_{g-2}(C)$ whose general element has a base point. Then a general element $\mathcal{O}(D)$ of Σ is contained in $\Sigma^1_e + W_{g-2-e}(C)$ for some e, where $4 \leq e \leq g-3$ and Σ^1_e is a sublocus of $W^1_e(C)$. Note that such a sublocus Σ^1_e cannot be contained in a component of $W^1_e(C)$ whose dimension is strictly less than $e-4$. For if it were,

$$g-6 \leq \dim\Sigma \leq \dim[\Sigma^1_e + W_{g-2-e}(C)] \leq (e-5) + (g-2-e) = g-7.$$

Thus Σ^1_e is an $(e-4)$-dimensional family of g^1_e's. Let e be an integer such that a general member if Σ^1_e has no base point. Then one can argue exactly as in the proof of Theorem 3.2 to conclude that $e = 4$ or 6; see [K, page 312–314]. But if $e = 6$, C is a double covering of a genus 2 curve and we are done by Theorem 3.5. Thus $e = 4$ and $\Sigma = \mathcal{O}(G) + W_{g-6}(C)$, where $G \in g^1_4$ which is unique on C.

Suppose that $W^1_{g-2}(C) = \Sigma = \mathcal{O}(G) + W_{g-6}(C)$, i.e. every pencil of degree $g-2$ has a base point. Then in particular,

$$\mathcal{O}(K-2G) - W_{g-8}(C) \subset \Sigma = \mathcal{O}(G) + W_{g-6}(C),$$

i.e. $\mathcal{O}(K-3G) - W_{g-8}(C) \subset W_{g-6}(C)$. But this is not possible; we have $h^0(C, \mathcal{O}(3G)) = 4$ and hence $h^0(C, \mathcal{O}(K-3G)) = g-8$. Therefore $h^0(C, \mathcal{O}(K-3G-p_1-\cdots-p_{g-8})) = 0$ for general choice of p_i's. ■

We now can answer the question which has been raised in the previous section.

Proposition 3.6 *For a non-hyperelliptic curve C of genus g, $d_0(C) = g-1$ if and only if C is bi-elliptic of odd genus.*

Proof. Assume C is bi-elliptic of odd genus with the bi-elliptic covering $\pi : C \longrightarrow E$. By Castelnuovo-Severi inequality, every base-point-free pencil on C of degree $d \leq g-2$ is composed with the bi-elliptic involution; i.e. any $g^1_d = \pi^*(g^1_{\frac{d}{2}})$ for some $g^1_{\frac{d}{2}}$ on E and d is even. Therefore any g^1_{g-2} on C must have a base point since g is odd, whence $d_0(C) = g-1$. Assume C is not bi-elliptic of odd genus. By Theorem 3.1, C has a base-point-free pencil of degree $g-2$ and it follows that $d_0(C) \leq g-2$. ■

On a double covering of a curve of genus $h = 2$, every pencil of degree $g-4 = g-2h$ or less is composed with the involution by the Castelnuovo-Severi inequality. Recall that on a bi-elliptic curve, we have seen that there exists a base-point-free pencil not composed with the bi-elliptic involution of degree d for every $d > g-2$, which is a very sharp result in view of Castelnuovo-Severi inequality; cf. Remark 2.11. Therefore one may wonder if there exists a base-point-free pencil of degree d for every $d > g-4$ on any double covering of a genus two curve. Putting the problem in a more general setting, we would like to pose the following question.

Question 3.7 *Given a double covering of a curve of genus h, does there exist a base-point-free pencil of degree d for every $d > g - 2h$ which is not composed with the involution ?*

We shall give two partial answers to the above question; one is an example which shows that the statement in Question 3.7 is a little bit too much to expect. On the other hand, we shall exhibit an example of a double covering of a curve of genus 2 on which the statement in Question 3.7 is indeed valid.

Example 3.8 (i) *There exists a curve C of genus 7 which is a double covering of a curve E of genus 2 with an extra pencil of degree 4 not composed with the given involution.*
(ii) *There also exists a curve C of genus 7 which is a double covering of a curve E of genus 2 with no extra pencil of degree 4 other than the one composed with the given involution.*

Proof. Let $C \xrightarrow{\pi} E$ be a double covering of a curve E with $g(E) = 2$ and $g(C) = 7$. Let $R \subset E$ be the branch locus of π. By the Riemann-Hurwitz relation (3.5.2) and base-point-free pencil trick, we have $\mathcal{S} \cong \mathcal{O}_E(-g^1_2 - p - q)$. Let $s \in H^0(E, \mathcal{O}_E(2g^1_2 + 2p + 2q))$ be a section with the zero locus $(s)_0 = R$. Let V be a subspace of $H^0(E, \mathcal{O}_E(2g^1_2 + 2p + 2q))$ which is the image of the cup product map

$$H^0(E, \mathcal{O}_E(g^1_2 + p + q))^{\otimes 2} \xrightarrow{\delta} H^0(E, \mathcal{O}_E(2g^1_2 + 2p + 2q))$$

and consider a natural morphism

$$H^0(C, \pi^*\mathcal{O}_E(g^1_2 + p + q))^{\otimes 2} \xrightarrow{\alpha} H^0(C, \pi^*\mathcal{O}_E(2g^1_2 + 2p + 2q)).$$

Using the identifications

$$\begin{aligned} H^0(C, \mathcal{O}_C(\pi^*(g^1_2 + p + q))) &\cong H^0(E, \pi_*\pi^*\mathcal{O}_E(g^1_2 + p + q)) \\ &\cong H^0(E, \mathcal{O}_E(g^1_2 + p + q)) \oplus H^0(E, \mathcal{S} \otimes \mathcal{O}_E(g^1_2 + p + q)) \\ &\cong H^0(E, \mathcal{O}_E(g^1_2 + p + q)) \oplus H^0(E, \mathcal{O}_E) \end{aligned}$$

and

$$\begin{aligned} &H^0(C, \mathcal{O}_C(\pi^*(2g^1_2 + 2p + 2q))) \\ &\qquad \cong H^0(E, \mathcal{O}_E(2g^1_2 + 2p + 2q)) \oplus H^0(E, \mathcal{O}_E(g^1_2 + p + q)), \end{aligned} \tag{3.8.1}$$

we see that the morphism α induces the map

$$\psi : H^0(E, \mathcal{O}_E) \otimes H^0(E, \mathcal{O}_E) \cong k \otimes k \to H^0(E, \mathcal{O}_E(2g^1_2 + 2p + 2q))$$

with $\psi(1) = s$ by the algebra structure of $\pi_*(\mathcal{O}_C)$. We also see that

$$\operatorname{Im}\alpha = \operatorname{Span}\{V, s\} \oplus H^0(E, \mathcal{O}_E(g_2^1 + p + q)), \tag{3.8.2}$$

under the identification (3.8.1).

Claim 1. Suppose that C is not bi-elliptic and put $g_4^1 = \pi^*(g_2^1)$. Then we have
(i) $h^0(C, \mathcal{O}_C(2g_4^1)) \geq 4$ if and only if $\mathcal{S} \cong \mathcal{O}_E(-2g_2^1)$,
(ii) Suppose $\mathcal{S} \not\cong \mathcal{O}_E(-2g_2^1)$. Then C has at least two complete pencil of degree 4 if and only if $s \in V$.

Proof of Claim 1. (i) follows directly from the following equality;

$$\begin{aligned} h^0(C, \mathcal{O}_C(2g_4^1)) &= h^0(C, \mathcal{O}_C(\pi^*(2g_2^1))) = h^0(E, \pi_*\mathcal{O}_C(\pi^*(2g_2^1))) \\ &= h^0(E, \mathcal{O}_E(2g_2^1)) + h^0(E, \mathcal{O}_E(2g_2^1) \otimes \mathcal{S})). \end{aligned}$$

(ii) Since $\mathcal{S} \not\cong \mathcal{O}_E(-2g_2^1)$, $\mathcal{S} \cong \mathcal{O}_E(-g_2^1 - p - q)$ for $p+q \notin g_2^1$ and hence the morphism $E \to \mathbb{P}^2$ induced by $|g_2^1 + p + q|$ is birationally very ample. Thus we have an exact sequence

$$0 \to \operatorname{Sym}^2 H^0(E, \mathcal{O}_E(g_2^1 + p + q)) \xrightarrow{\tilde{\delta}} H^0(E, \mathcal{O}_E(2g_2^1 + 2p + 2q)),$$

and we see that

$$\dim V = \dim \operatorname{Im}\delta = \dim \operatorname{Im}\tilde{\delta} = \dim \operatorname{Sym}^2 H^0(E, \mathcal{O}_E(g_2^1 + p + q)) = 6.$$

Therefore it follows that α is surjective if and only if $s \notin V$ by (3.8.2).

Assume that C has a pencil $h_4^1 \neq g_4^1$ and put $k_4^1 := |K_C - g_4^1 - h_4^1|$. Since C is not bi-elliptic, the morphism $\zeta : C \to \mathbb{P}^3$ induced by $|K_C - g_4^1| = |h_4^1 + k_4^1|$ birational onto its image and $\zeta(C) \subset \mathbb{P}^3$ lies on a quadric surface. By the Riemann-Hurwitz relation $|K_C| = |\pi^*(K_E + \mathcal{S}^{-1})|$, we have $|\pi^*(g_2^1 + p + q)| = |h_4^1 + k_4^1|$. Since

$$\dim \operatorname{Sym}^2 H^0(C, \pi^*\mathcal{O}_E(g_2^1 + p + q)) = h^0(C, \pi^*\mathcal{O}_E(2g_2^1 + 2p + 2q) = 10,$$

it follows that the map

$$\operatorname{Sym}^2 H^0(C, \pi^*\mathcal{O}_E(g_2^1 + p + q)) \longrightarrow H^0(C, \pi^*\mathcal{O}_E(2g_2^1 + 2p + 2q))$$

is not surjective. Hence the map α is not surjective and we have $s \in V$.

Conversely, assume that $s \in V$. Then the morphism α is not surjective and hence the image of the morphism $C \to \mathbb{P}^3$ induced by the birationally very ample $|\pi^*(g_2^1 + p + q)|$ is contained in a quadric surface $S \subset \mathbb{P}^3$. In case S is a cone, there is a pencil h_4^1 such that $h^0(C, \mathcal{O}_C(2h_4^1)) = 4$. From the assumption $\mathcal{S} \not\cong \mathcal{O}_E(-2g_2^1)$, it follows that $h^0(C, \mathcal{O}_C(2g_4^1)) = 3$ by (i) and hence $g_4^1 \neq h_4^1$. In case S is a non-singular quadric, we also have two pencils of degree 4 corresponding to the rulings of S and this finishes the proof of Claim 1.

Note that a curve C of genus $g \leq 7$ which is a double covering of a curve of genus 2 may be also bi-elliptic, whereas a curve of genus $g \geq 8$ cannot be both bi-elliptic and a double covering of a curve of genus 2 by Castelnuovo-Severi inequality. The following Claim 2 provides a simple criteria for a double covering of a curve of genus 2 being bi-elliptic.

Claim 2. Let C be a curve of genus $g = 7$ which is a double covering of a curve of genus 2 with an involution ι induced by the covering. If C is also bi-elliptic, then the bi-elliptic involution τ commutes with ι; i.e., $\iota\tau = \tau\iota$.

Proof of Claim 2. Let $\pi_1 : C \to E_1$ be a double covering, where E_1 is an elliptic curve and let τ be the involution induced by π_1. Consider the double covering $\pi_2 : C \to E_2$ induced by $\iota^{-1}\tau\iota$. We remark that $Q \in C$ is invariant under $\iota^{-1}\tau\iota$ if and only if $\iota(Q)$ is invariant under τ. Thus, if R is the ramification locus of π_1 then $\iota(R)$ is the ramification locus of π_2. It follows that $\iota^{-1}\tau\iota$ is also a bi-elliptic involution by the Riemann-Hurwitz formula. By Castelnuovo-Severi inequality, bi-elliptic involution of C is unique, whence $\iota\tau = \tau\iota$ and we are done with the proof of Claim 2.

By the Claim 2, it follows easily that if a double covering $C \xrightarrow{\pi} E$ is also bi-elliptic then there is an automorphism on E which lifts to the bi-elliptic involution τ via π.

Finally, we take a curve E of genus 2 such that $\mathrm{Aut}(E) = \{\sigma, 1_E\}$ where σ is the hyperelliptic involution on E. Let

$$C = \mathbf{Spec}(\mathcal{O}_E \oplus \mathcal{O}_E(-g_2^1 - p - q))$$

with $p + q \notin g_2^1$ where $p, q \in E$ are not fixed points of σ. If C is bi-elliptic with a bi-elliptic involution τ, then σ lifts to τ; note that 1_E does not lift to τ via π. Then it follows that $\sigma^*\mathcal{O}_E(-g_2^1 - p - q) \cong \mathcal{O}_E(-g_2^1 - p - q)$, which implies $\sigma(p) + \sigma(q) = p + q$ and hence $\sigma(p) = p$, $\sigma(q) = q$ contrary to the choice of $p, q \in E$ as non-fixed points of σ. Therefore C cannot be bi-elliptic.

(i) Choose $t_1, t_2 \in H^0(E, \mathcal{O}_E(g_2^1 + p + q))$ such that $(t_1)_0$, $(t_2)_0$ are reduced and $(t_1)_0 \cap (t_2)_0 = \emptyset$. Then $R = (s)_0$, where $s = t_1 t_2$, is also reduced. We note that $s \in V$. Therefore the curve $C = \mathbf{Spec}(\mathcal{O}_E \oplus \mathcal{O}_E(-g_2^1 - p - q))$ with the branch locus R is non-singular, which has at least two pencils of degree 4 by Claim 1.

(ii) Recall that $V \subsetneq H^0(E, \mathcal{O}_E(2g_2^1 + 2p + 2q))$ and take $s \notin V$ such that $R = (s)_0$ is reduced. Then the curve C with branch locus R has no extra pencil of degree 4 other than $\pi^*(g_2^1)$ by Claim 1. ■

It should be remarked that Example 3.8 has been constructed and used for another purpose in a joint work with Dr. K. Cho and Professor Akira Ohbuchi; cf.

[CKO2].

Example 3.9 *Given a curve E of genus $h \geq 1$ and an integer $g \geq 8h$, there exists curve C of genus g which is a double covering of E possessing a base-point-free pencil g^1_d not posed with the involution for every $d \geq g - 2h + 1$.*

Proof. On a ruled surface $S := E \times \mathbb{P}^1$ we consider a linear system

$$|\Gamma| = |D \times \mathbb{P}^1 + E \times 2p_0|$$

where $D \in E_{g-2h+1}$ and $p_0 \in \mathbb{P}^1$. Since $g >> h$, $|\Gamma|$ is very ample. Therefore there exists a smooth irreducible curve $C \in |\Gamma|$ by a theorem of Bertini. Using adjunction formula, one computes easily that $p_a(C) = g$. Denoting π_i the restrictions to $C \subset S$ of the projection morphisms from S onto its factors, we see that $\deg \pi_1 = 2$ and $\deg \pi_2 = g - 2h + 1$. It is now clear that the morphism π_1 is a double covering of E and π_2 induces a base-point-free pencil of degree $g - 2h + 1$ which is not composed with the involution. Thus we are done for $d = g-2h+1$. For the case $d > g-2h+1$, we leave as an exercise to the interested reader. [Hint; use a method used in the proof of Proposition 5.1, or use Lemma 5.2] ■

4 Double covering of a general curve of higher genus.

In previous sections, we mainly dealt with double coverings of a curve genus $h \leq 2$. For a double covering of a curve of genus $h \geq 3$, the situation becomes quite complicated and we will only present results which are relatively weaker. The following theorem may be considered as a generalization of our results in previous sections except the fact that the base curve of the double covering being considered in this section is a general curve of genus h.

Theorem 4.1 (Ballico-Keem, [BK1, Theorem 0.1]) *Let C be a smooth algebraic curve of genus g which admits a two sheeted covering onto a general curve E of genus h, $g \geq 5h + 1$. Then there exists a base-point-free pencil of any degree $d \geq g - h$ which is not composed with the given involution.*

A proof of Theorem 4.1 requires several preparatory results and we begin with the following theorem due to Matelski [Mat] ; see also [Ho2, Corollary 1].

Lemma 4.2 *Let C be a smooth algebraic curve of genus $g \geq 4j + 3, j \geq 0$. If* $\dim W^1_d(C) = d-2-j$ *for some d such that $j+2 \leq d \leq g-1-j$, then* $\dim W^1_{2j+2}(C) = j$.

Lemma 4.3 *Let C be a curve of genus g which is a double covering of a general curve E of genus h. If $g \geq 4h$ then $W^1_{g-h}(C)$ has the expected dimension $\rho(g-h, g, 1) = g - 2h - 2$.*

Proof. Suppose $W^1_{g-h}(C)$ does not have the expected (and minimal) dimension $g - 2h - 2$. Set

$$\text{dim}W^1_{g-h}(C) = (g-h) - 2 - j \geq g - 2h - 1 \tag{4.3.1}$$

Then by substituting $d = g - h$ in Lemma 4.2, we easily see that the numerical hypothesis of Lemma 4.2 are satisfied. Thus by Lemma 4.2, we have $\text{dim}W^1_{2j+2}(C) = j$ and hence

$$\text{dim}W^1_{2h}(C) \geq 2h - (2j+2) + j = 2h - j - 2.$$

On the other hand, by Castelnuovo–Severi inequality and by the hypothesis $g \geq 4h$, every component of $W^1_{2h}(C)$ is of the form $\pi^*(\Sigma^1_n(E)) + W_{2h-2n}(C)$ for some n, where $\pi : C \to E$ is the double covering and $\Sigma^1_n(E)$ is a component of $W^1_n(E)$. But

$$\text{dim}\pi^*(\Sigma^1_n(E)) + W_{2h-2n}(C) = \rho(n, h, 1) + 2h - 2n = h - 2$$

since the base curve E is general. Consequently we have

$$\text{dim}W^1_{2h}(C) = h - 2 \geq 2h - j - 2$$

which is contradictory to (4.3.1). ■

Lemma 4.4 *Let C be a curve of genus g which is a double covering of a general curve E of genus h. Let $g \geq 5h+1$. Then the components of $W^1_{g-h}(C)$ whose general elements have base points are of the form $\pi^*(\Sigma^1_{n/2}(E)) + W_{g-h-n}(C)$ with $\Sigma^1_{n/2}(E)$ a component of $W^1_{n/2}(E)$, where n is even and $2[\frac{h+3}{2}] \leq n \leq 2h+2$ and $\pi : C \to E$ is the double covering.*

Proof. Let Σ be a component of $W^1_{g-h}(C)$ whose general element has a base point. Then $\Sigma = \Sigma^1_n + W_{g-h-n}(C)$ for some $n, 0 \leq n \leq g - h - 1$, where Σ^1_n is a subvariety of $W^1_n(C)$ and a general element of Σ^1_n is base-point-free .

We will first argue that n is relatively small compared to g. First note that Σ has dimension $g - 2h - 2$ by Lemma 4.3. We also have $\text{dim}\Sigma^1_n = n - h - 2$, otherwise

$$g - 2h - 2 = \text{dim}\Sigma = \text{dim}(\Sigma^1_n + W_{g-h-n}(C)) \neq (n-h-2) + (g-h-n) = g - 2h - 2.$$

Let L be a general element of Σ^1_n. By the description of the Zariski tangent space to the variety W^r_d in general, we have

$$\text{dim}(\text{Im}\,\mu_0)^\perp = \text{dim}T_L(\Sigma^1_n) \geq \text{dim}\Sigma^1_n \geq n - h - 2$$

where $\mu_0 : H^0(C,L) \otimes H^0(C,K_CL^{-1}) \to H^0(C,K_C)$ is the usual cup product map. By the base-point-free pencil trick, we have

$$\begin{aligned}\dim(\operatorname{Im}\mu_0)^\perp &= g - \dim(\operatorname{Im}\mu_0) = g - h^0(C,L)h^1(C,L) + \dim(\operatorname{Ker}\mu_0)\\ &= g - 2(g-n+1) + h^0(C,K_CL^{-2}) = h^0(C,L^2) - 3 \geq n-h-2.\end{aligned}$$

Hence $h^0(C,L^2) \geq n-h+1$ which implies $W_{2n}^{n-h}(C) \geq n-h-2$. By reducing to pencils we have

$$\dim W^1_{n+h+1}(C) = \dim W^1_{2n-(n-h-1)}(C) \geq n-h-2+(n-h-1) = 2(n-h)-3.$$

Note that $n \leq g-h-1$, thus $n+h+1 \leq g$. We consider the following two cases:

(1) If $n+h+1 = g$, then by passing to residual series

$$\dim W^1_{n+h+1}(C) = \dim W_{g-2}(C) = g-2 \geq 2(n-h)-3 \Leftrightarrow g \leq 4h+3,$$

contradictory to the genus bound $g \geq 5h+1$.
(2) If $n+h+1 \leq g-1$, we have

$$2(n-h)-3 \leq \dim W^1_{n+h+1}(C) \leq (n+h+1)-2-1 \Leftrightarrow n \leq 3h+1$$

by H.Martens' well-known theorem; Theorem 1.5. Thus $n \leq 3h+1 \leq g-2h$ by the genus bound $g \geq 5h+1$, and again by the Castelnuovo–Severi inequality every element of Σ^1_n is a pull-back of a $g^1_{n/2}$ on E, i.e. $\Sigma^1_n = \pi^*(\Sigma^1_{n/2}(E))$, where $\Sigma^1_{n/2}(E)$ is a component of $W^1_{n/2}(E)$. Since

$$\dim\Sigma^1_n = \dim\pi^*(\Sigma^1_{n/2}(E)) = \dim(\Sigma^1_{n/2}(E)) = \dim(W^1_{n/2}(E)) = n-h-2 \leq h,$$

we have $n \leq 2h+2$, and $[\frac{h+3}{2}] \leq \frac{n}{2}$ since E is general. ■

Proof of Theorem 4.1. We use the same notations and conventions explained in the beginning of Section 2; $u : C_d \to J(C)$ is the abelian sum map, θ is the class of the theta divisor in $J(C)$, $u^* : H^*(J(C),\mathbb{Q}) \to H^*(C_{g-h},\mathbb{Q})$ is the homomorphism induced by u. Also recall that by fixing a point p on C, one has the map $\iota : C_{d-1} \to C_d$ defined by $\iota(D) = D+p$. We denote the class of $\iota(C_{d-1})$ in C_d by x.

Let $\pi : C \to E$ be the double covering. By Lemma 4.4, the only component of C^1_{g-h} ($\subset C_{g-h}$) consisting of divisors whose complete linear series have base points are the components of the equi-dimensional locus $\pi^*(E^1_{n/2}) + C_{g-h-n}$, with $[\frac{h+3}{2}] \leq n/2 \leq h+1$, $n =$ even. We denote α_n and $\tilde{\alpha}_n$ by the class of $\pi^*(E^1_{n/2}) + C_{g-h-n}$ in C_{g-h} and the class of $\pi^*(E^1_{n/2})$ in C_n respectively. Because C^1_{g-h} is of pure and expected dimension $\rho(g-h, g, 1) + 1$, the class x^1_{g-h} of C^1_{g-h} is well-known (cf. [ACGH, page 326]), namely

$$x^1_{g-h} = \frac{1}{h!(h+1)!}(h!\theta^{h+1} - (h+1)!x\theta^h). \tag{4.1.1}$$

As in Section 2, let

$$A_k : H^{2m}(C_d, \mathbb{Q}) \to H^{2m}(C_{d+k}, \mathbb{Q}), B_k : H^{2m}(C_d, \mathbb{Q}) \to H^{2m-2k}(C_{d-k}, \mathbb{Q})$$

be maps induced by the assignments

$$Z \mapsto A_k(Z) := \{E \in C_{d+k} : E - D \geq 0 \quad \text{for some} \quad D \in Z\}$$

$$Z \mapsto B_k(Z) := \{E \in C_{d-k} : D - E \geq 0 \quad \text{for some} \quad D \in Z\}$$

given a cycle Z in C_d.

Thus by the Formula 2.1 and Formula 2.4 (Push-pull formulae for symmetric products), we have

$$B_{g-h-n}(x^{g-2h-1}) = \binom{g-2h-1}{g-h-n} \cdot x^{n-h-1}$$

and

$$\begin{aligned}(\alpha_n \cdot x^{g-2h-1})_{C_{g-h}} &= (A_{g-h-n}(\tilde{\alpha}_n) \cdot x^{g-2h-1})_{C_{g-h}} \\ &= (\tilde{\alpha}_n \cdot B_{g-h-n}(x^{g-2h-1}))_{C_n} = \binom{g-2h-1}{g-h-n}\end{aligned}$$

by noting the fact that $(\tilde{\alpha}_n \cdot x^{n-h-1})_{C_n} = 1$

On the other hand, by (4.1.1)

$$\begin{aligned}(x^1_{g-h} \cdot x^{g-2h-1})_{C_{g-h}} &= \frac{1}{h!(h+1)!}((h!\theta^{h+1} - (h+1)!x\theta^h) \cdot x^{g-2h-1})_{C_{g-h}} \\ &= \frac{g!}{(g-h-1)!} \cdot \frac{1}{(h+1)!} - \frac{g!}{(g-h)!} \cdot \frac{1}{h!}\end{aligned}$$

where the last equality comes from the fact that $(x^{d-\beta} \cdot \theta^{\beta})_{C_d} = g!/(g-\beta)!$, which is a consequence of the Poincarè formula; Formula 2.3.

Finally a messy calculation yields

$$\sum_{\substack{2[\frac{h+3}{2}] \leq n \leq 2h+2 \\ n-even}} \binom{g-2h-1}{g-h-n} < \frac{g!}{(g-h-1)!} \cdot \frac{1}{(h+1)!} - \frac{g!}{(g-h)!} \cdot \frac{1}{h!}$$

and this shows that there exists a component in C^1_{g-h} other than those of the form $\pi^*(E^1_{n/2}) + C_{g-h-n}$ which in turn proves that there exists a divisor of degree $g-h$ on C which moves in a base-point-free pencil and whose complete linear series is not composed with the given double covering.

By the excess linear series result [ACGH, page 329], one has

$$\dim W^r_{d-1}(C) \geq \dim W^r_d(C) - r - 1.$$

Because $\dim W^1_{g-h}(C) = \rho(g-h, g, 1)$, for any $d \geq g-h+1$, we have

$$\rho(d,g,1) \leq \dim W^1_d(C) \leq \dim W^1_{d-1}(C) + 2 \leq \rho(d-1,,g,1) + 2 = \rho(d,g,1).$$

Thus

$$\dim W^1_d(C) = \rho(d,g,1)$$

for any $d \geq g-h$, which implies that there exists a complete base-point-free pencil of degree d which is not composed with the double covering. ∎

5 Irreducibility of $W^1_d(C)$ for double coverings.

In Theorem 4.1, we showed not only the existence of a base-point-free pencil of degree $d \geq g-h$ which is not composed with the given involution but also the reducibility of $W^1_{g-h}(C)$ for a double covering of a general curve of genus h. Therefore one may wonder what would be the range of d in terms of g and h for the irreducibility of $W^1_d(C)$. It turns out that Theorem 4.1 is quite a sharp result; i.e. $W^1_d(C)$ is irreducible for every $d \geq g-h+1$ for a double covering of a general curve of genus h. We first prove the following proposition, which is an intermediate step toward the proof of the irreducibility of $W^1_d(C)$ in the range $d \geq g-h+1$ for a double covering of a general curve of genus h.

Proposition 5.1 (Ballico-Keem, [BK2, Proposition 1.5]) *Let C be an irreducible smooth algebraic curve of genus g which admits a two sheeted covering $\pi : C \to E$ onto a general curve E of genus $h > 0$, $g \geq 5h-2$. Then the variety $W^1_d(C)$ of pencils of degree $d \geq g-h+1$ on C is generically reduced and a general element of any component of $W^1_d(C)$ is base-point-free.*

We need the following lemma due to Coppens which is an application of the so-called excess linear series argument.

Lemma 5.2 (M. Coppens [C, Theorem 4]) *Let C be an algebraic curve of genus g. Suppose that $W^r_d(C)$ has the expected dimension, i.e.* $\dim W^r_d(C) = \rho(d,g,r) := g-(r+1)(g-d+r)$. *Then* $\dim W^r_{d+1}(C) = \rho(d+1,g,r)$ *and $W^r_{d+1}(C)$ is irreducible (resp. reduced) if $W^r_d(C)$ is irreducible (resp. reduced).*

The following is a well-known criteria for the irreducibility of $W^r_d(C)$ which follows from [FL, Remark 1.8].

Lemma 5.3 *Let C be a smooth algebraic curve of genus g. Suppose that $W^r_d(C)$ has the expected dimension $\rho(d,g,r) > 0$ and that the codimension of the singular locus* $\operatorname{Sing} W^r_d(C)$ *is at least two. Then $W^r_d(C)$ is irreducible.*

We also need the following dimension theoretic statement for W^2_d; [CKM1, Theorem 3.3.1].

Lemma 5.4 *Let C be a smooth algebraic curve of genus g. Let $n \in \mathbb{N}$, $g \geq 2(n+1)^2$ and* $\dim W^1_{n+3}(C) < 1$. *Then* $\dim W^2_d(C) \leq 2d-6-g$ *for $g-n < d \leq g$.*

Proof of Proposition 5.1. By Lemma 5.2, it is enough to prove Proposition 5.1 for $d = g-h+1$. Let Σ be a component of $W^1_{g-h+1}(C)$ whose general element has a base point. Set $\Sigma = \Sigma^1_n + W_{g-h+1-n}(C)$ for some $n \leq g-h$, where Σ^1_n is a subvariety of $W^1_n(C)$ whose general element is base-point-free. Hence

$$\dim \Sigma^1_n = \dim \Sigma - (g-h+1-n) = n-h-1.$$

Let $L \in \Sigma^1_n$ be general. Then $\dim T_L\Sigma^1_n \geq \dim \Sigma^1_n \geq n-h-1$, and hence by the base-point-free pencil trick,

$$\begin{aligned}\dim(\operatorname{Im}\mu_0)^\perp &= g - h^0(C,L)h^1(C,L) + \dim \operatorname{Ker}\mu_0 \\ &= g-2(g-n+1)+h^0(C,KL^{-2}) = h^0(C,L^2)-3 \geq n-h-1.\end{aligned}$$

Then $h^0(C,L^2) = n-h+2$ for general $L \in \Sigma^1_n$ and hence $\dim W^{n-h+1}_{2n}(C) \geq n-h-1$. By taking off $(n-h)$ general points on C, we have

$$\dim W^1_{n+h}(C) \geq 2(n-h)-1. \tag{5.1.1}$$

Note that $n+h \leq g$ and we distinguish the following two cases.

(i) If $n + h = g$,

$$\dim W^1_g(C) = \dim W_{g-2}(C) = g - 2 \geq 2(n-h) - 1 = 2(g-2h) - 1,$$

which is contradictory to the genus bound $g \geq 5h - 2$.
(ii) If $n + h \leq g - 1$, $2(n-h) - 1 \leq \dim W^1_{n+h}(C) \leq n + h - 3$ by (5.1.1) and by H. Martens theorem. Then by the genus bound $g \geq 5h - 2$, $n \leq 3h - 2 \leq g - 2h$ and hence by Castelnuovo-Severi inequality, one has $\Sigma^1_n \subset \pi^*(W^1_{\frac{n}{2}}(E))$. On the other hand $\dim W^1_{\frac{n}{2}}(E) = n - h - 2$ since E is general. Hence

$$n - h - 1 = \dim \Sigma^1_n \leq \dim \pi^*(W^1_{\frac{n}{2}}(E)) = n - h - 2$$

which is a contradiction. And this proves that a general element of any component of $W^1_d(C)$ is base-point-free.

For the generically reducedness of $W^1_{g-h+1}(C)$, we only need to compute the dimension of the Zariski tangent space $T_L W^1_{g-h+1}(C)$ at a general L. Suppose

$$\dim T_L W^1_{g-h+1}(C) = \dim(\operatorname{Im} \mu_o)^\perp > \dim W^1_{g-h+1}(C) = g - 2h$$

for a general $L \in W^1_{g-h+1}(C)$. L being base-point-free, it easily follows that $h^0(C, KL^{-2}) \geq 1$ for general $L \in W^1_{g-h+1}(C)$ by the base-point-free pencil trick. Then we have

$$g - 2h \leq \dim W^1_{g-h+1}(C) \leq \dim W_{2h-4}(C) = 2h - 4,$$

which is contradictory to the genus bound $g \geq 5h - 2$. ■

We now ready to prove the main theorem in this section.

Theorem 5.5 (Ballico-Keem, [BK2, Theorem 1.1]) *Let C be a smooth algebraic curve of genus g which admits a two sheeted covering $\pi : C \to E$ onto a general curve E of genus $h > 0$, $g \geq \max\{2h^2, 5h+3\} =: \varepsilon(h)$. Then the variety $W^1_d(C)$ of pencils of degree d on C is generically reduced and irreducible with the expected dimension for all $d \geq g - h + 1$.*

Proof. We first claim that $W^1_{g-h+1}(C)$ is equi-dimensional of the expected dimension $\rho(g-h+1, g, 1) = g - 2h$. Indeed, in Lemma 4.3, it is proved that $W^1_{g-h}(C)$ has the expected dimension if $g \geq 4h$. Hence the same is true for $W^1_{g-h+1}(C)$ by Lemma 5.2. Therefore, by 5.2 it is sufficient to prove the theorem only for $W^1_{g-h+1}(C)$.

By a result of Mayer [M], one has

$$\mathrm{Sing}(W^1_{g-h+1}(C)) \supset W^2_{g-h+1}(C).$$

We now claim that

$$\dim W^2_{g-h+1}(C) \le \rho(g-h+1, g, 1) - 2 = g - 2h - 2$$

Suppose $h = 2e+1$ is odd and take $n = h-1 = 2e$. Then by the Castelnuovo-Severi inequality, one has

$$W^1_{n+3}(C) = W^1_{2e+3}(C) = \pi^* W^1_{e+1}(E) + W_1(C).$$

Because E is general, $\dim W^1_{e+1}(E) = \rho(e+1, h, 1) = -1$. Therefore $W^1_{n+3}(C) = \emptyset$ and hence $\dim W^1_{n+3}(C) < 1$. By taking $d = g-h+2$ in Lemma 5.4, one has

$$\dim W^2_{g-h+1}(C) \le \dim W^2_{g-h+2}(C) \le 2(g-h+2) - 6 - g = g - 2h - 2.$$

Suppose $h = 2e$ and take $n = h-1$. Again by Castelnuovo-Severi inequality, one has

$$W^1_{n+3}(C) = W^1_{2e+2}(C) = \pi^* W^1_{e+1}(E).$$

Since E is general $\dim W^1_{e+1}(E) = \rho(e+1, h, 1) = 0$ and hence $\dim W^1_{n+3}(C) = 0 < 1$. By taking $d = g-h+2$ in Lemma 5.4, one also has

$$\dim W^2_{g-h+1}(C) \le \dim W^2_{g-h+2}(C) \le 2(g-h+2) - 6 - g = g - 2h - 2$$

and this finishes the proof of the claim.

We now suppose that the singular locus $\mathrm{Sing}\, W^1_{g-h+1}(C)$ has codimension at most one in $W^1_{g-h+1}(C)$. By the above claim we may also assume that

$$\mathrm{Sing}\, W^1_{g-h+1}(C) \supset W^2_{g-h+1}(C) \cup A,$$

where A is an irreducible closed subvariety of $W^1_{g-h+1}(C)$ such that $\dim A \ge g-2h-1$ and $A \not\subset W^2_{g-h+1}(C)$. We break up the proof into the following two cases.

(i) Assume that a general element of A has no base point, and choose $L \in A$ a general element. Since $L \in \mathrm{Sing}\, W^1_{g-h+1}(C)$ and by the base-point-free pencil trick, one has

$$\begin{aligned}\dim T_L W^1_{g-h+1}(C) &= \dim(\mathrm{Im}\,\mu_0)^\perp = g - 2h + \dim \mathrm{Ker}\,\mu_0 \\ &= g - 2h + h^0(C, KL^{-2}) > \dim W^1_{g-h+1}(C) = g - 2h\end{aligned}$$

where $\mu_0 : H^0(C,L) \otimes H^0(C,KL^{-1}) \to H^0(C,K)$ is the usual cup-product map; this follows from a general theory of special linear series (cf. [ACGH, Prop.(4.2), page 189]). Therefore $h^0(C,KL^{-2}) > 0$ and hence $KL^{-2} \in W_{2h-4}(C)$ for general $L \in A$. We then havc

$$g - 2h - 1 \leq \dim A \leq \dim W_{2h-4}(C) = 2h - 4,$$

which is contradictory to the genus bound $g \geq \varepsilon(h)$.

(ii) Assume that $A \subset W^1_{g-h}(C) + W_1$, i.e. a general element of A has a base point. Note that $\dim A = g - 2h - 1$ since $W^1_{g-h+1}(C)$ is generically reduced by Proposition 5.1. Because $\dim W^1_{g-h}(C) = g - 2h - 2$, there exists a component Y of $W^1_{g-h}(C)$ such that $A = Y + W_1(C)$, $\dim Y = g - 2h - 2$.

(ii-a) Suppose Y is not of the form $Y' + W_1(C)$ for some $Y' \subset W^1_{g-h-1}(C)$. Then a general $M \in Y$ is base-point-free and $M \otimes \mathcal{O}(p)$, $p \in C$ general, has only one base point p. By the base-point-free pencil trick applied to the cup-product map

$$\mu_0 : H^0(C, M \otimes \mathcal{O}(p)) \otimes H^0(C, KM^{-1} \otimes \mathcal{O}(-p)) \to H^0(C,K),$$

$\operatorname{Ker} \mu_0 \cong H^0(C, KM^{-2} \otimes \mathcal{O}(-p)) \neq 0$ since $M \otimes \mathcal{O}(p) \in A \subset \operatorname{Sing} W^1_{g-h+1}(C)$. Therefore we have $KM^{-2} \otimes \mathcal{O}(-p) \in W_{2h-3}(C)$ for general $M \in Y$ and $p \in C$. From this we get an inequality,

$$g - 2h - 1 = \dim A \leq \dim W_{2h-3}(C) + 1 = 2h - 2,$$

which is contradictory to the assumption that $g \geq \varepsilon(h)$.

(ii-b) Suppose Y is of the form $Y' + W_1(C)$ for some $Y' \subset W^1_{g-h-1}(C)$. We claim that Y is of the form $\pi^*(\Sigma^1_{n/2}(E)) + W_{g-h-n}(C)$ with $\Sigma^1_{n/2}(E)$ a component of $W^1_{n/2}(E)$, where n is even and $2[\frac{h+3}{2}] \leq n \leq 2h + 2$.

Proof of Claim. Because Y is of the form $Y' + W_1(C)$ for some $Y' \subset W^1_{g-h-1}(C)$, Y is a component of $W^1_{g-h}(C)$ whose general element has a base point. Then $Y = \Sigma^1_n + W_{g-h-n}(C)$ for some $n, 0 \leq n \leq g - h - 1$, where Σ^1_n is a subvariety of $W^1_n(C)$ and a general element of Σ^1_n is base-point-free. We will first argue that n is relatively small compared to g. Because Y has dimension $g-2h-2$, one has $\dim \Sigma^1_n = n-h-2$, otherwise

$$g-2h-2 = \dim Y = \dim(\Sigma^1_n + W_{g-h-n}(C)) \neq (n-h-2)+(g-h-n) = g-2h-2.$$

Let L be a general element of Σ^1_n. By the standard description of the Zariski tangent space to the variety W^r_d, we have

$$\dim(\operatorname{Im} \mu_0)^{\perp} = \dim T_L(\Sigma^1_n) \geq \dim \Sigma^1_n \geq n - h - 2$$

where $\mu_0 : H^0(C, L) \otimes H^0(C, KL^{-1}) \to H^0(C, K)$ is the usual cup-product map. By the base-point-free pencil trick, we have

$$\begin{aligned}\dim(\operatorname{Im}\mu_0)^\perp &= g - \dim(\operatorname{Im}\mu_0) = g - h^0(C, L)h^1(C, L) + \dim(\operatorname{Ker}\mu_0)\\ &= g - 2(g - n + 1) + h^0(C, KL^{-2}) = h^0(C, L^2) - 3 \geq n - h - 2.\end{aligned}$$

Hence $h^0(C, L^2) \geq n - h + 1$ which implies $W_{2n}^{n-h}(C) \geq n - h - 2$. By reducing to pencils we have

$$\dim W_{n+h+1}^1(C) = \dim W_{2n-(n-h-1)}^1(C) \geq n - h - 2 + (n - h - 1) = 2(n - h) - 3.$$

Note that $n \leq g - h - 1$, thus $n + h + 1 \leq g$. We consider the following two cases:
(1) If $n + h + 1 = g$, then by passing to residual series

$$\dim W_{n+h+1}^1(C) = \dim W_{g-2}(C) = g - 2 \geq 2(n - h) - 3 \Leftrightarrow g \leq 4h + 3,$$

contradictory to the genus bound $g \geq \varepsilon(h)$.
(2) If $n + h + 1 \leq g - 1$, we have

$$2(n - h) - 3 \leq \dim W_{n+h+1}^1(C) \leq (n + h + 1) - 2 - 1 \Leftrightarrow n \leq 3h + 1$$

by H.Martens' theorem; cf. Theorem 1.5. Thus $n \leq 3h + 1 \leq g - 2h$ by the genus bound $g \geq \varepsilon(h)$, and again by the Castelnuovo-Severi inequality every element of Σ_n^1 is a pull-back of a $g_{n/2}^1$ on E, i.e. $\Sigma_n^1 = \pi^*(\Sigma_{n/2}^1(E))$, where $\Sigma_{n/2}^1(E)$ is a component of $W_{n/2}^1(E)$. Since

$$\dim \Sigma_n^1 = \dim \pi^*(\Sigma_{n/2}^1(E)) = \dim(\Sigma_{n/2}^1(E)) = \dim(W_{n/2}^1(E)) = n - h - 2 \leq h,$$

we have $n \leq 2h + 2$, and $[\frac{h+3}{2}] \leq \frac{n}{2}$ since E is general. This finishes the proof of the claim.

We next claim that

$$h^0(C, (\pi^* N)^{\otimes 2}) = h^0(E, N^{\otimes 2}) = n - h + 1$$

for $N \in \Sigma_{n/2}^1(E)$ general, $2[\frac{h+3}{2}] \leq n \leq 2h + 2$: Since E is general, $W_{n/2}^1(E)$ is reduced at a general point $N \in \Sigma_{n/2}^1(E)$, hence $h^0(E, N^{\otimes 2}) = n - h + 1$. Suppose that $h^0(C, (\pi^* N)^{\otimes 2}) > h^0(E, N^{\otimes 2})$, i.e. $\pi^* H^0(E, N^{\otimes 2}) \subsetneq H^0(C, (\pi^* N)^{\otimes 2})$. Then it follows that the complete linear system $(\pi^* N)^{\otimes 2}$ is not composed with π. Thus C has a base-point-free g_x^1 which is not composed with π,

$$x \leq \deg(\pi^* N)^{\otimes 2} - (n - h) = 2n - (n - h) = n + h \leq 3h + 2$$

by subtracting $n-h$ generically chosen points on C. But this is contradictory to the Castelnuovo-Severi inequality since $g \geq \varepsilon(h)$.

Now consider a general $M \in Y = \pi^*(\Sigma^1_{n/2}(E)) + W_{g-h-n}(C)$, and $M \otimes \mathcal{O}(p) \in A = Y + W_1(C), p \in C$ general. Then

$$M = \pi^*N \otimes \mathcal{O}(p_1 + \cdots + p_{g-h-n}), M \otimes \mathcal{O}(p) = \pi^*N \otimes \mathcal{O}(p_1 + \cdots + p_{g-h-n} + p),$$

where $N \in \Sigma^1_{n/2}(E)$. Applying the base-point-free pencil trick to the cup-product map $\mu_0 : H^0(C, M \otimes \mathcal{O}(p)) \otimes H^0(C, KM^{-1} \otimes \mathcal{O}(-p)) \to H^0(C, K)$, one has

$$\operatorname{Ker} \mu_0 \cong H^0(C, K \otimes (\pi^*N)^{\otimes -2} \otimes \mathcal{O}(-p_1 - \cdots p_{g-h-n} - p)).$$

On the other hand, from the previous claim $h^0(C, (\pi^*N)^{\otimes 2}) = n - h + 1$ and hence

$$h^0(C, K \otimes (\pi^*N)^{\otimes -2}) = g - n - h.$$

Since $p_1, \cdots, p_{g-h-n}, p \in C$ have been chosen generically, we have

$$\dim \operatorname{Ker} \mu_0 = h^0(C, K \otimes (\pi^*N)^{\otimes -2} \otimes \mathcal{O}(-p_1 - \cdots p_{g-h-n} - p)) = 0.$$

But this is contradictory to the fact that $M \otimes \mathcal{O}(p) \in \operatorname{Sing} W^1_{g-h+1}(C)$, i.e. $\dim \operatorname{Ker} \mu_0 > 0$.

So far, we have shown that $\operatorname{Sing} W^1_{g-h+1}(C)$ has codimension at least two in $W^1_{g-h+1}(C)$. By [FL, Remark 1.8], we finally conclude that $W^1_{g-h+1}(C)$ is irreducible. ■

6 Epilogue; an odd end with general $k-$gonal curves.

In previous sections we mainly played with double coverings and saw that not only the Lüroth's semigroup $S(C)$ but also the invariant $d_0(C)$ may depend on the genus of the base curve. Therefore one may wonder what it would be like if we consider a general $k-$gonal curve instead of a double covering. Fortunately, we already have sufficient knowledge to answer this question.

Theorem 6.1 (Arbarrello-Cornalba [AC, Theorem 2.6]) *Let C be a general k-gonal curve of genus g, and $k < [(g+3)/2]$, i.e. $\rho(g,1,k) < 0$. Then C has a a unique g^1_k, and each base-point-free (incomplete) pencil g^1_d on C with $\rho(g,1,d) < 0$ is compounded of the g^1_k. Furthermore, $W^1_d(C) = W^1_k(C) + W_{d-k}$ for $k \leq d$ and $\rho(g,1,d) < 0$. In particular $W^1_d(C)$ is irreducible of dimension $d-k$ if $\rho(g,1,d) < 0$.*

According to Theorem 6.1, any complete and base-point-free pencil on a general k-gonal curve C different from the g^1_k must have non-negative Brill-Noether number. However Theorem 6.1 does not necessarily guarantee the existence of such pencils. The following Brill-Noether type theorem provides the existence of complete and base-point-free pencils.

Theorem 6.2 (Coppens-Keem-G.Martens [CKM2, Theorem 2.2]) *Let C be a general k-gonal curve of genus g, $3 \leq k < [(g+3)/2]$, and let $d \leq g$ be an integer such that the Brill-Noether number $\rho(g,1,d) = 2(d-1) - g$ is non-negative; i.e. $n \geq (g/2)+1$. Then*

$$W^1_d(C) \neq W^1_k(C) + W_{d-k}(C)$$

and each irreducible component Z of $W^1_d(C)$ different from $W^1_k(C) + W_{d-k}(C)$ has the minimal possible dimension $\dim Z = \rho(g,1,d)$.

Theorem 6.2 easily yields the following corollary.

Corollary 6.3 *Let C be a general k-gonal curve of genus g, $3 \leq k < [(g+3)/2]$. Then for any $d \in \mathbb{N}$ with $(g/2)+1 \leq d \leq g$ there is a complete and base-point-free pencil g^1_d on C such that $2g^1_d$ is non-special.*

Corollary 6.3 together with Theorem 6.1 easily yields the following result which completely determines the invariant $d_0(C)$ for a general k-gonal curve.

Proposition 6.4 *Let C be a general k-gonal curve of genus g. Assume that $3 \leq k < [(g+3)/2]$. If $[\frac{g+1}{2}]$ is a multiple of k, then $d_0(C) = [\frac{g+1}{2}]$. If $[\frac{g+1}{2}]$ is not a multiple of k, then $d_0(C) = [\frac{g+3}{2}]$.*

REFERNCES

[A] R. Accola, *Topics in the theory of Riemann surfaces*, Lecture Notes in Mathematics, **1595** (1994), Springer Verlag.

[AC] Arbarello, E and Cornalba, M., *Footnotes to a paper of B. Segre*, Math. Ann. **256** (1981), 341-362.

[ACGH] Arbarello, E., Cornalba, M., Griffiths P. A. and Harris J., Geometry of Algebraic Curves I, Springer Verlag, 1985.

[BK1] E. Ballico and C. Keem, *On multiple covering of irrational curves*, Arch. Math. **65** (1995), 151-160.

[BK2] E. Ballico and C. Keem, *Variety of linear systems on double covering curves*, Journal of Pure and Applied Algebra. **128** (1998), 213-224.

[C] Coppens, M., *Some remarks on the schemes* W_d^r, Annali di Matematica pura ed applicata (IV) **157** (1990), 183–197.

[CKM1] Coppens, M., Keem, C. and Martens, G., *Primitive linear series on curves*, Manuscripta Math. **77** (1992), 237-264.

[CKM2] Coppens, M., Keem, C. and Martens, G., *Primitive length of a general k-gonal curve*, Indag. Mathem., **N.S. 5(2)** (1994), 145-159.

[CKO1] Cho, K., Keem C. and Ohbuchi A., *On the variety special linear systems of degree* $g-1$ *on smooth algebraic curves*, International Journal of Mathematics, to appear.

[CKO2] Cho, K., Keem C. and Ohbuchi A., *Variety of nets of degree* $g-1$ *on smooth curves of low genus*, Journal of the Mathematical Society of Japan, to appear.

[FL] Fulton W. and Lazarsfeld R., *On the connectedness of degeneracy loci and special divisors*, Acta Math. **146** (1981), 271–283

[GH] Griffiths, P. and Harris, J., *The dimension of the variety of special linear systems on a general curve*, Duke Math. J. **47** (1980), 233–272.

[Ho1] R. Horiuchi, *On the existence of meromorphic functions with certain low order on non-hyperelliptic Riemann surfaces*, J. Math. Kyoto Univ. **21-2** (1981), 397-416.

[Ho2] Horiuchi, R. *Gap orders of meromorphic functions on Riemann surfaces*, J. reine angew. Math. **336** (1982), 213-220.

[K] Keem, C., *On the variety of special linear systems on an algebraic curve*, Math. Annalen **288** (1990), 309–322.

[KL] Kleiman, S. and Laksov, D., *On the existence of special divisors*, Am. J. Math. **94** (1972), 431–436.

[M] Mayer, A.L., *Special divisors and the Jacobian variety*, Math. Ann. **153** (1964), 163–167.

[Mat] Matelski J.P., On geometry of algebraic curves, Ph.D. Thesis, Princeton (1978).

[MH] Martens H., *On the varieties of special divisors on a curve*, J. Reine Angew.Math. **233** (1967), 111–120.

[P] Park, S., Special linear series on 5-gonal curves and bi-elliptic curves, and very ample series on general k-gonal curves, Ph.D. Thesis, Seoul National University (2002).

[Sh] V. Shokurov, *Distinguishing Prymians from Jacobians*, Invent. math. **65** (1981), 209-219.

Department of Mathematics
Seoul National University
Seoul 151-742 South Korea
email:ckeem@math.snu.ac.kr

ALGEBRAIC SURFACES WITH QUOTIENT SINGULARITIES - INCLUDING SOME DISCUSSION ON AUTOMORPHISMS AND FUNDAMENTAL GROUPS

JONGHAE KEUM AND DE-QI ZHANG

INTRODUCTION

We work over the complex numbers field **C**. In the present survey, we report some recent progress on the study of varieties with mild singularities like log terminal singularities (which are just quotient singularities in the case of dimension 2; see [KMM]). Singularities appear naturally in many ways. The minimal model program developed by Mori et al shows that a minimal model will inevitably have some terminal singularities [KMo]. Also the degenerate fibres of a family of varieties will have some singularities.

We first follow Iitaka's strategy to divide (singular) varieties Y according to the logarithmic Kodaira dimension $\kappa(Y^0)$ of the smooth locus Y^0 of Y. One key result **(2.3)** says that for a relatively minimal log terminal surface Y we have either nef K_Y or dominance of Y^0 by an affine-ruled surface. It is conjectured to be true for any dimension [KMc].

In smooth projective surfaces of general type case, we have Miyaoka-Yau inequality $c_1^2 \leq 3c_2$ and Noether inequalities: $p_g \leq (1/2)c_1^2 + 2$, $c_1^2 \geq (1/5)c_2 - (36/5)$. Similar inequalities are given for Y^0 in Section 4; these will give effective restriction on the region where non-complete algebraic surfaces of general type exist.

In Kodaira dimension zero case, an interesting conjecture **(3.12)** (which is certainly true when Y is smooth projective by the classification theory) claims that for a relatively minimal and log terminal surface Y of Kodaira dimension $\kappa(Y^0) = 0$, one has either $\pi_1(Y^0)$ finite, or an etale cover $Z^0 \to Y^0$ where Z^0 is the complement of a finite set in an abelian surface Z. Some partial answers to **(3.12)** are given in Section 3.

The topology of Y^0 is also very interesting. We still do not know whether π_1 of the complement of a plane curve is always residually finite or not. Conjecture **(2.4)** proposed in [Z7] claims that the smooth locus of a log terminal Fano variety has finite topological fundamental group. This is confirmed when the dimension is two and now there are three proofs: [GZ1, 2] (using Lefschetz hyperplane section theorem and Van Kampen theorem), [KMc] (via rational connectivity), [FKL] (geometric).

The other interesting topic covered is the automorphism groups. Recent progress in K3 surface case is treated in Section 5. For generic rational surface X of degree ≤ 5, it is classically known that $|\mathrm{Aut}(X)|$ divides 5!. However, when Y is a log terminal del Pezzo singular surface of Picard number 1, it is very often that $\mathrm{Aut}(Y)$ contains $\mathbf{Z}/(p)$ for all prime $p \geq 5$ (see [Z9] or **(6.2)**).

Terminology and Notation

(1). For a variety V we denote by $V^0 = V - \mathrm{Sing}\ V$ the smooth locus.
(2). A $(-n)$-curve C on a smooth surface is a smooth rational curve with $C^2 = -n$.
(3). For a divisor D, we denote by $\#D$ the number of irreducible components of Supp D.
(4). For a variety V, the $e(V)$ is the Euler number.

Section 1. Preliminaries

(1.1). Let V^0 be a nonsingular variety and let V be a *smooth completion* of V^0, i.e., V is nonsingular projective and $D := V \setminus V^0$ is a divisor with simple normal crossings. If $H^0(V, m(K_V + D)) = 0$ for all $m \geq 1$, we define the *Kodaira (logarithmic) dimension* $\kappa(V^0) = -\infty$. Otherwise, $|m(K_V + D)|$ gives rise to a rational map φ_m for some m and we define the *Kodaira dimension* $\kappa(V^0)$ as the maximum of $\dim(\varphi_m(V^0))$.

The Kodaira dimension of V^0 does not depend on the choice of the completion V [I3, §11.2]. Also $\kappa(V^0)$ takes value in $\{-\infty, 0, 1, \dots, \dim V^0\}$.

V^0 is of *general type* if $\kappa(V^0) = \dim V^0$.

$p_g(V^0) = h^0(V, K_V + D)$ is called the *logarithmic geometric genus* which does not depend on the choice of the completion V [ibid.].

(1.2).
(a) Let $G \subseteq GL_2(\mathbf{C})$ be a non-trivial finite group with no reflection elements. Then $\mathbf{C}^2/G$ has a unique singularity at $\overline{O}$ (the image of the origin of the affine plane $\mathbf{C}^2$). A singularity Q of a normal surface Y is a *quotient singularity* if locally the germ (V, Q) is analytically isomorphic to $(\mathbf{C}^2/G, \overline{O})$ for some G. Quotient singularities are classified in [Br, Satz 2.11].
(b) When $\dim Y = 2$, the Q in Y is a quotient singularity if and only if it is a log terminal singularity [Ka2, Cor 1.9].

Q is a *Du Val* (or *rational double*, or *Dynkin type* ADE, or *canonical*, or *rational Gorenstein* in other notation) singularity if $G \subseteq SL_2(\mathbf{C})$ (see [Du], [Re1]).

In **(1.3) - (1.6)** below, we assume that Y is a normal projective surface with at worst quotient singularities.

(1.3). Let $f : \widetilde{Y} \to Y$ be the minimal resolution and D the exceptional divisor. We can write $f^*K_Y = K_{\widetilde{Y}} + D^*$ where D^* is an effective $\mathbf{Q}$-divisor with support in D. Write $D = \sum_{i=1}^n D_i$ with irreducible D_i and $D^* = \sum_{i=1}^n d_i D_i$.

Lemma.
(1) Each D_i is a $(-n_i)$-curve for some $n_i \geq 2$.
(2) $0 \leq d_i < 1$.
(3) $d_i = 0$ holds if and only if the connected component of D containing D_i is contracted to a Du Val singularity on Y (i.e., $f(D_i)$ is a Du Val singularity).

Proof. (2) follows from the fact that a quotient singularity is just a log terminal singularity [Ka2, Cor 1.9]. For (1) and (3), see [Br, Satz 2.11] and [Ar1, Theorem 2.7].

(1.4). Y is a *log del Pezzo* surface if the anti-canonical divisor $-K_Y$ is $\mathbf{Q}$-ample. Y is a *Gorenstein (log) del Pezzo* surface if further Y has at worst Du Val singularities (see [Mi, Ch II, 5.1], [MZ1]).

A normal variety V is *Fano* if $-K_V$ is $\mathbf{Q}$-ample.

A log del Pezzo surface is nothing but a log terminal Fano surface, and a Gorenstein del Pezzo surface is nothing but a canonical Fano surface.

(1.5). Y is a *log Enriques surface* if the irregularity $q(Y) = h^1(Y, \mathcal{O}_Y) = 0$ and if $mK_Y \sim 0$ (linear equivalence) for some positive integer m. The smallest m is called the index of Y and denoted by $I(Y)$ [Z4, Part I, Definition 1.1].

A log Enriques surface of index 1 is nothing but a K3 surface possibly with Du Val singularities. A non-rational log Enriques surface is of index 2 if and only if it is an Enriques surface possibly with Du Val singularities. The case of Y with a unique singularity is classified by Tsunoda [Ts, Proposition 2.2] (see also [Z4, Part I, Proposition 1.6]).

Proposition. *Let Y be a rational log Enriques surface with* $\#(\mathrm{Sing}\, Y) = 1$. *Then* $I(Y) = 2$ *and the unique singularity is of type* $(1/4n)(1, 2n-1)$ *for some* $n \geq 1$.

(1.6). The surface Y is *relatively minimal* if for every curve C, we have either

$K_Y.C \geq 0$ or $C^2 \geq 0$.

Suppose that Y is not relatively minimal. Then there is a curve C such that $K_Y.C < 0$ and $C^2 < 0$. By [MT1, Lemma 1.7 (2)], we see that there is a contraction $Y \to Z$ of the curve C to a smooth or quotient singularity such that the Picard number $\rho(Z) = \rho(Y) - 1$. So every projective surface with at worst quotient singularities has a relatively minimal model.

Y is *strongly minimal* if it is relatively minimal and if there is no curve C with $C^2 < 0$ and $C.K_Y = 0$ [Mi, Ch II, **(4.9)**].

(1.7). A smooth projective rational surface X is a *Coble* surface if $|-K_X| = \emptyset$ while $|-2K_X| \neq \emptyset$. A Coble surface is *terminal* if it is not the image of any birational but not biregular morphism of Coble surface.

Coble surfaces are classified in [DZ]. Here is an example. Let Z be a rational elliptic surface with a multiplicity-2 fibre F_0 and a non-multiple fibre F_1 of type I_n (see [CD] for classification of Z). Let $X \to Z$ be the blow up of all n intersection points in F_1. Then X is a terminal Coble surface. Coble surfaces and log Enriques surfaces are closely related.

Proposition [DZ, Proposition 6.4].
(1) *The minimal resolution X of a rational log Enriques surface Y of index 2 is a Coble surface with $h^0(X, -2K_X) = 1$ and the only member D in $|-2K_X|$ is reduced and a disjoint union of D_i, where D_i is either a single (-4)-curve or a linear chain with the dual graph below (each D_i is contractible to a singularity of type $(1/4n_i)(1, 2n_i - 1)$ with $n_i = \#D_i$):*

$$(-3) - -(-2) - - \cdots - -(-2) - -(-3).$$

If we let $X_{te} \to X$ be the blow up of all intersection points in D, then X_{te} is a terminal Coble surface with $h^0(X_{te}, -2K_{X_{te}}) = 1$ and the only member in $|-2K_{X_{te}}|$ is a disjoint union of n of (-4)-curves with $n = \sum_i n_i$.
(2) *Conversely, a terminal Coble surface X has a unique member D in $|-2K_X|$, and D is reduced and a disjoint union of (-4)-curves.*

(1.8). A smooth affine surface S is a **Q**-*homology plane* if $H_i(S, \mathbf{Q}) = 0$ for all $i > 0$. Similarly we can define a **Z**-*homology plane* and **Q**-*homology plane with quotient singularities.* The following very important theorem is proved by Gurjar, Pradeep and Shastri in their papers [GS], [PS], [GPS] and GPr].

Theorem 1.9. *A* **Q**-*homology plane with at worst quotient singularities is a rational surface.*

Theorem 1.10 [Mi, Theorem 4.10]. *Let Y be a $\mathbf{Q}$-homology plane. Let ν be the number of topologically contractible curves in Y. Then we have:*
(1) *Every topologically contractible curve is isomorphic to the affine line.*
(2) *The Kodaira dimension $\kappa(Y^0) = 2, 1$ or $0, -\infty$ if and only if $\nu = 0$, finite, ∞, respectively.* (Gurjar and Parameswaran [GPa] have determined the number ν when $\kappa(Y^0) = 0$).
(3) *Suppose that Y is a homology plane. Then $\kappa(Y^0) = 2, 1, -\infty$ if and only if $\nu = 0, 1, \infty$, respectively* (see **(3.17)**).

SECTION 2. NORMAL ALGEBRAIC SURFACES Y WITH KODAIRA DIMENSION $\kappa(Y^0) = -\infty$ AND FANO VARIETIES

In this section we consider projective varieties with at worst log terminal singularities and Kodaira dimension $\kappa(Y^0) = -\infty$, where $Y^0 = Y - \operatorname{Sing} Y$.

The following result is a special case of [MT1, Theorem 2.11]. We will sketch a different and direct proof here by making use of [KMM].

Theorem 2.1. *Let Y be a relatively minimal surface with at worst quotient singularities. Then one of the following occurs.*
(1) *The Kodaira dimension $\kappa(Y^0) \geq 0$ and K_Y is numerically effective.*
(2) *$\kappa(Y^0) = -\infty$ and K_Y is not numerically effective. To be precise, either*
(2a) *Y^0 is ruled, i.e. Y^0 has a Zariski open set of the form $\mathbf{P}^1 \times C$ with a curve C, or*
(2b) *Y is a log del Pezzo surface of Picard number* 1.

Proof. We may assume that K_Y is not nef. Then by [KMM, Theorems 4-2-1 and 3-2-1], there is an extremal ray $\mathbf{R}_{>0}[C]$ with C a rational curve, and a corresponding morphism $\Phi : Y \to Z$ with connected fibres such that a curve E is mapped to a point by Φ if and only if the class of E is in $\mathbf{R}_{>0}[C]$.

Case $\dim Z = 2$**.** Then Z has at worst log terminal singularities (= quotient singularities) by [KMM, Proposition 5-1-6]. This contradicts the relative minimality of Y.

Case $\dim Z = 0$**.** Then $\operatorname{Pic} Y$ is generated over $\mathbf{Q}$ by C and hence Picard number $\rho(Y) = 1$ and C is $\mathbf{Q}$-ample. Since $K_Y.C < 0$, we have $K_Y = aC$ (numerically) with $a < 0$. So the case (2b) occurs.

Case $\dim Z = 1$**.** Then a general fibre F of Φ is $\mathbf{P}^1$ because $F^2 = 0$ and $K_Y.F < 0$ (pull back to $\widetilde{Y}$ and use genus formula). Clearly the case (2a) occurs. This proves the theorem.

Theorem 2.2 ([KMc, Cor. 1.6], [Mi, Ch II, Theorems 2.1 and 2.17]).
Let Y be a log del Pezzo surface. Then there is a dominant morphism $X^0 \to Y^0$ such that X^0 is an affine-ruled surface (i.e., X^0 contains a Zariski open set of the form $\mathbf{A}^1 \times C$ for some curve C).

When Y is Gorenstein, Theorem 2.2 was proved in [Z2, Theorem 3.6]; the general case of Theorem 2.2 was proved in a lengthy book [KMc, Cor 1.6]; Theorems 2.1 and 2.2 together give the proof of the following result, which is the quotient surface case of Miyanishi Conjecture.

Theorem 2.3. *Let Y be a projective surface with at worst quotient singularities. Suppose that Y is relatively minimal. Then the following are equivalent.*
(1) *K_Y is not nef.*
(2) $\kappa(Y^0) = -\infty$.
(3) *There is a dominant morphism $X^0 \to Y^0$ such that X^0 is an affine-ruled surface.*

Proof. The equivalence of (1) and (2) is proved in [MT1, Theorem 2.11]. (2) implies (3) by Theorems 2.1 and 2.2. Assume (3). Now $\kappa(X^0) = -\infty$ is clear by considering a ruled surface as a completion of X^0 with the boundary equal to the union of a section (or empty set) and a few fibre components. Since $\kappa(X^0) \geq \kappa(Y^0)$, (2) follows.

Now we turn to the topology of smooth locus of a variety. We proposed the following in [Z7].

Conjecture 2.4. *Let V be a Fano variety with at worst log terminal singularities. Then the topological fundamental group $\pi_1(V^0)$ is finite.*

The affirmative answer to **(2.4)** would imply the following which was conjectured in [KZ] and is now a theorem of S. Takayama [Ta]. Indeed, **(2.4)** would imply that $\pi_1(V)$ is finite and we let $U \to V$ be the universal cover. Then $\chi(\mathcal{O}_U) = n\chi(\mathcal{O}_V)$, where $n = |\pi_1(V)|$. The Kawamata-Viehweg vanishing implies that $\chi(\mathcal{O}_X) = h^0(X, \mathcal{O}_X)$ $(= 1)$ for both $X = U$ and V. Hence $n = 1$.

Theorem 2.5. *Let V be a Fano variety with at worst log terminal singularities. Then $\pi_1(V) = (1)$.*

The result **(2.5)** would also follow from the following conjecture which is still open for dimension 4 or higher. It is proved in 3-fold case by [Ca] and [KoMiMo]. Recently, Graber, Harris and Starr [GHS] have proved that any complex algebraic variety having a fibration with rationally connected general fibres and image (or base), is again rationally connected.

Conjecture 2.6. *Let V be a Fano variety with log terminal singularities. Then V is rationally connected, i.e., any two general points are connected by an irreducible rational curve.*

Partial answers to **(2.4)** are given in **(2.7)** $\sim$ **(2.10)**.

(2.7). When dimension is 2, Conjecture 2.4 was proved in affirmative by [GZ1, 2]; for a differential geometric proof, see [FKL]. In [KMc, Cor 1.6], it was proved that for a log del Pezzo surface Y, the Y^0 is rationally connected and hence has finite $\pi_1(Y^0)$ (see [Ca] and [KoMiMo]).

Theorem 2.8 [Z7, Theorem 2].
Conjecture 2.4 *is true if one of the following occurs.*
(1) $\dim V \leq 2$.
(2) *The Fano index* $r(V) > \dim V - 2$.
(3) V has only isolated singularities and $r(V) = \dim V - 2 = 1$.

Theorem 2.9 [Z7, Theorem 2].
Let V be a Fano variety of Fano index $r(V) > \dim V - 2$ and with at worst canonical singularities. Then $\pi_1(V^0)$ is abelian of order ≤ 9.

Theorem 2.10 [Z7, Theorems 1 and 2].
Let V be a Fano variety. Then $\pi_1(V^0) = (1)$ if one of the following occurs:
(1) *The Fano index* $r(V) > \dim V - 1$.
(2) $\dim V = 3$ *and V has only Gorenstein isolated singularities.*

(2.11). The following gives a concrete upper bound for $\pi_1(V^0)$ in certain case. A relation $m(K_V + H) \sim 0$ in the theorem below occurs when V has Fano index 1 and Cartier index m. It is conjectured that $m = 1, 2$. To prove the theorem below, we show first that there is a natural surjective map $\pi_1(H^0) \to \pi_1(V^0)$ and also use the fact that H is Du Val K3 or Enriques. Now the theorem follows from the results on H in [KZ, Theorems 1 and 2]. For each of the three exceptional cases of (p, c) below, we note that there is a Du Val K3 or Enriques surface Y with $\operatorname{Sing} Y = cA_{p-1}$ and $\pi_1(Y^0)$ infinity [ibid.].

Theorem [KZ, Theorem 3].
Let p be a prime number. Let V be a log terminal Fano 3-fold with a Cartier divisor

H such that $m(K_V + H) \sim 0$ (linear equivalence) for $m = 1$ or 2. Suppose that a member H of $|H|$ is irreducible normal and has c singularities of type A_{p-1} and no other singularities.
Then $\pi_1(V^0)$ is soluble; and if $(p, c) \neq (2, 8), (2, 16), (3, 9)$, then $|\pi_1(V^0)| \leq 2p^k$ for some $k \leq 4$.

Remark 2.12.
(1) In **(2.4)** if we replace "log terminal" by "log canonical", then **(2.4)** has counter-examples; more precisely, if V is a normal Fano surface with at worst rational log canonical singularities, then $\pi_1(V^0)$ contains a finite-index abelian subgroup of rank k $(k = 0, 2)$ [Z8, Theorem 2.3].
(2) In **(2.9)**, the upper bound is optimum [MZ1, Lemma 6]; also "canonical" can not be replaced by "log terminal" [Z3, Appendix].

(2.13). In [Kj2], log del Pezzo surfaces with a unique singularity are classified (including the existence part). The classification of log del Pezzo surface of Cartier index ≤ 2 were announced in [AN]. In [Kj4], Kojima classified Picard number 1 log del Pezzo surfaces Y of index 2, in a way different from [AN]: there are exactly 18 types of $\mathrm{Sing}\, Y$ and the $\pi_1(Y^0) \leq 8$; the $\pi_1(Y^0) = (1)$ holds if and only if Y contains the affine plane as a Zariski open set.

(2.14). In [Ni3], the Picard number $\rho(\widetilde{Y})$ of the minimal resolution $\widetilde{Y}$ of a log del Pezzo surface Y is bounded from above in terms of the maximum of multiplicities of Y.

SECTION 3. NORMAL ALGEBRAIC SURFACES Y WITH KODAIRA DIMENSION $\kappa(Y^0) = 0$

In this section we consider projective surfaces Y with at worst quotient singularities and Kodaira dimension $\kappa(Y^0) = 0$, where $Y^0 = Y -$ Sing Y.

Theorem 3.1 ([Ka1, Theorem 2.2], [Mi, Ch II, 6.1.3]).
Let Y be a projective surface with at worst quotient singularities. Then the following are equivalent:
(1) Y is relatively minimal with Kodaira dimension $\kappa(Y^0) = 0$.
(2) There is a positive integer m such that $mK_Y \sim 0$ (linear equivalence).

(3.2). The smallest positive integer m with $mK_Y \sim 0$ is called the *index* of Y and denoted by $I = I(Y)$.

Proposition 3.3. *Let Y be a projective surface with at worst quotient singularities* *and* $IK_Y \sim 0$, *where* $I > 0$ *is the index of* Y. *Suppose that* Y *is irrational. Then one of the following occurs.*

(1) Y *is a (smooth) abelian surface* ($I = 1$) *or a hyperelliptic surface* ($I = 2, 3, 4, 6$).

(2) Y *has at worst Du Val singularities. The minimal resolution of* Y *is either a* $K3$ *surface* ($I - 1$) *or an Enriques surface* ($I = 2$).

Proof. In notation of **(1.3)**, we have $I(K_{\widetilde{Y}} + D^*) \sim 0$. So $\kappa(\widetilde{Y}) \le 0$. If $\kappa(\widetilde{Y}) = 0$, Then **(3.3)** follows from the classification of smooth surfaces. If $\kappa(\widetilde{Y}) = -\infty$, then $\widetilde{Y}$ is an irrational ruled surface over a base curve of genus ≥ 1. However, $K_{\widetilde{Y}} + D^* = 0$ (numerically) implies that D^* contains some horizontal components (this can be seen by going to a relative minimal model of $\widetilde{Y}$) which dominates the base curve and hence is irrational, contradicting the fact that D consists of rational curves only **(1.3)**. So $\kappa(\widetilde{Y}) = -\infty$ is impossible. This proves the proposition.

In view of **(3.3)**, to classify those Y with at worst quotient singularities and $\kappa(Y^0) = 0$, we need only to consider rational surfaces Y with $mK_Y \sim 0$ for some integer $m \ge 2$ (see **(1.5)**). These are precisely rational log Enriques surfaces **(1.5)**.

(3.4). Let Y be a rational log Enriques surface. Then the index $I = I(Y) \ge 2$. Since $IK_Y \sim 0$, there is a *canonical* $\mathbf{Z}/(I)$-Galois cover $\pi : X = Spec \oplus_{j=0}^{I-1} \mathcal{O}(-jK_Y) \to Y$ which is unramified over $(Y \setminus \{\text{non-Du Val singularities}\}) \supseteq Y^0$ and satisfies $K_X \sim 0$. Therefore, either X is a (smooth) abelian surface or a K3 surface possibly with some Du Val singularities.

Recently, Suzuki [Su] has proved Morrison's cone conjecture for rational log Enriques surfaces Y : there is a finite rational polyhedral cone which is a fundamental domain for the action of $\mathrm{Aut}(Y)$ on the rational convex hull of its ample cone.

Theorem 3.5. *Suppose that* Y *is a rational log Enriques surface of index* I *and that the canonical* $\mathbf{Z}/(I)$*-cover* X *of* Y *is an abelian surface. Then we have:*

(1) $I = 3$ *or* 5. *If* $I = 3$, *then* Sing Y *consists of* 9 *singularities of type* $(1/3)(1, 1)$; *if* $I = 5$, *then* Sing Y *consists of* 5 *singularities of type* $(1/5)(1, 2)$; see [Re1] for notation.

(2) *For each* $I = 3, 5$, *there is a unique log Enriques surface* Y_I *with* X_I *abelain and* $I(Y_I) = I$. To be precise, $Y_I = X_I/\langle g_I \rangle$, where $X_3 = E_{\zeta_3} \times E_{\zeta_3}$ with $E_{\zeta_3} = \mathbf{C}/(\mathbf{Z} + \mathbf{Z}\zeta_3)$ an elliptic curve of period $\zeta_3 = \exp(2\pi\sqrt{-1}/3)$, $g_3 = \mathrm{diag}(\zeta_3, \zeta_3)$, X_5 is the Jacobian surface of the genus-2 curve : $y^2 = x^5 - 1$, and g_5 in $\mathrm{Aut}(X_5)$ is induced by the curve automorphism : $(x, y) \mapsto (\zeta_5 x, y)$ (see [Bl, Example 1.2], [Su, Proposition 1.2] and [Z4, Example 4.2]).

Proof. (1) is proved in [Z4, Theorem 4.1]. (2) is proved in [Bl, Su].

Theorem 3.6. *Let Y be a log Enriques surface of index I. Then $I \leq 21$.*

Remark 3.7.
(1) It is easy to see that the Euler function $\varphi(I) \leq 21$ and hence $I \leq 66$ [Z4, Part I, Lemma 2.3]. In [Bl, Theorem C], it is proved that $I \leq 21$.
(2) Examples of Y with prime $I(Y)$ are constructed in [Z4, Example 5.3-5.8; Part II, Example 7.3] and [Bl, Example 4.1].

(3.8). A log Enriques surface Y is *maximum* if any birational morphism $Y \to Z$ to another log Enriques surface is an isomorphism. By [Z4, Part II, Theorem 2.11'], for every log Enriques surface Y of prime index, there is a unique maximum log Enriques surface Y_{max} with $I(Y_{max}) = I(Y)$ and a birational morphism $Y_{max} \to Y$; each singularity (if exists) of the canonical cover of Y_{max} is of type A_1.

The surface $Y(A_{19})$ in the assertion(3) below is not isomorphic to the unique (modulo projective transformation) quartic $K3$ surface in $\mathbf{P}^3$ with a Dynkin type A_{19} singularity; neither can $Y(D_{19})$ be embedded in $\mathbf{P}^3$ [KN]. The $Y(D_{19})$ is constructed in two different ways in [Z4, Example 6.11: the $\overline{V}'$] and [OZ1, Example 1], and $Y(A_{19})$ in [Z4, Example 3.2 : the $\overline{V}$] and [OZ1, Example 2]. The uniqueness problem of $Y(D_{19})$ was initiated by [Re2, Round 3, Example 6].

Theorem [OZ4, Corollary 4; OZ3, Corollary in §1; OZ1, Theorems 1 and 2].
(1) *For each $I = 13, 17, 19$, there is a unique maximum log Enriques surface Y with $I(Y) = I$.*
(2) *All maximum log Enriques surfaces of index* 11 *form a family of dimension* 1 *and are all given in* [OZ3, §1, Corollary].
(3) *For each D in $\{D_{19}, A_{19}\}$, there is a unique rational log Enriques surface $Y(D)$ whose canonical cover has a singularity of Dynkin type D. The index $I(Y)$ equals* 3 *(resp.* 2*) when Y equals $Y(D_{19})$ (resp. $Y(A_{19})$).*

Next we will investigate the behaviour of $\pi_1(Y^0)$ and propose a conjecture **(3.12)** generalizing the one in [CKO]. Note that $\mathbf{Z}/(I)$ is the image of $\pi_1(Y^0)$ by a homomorphism.

Theorem 3.9 [Z4, Part II, Theorem 2.11', Cor 1]. *Let Y be a maximum log Enriques surface of odd prime index I. Let $X \to Y$ be the canonical $\mathbf{Z}/(I)$-cover. Then we have:*
(1) *X has at worst type A_1 singularities and* $\#(\mathrm{Sing}\, X) \leq 6$.
(2) $\pi_1(Y^0) = \mathbf{Z}/(I)$.

Proof. (1) is proved in [Z4, Part II, Cor. 1]. For (2), we have only to show that $\pi_1(X^0) = (1)$ because the inverse of Y^0 via the canonical map $X \to Y$ (unramified over Y^0) is X^0 with a few smooth points removed (the removal of smooth points in a complex surface does not change π_1). Since $\#(\operatorname{Sing} X) \leq 6$ by (1), we have $\pi_1(X^0) = (1)$ [KZ, Theorem 1]. This proves the theorem.

Theorem 3.10 [SZ, Proposition 4.1, Theorem 4.3, Cor 4.4]
(1) *Suppose that "$\pi_1^{alg}(X^0) = (1) \Rightarrow \pi_1(X^0) = (1)$ for all Du Val $K3$ surfaces X" holds (X satisfies this condition if (∗) in (2) holds for X). Let Y be a log Enriques surface of index I. Then either $\pi_1(Y^0)$ is finite or there is a finite morphism $Z \to Y$ from an abelian surface which is unramified over Y^0. In particular, $\pi_1(Y^0)$ contains a finite-index abelian subgroup of rank k ($k = 0, 4$).*
(2) *Let $X \to Y$ be the canonical cover of a log Enriques surface of index I, let $\widetilde{X} \to X$ be the minimal resolution and $D = \sum D_i$ the exceptional divisor. Then $\pi_1(Y^0) = \mathbf{Z}/(I)$ if the lattice $\Gamma = \mathbf{Z}[\cup D_i]$ is primitive in $H^2(\widetilde{X}, \mathbf{Z})$ and satisfies:*
(*) $r = rank(\Gamma) \leq 18$ and the discriminant group $\Gamma^\vee/\Gamma$ is generated by k elements with $k \leq min\{r, 20 - r\}$.

Proof. Let $X \to Y$ be the canonical $\mathbf{Z}/(I)$-cover. Now the conclusion in **(3.10)** (1) with Y replaced by X holds by [SZ, Proposition 4.1]. So (1) is true for $X \to Y$ is unramified over Y^0. For (2), [SZ, Theorem 4.3] implies $\pi_1(X^0) = (1)$. So (2) is true. This proves the theorem.

There is a concrete upper bound of $\pi_1(Y^0)$ for certain Y.

Theorem 3.11 [GZ3, Theorem 1]. *Let Y be a rational log Enriques surface of index 2. Assume that Y has no Du Val singularities. Then $\pi_1(Y^0)$ is a soluble group of order $n_1 n_2$ with $n_i \leq 16$.*

The results **(3.9)** $\sim$ **(3.11)** support the following which is just the conjecture in [CKO] when Y is a Du Val $K3$ surface, i.e., when $I(Y) = 1$ and $q(Y) = 0$.

Conjecture 3.12. *Let Y be a log Enriques surface. Then the universal cover U of Y^0 is a big open set (= the complement of a discrete subset) of either a Du Val $K3$ surface or of $\mathbf{C}^2$; in the latter case, $U \to Y^0$ factors through a finite etale cover $Z^0 \to Y^0$, where Z^0 is a big open set of an abelian surface Z.*

(3.13). Since the canonical cover $X \to Y$ of a log Enriques surface is unramified over Y^0, we have $\pi_1(Y^0)/\pi_1(X^0) = \mathbf{Z}/(I)$. So **(3.12)** is, in most cases, reduced to the problem of $\pi_1(X^0)$ for a Du Val K3 surface X (see **(3.5)**). We have the following:

Theorem [KZ, Theorems 1 and 2]. *Conjecture* 3.12 *is true if* Y *is a Du Val* $K3$ *or Enriques surface and has several singularities of type* A_{p-1} *and no other singularities; here* p *is a prime number.*

(3.14). The following results contributes towards an answer in affirmative to **(3.12)**. These are just applications of [CKO, Theorems A and B] to the canonical cover X of Y. Also the upper bound $\#D \leq 15$ is an optimum condition for $\pi_1(X^0)$ to be finite by considering Kummer surfaces.

Theorem. (1) *Let* Y *be a log Enriques surface with an elliptic fibration. Then either* $\pi_1(Y^0)$ *is finite or there is a finite cover of* $Z \to Y$ *from an abelian surface which is unramified over* Y^0.
(2) *Let* Y *be a log Enriques surface,* $X \to Y$ *the canonical cover and* $\widetilde{X} \to X$ *the minimal resolution with* D *the exceptional divisor. Suppose that* $\#D \leq 15$. *Then* $\pi_1(Y^0)$ *is finite.*

(3.15). In [Oh], pairs (S, Δ) of normal surface S and a **Q**-divisor Δ satisfying $K_S + \Delta \equiv 0$ (numerically) are considered. These pairs appear naturally as degenerate fibres in log degeneration; for many interesting cases, he completed the classification of these pairs.

(3.16). In [Kj1, Theorem 0.1], strongly minimal smooth affine surface S with $\kappa(S) = 0$ is classified and its invariants are classified (*strongly minimal* means almost minimal and having no exceptional curve of the second kind [Mi, Ch II, **(4.9)**]). In particular, the minimal $m > 0$ with log pluri-genus $P_m(S) > 0$, the log irregularity $q(S)$ and the Euler number $e(S)$ satisfy the following ($\pi_1(S)$ is also calculated there, which is generated by at most two elements):

$$m|6, \quad q(S) \in \{0, 1, 2\}, \quad e(S) \in \{0, 1, 2, 3, 4\}.$$

(3.17). In [Fu, §8], all **Q**-homology planes of Kodaira dimension 0 are classified. It was also proved there that there is no **Z**-homology plane S of Kodaira dimension $\kappa(S) = 0$. The paper [Fu] is very important and also essentially used in [Kj1].

(3.18). Iitaka [I2] conjectured that an affine normal variety S is isomorphic to $(\mathbf{C}^*)^n$ if and only if $\kappa(S) = 0$ and $q(S) = \dim S$. In the same paper, he himself proved it when $\dim S = 2$.

According to [I1], a (possibly open) surface S is *logarithmic* $K3$ if the logarithmic invariants satisfy : $q(S) = 0$, $p_g(S) = 1$, $\kappa(S) = 0$. In [I1] log K3 surfaces were classified. In [Z1], one defines the *Iitaka surface* as a pair $(V, A+N)$ of smooth projective rational surface V and reduced divisor $A+N$ with $A + K_V \sim 0$ and N contractible to Du Val singularities, and the classification of such pairs were done there.

(3.19). In [Kj3], Kojima studies complements S of reduced plane curves with $\kappa(S) = 0$; in particular he proves that the logarithmic geometric genus $p_g(S) = 1$.

Section 4. Normal algebraic surfaces Y with Kodaira dimension $\kappa(Y^0) = 1, 2$

In this section we consider projective surface Y with at worst quotient singularities and $\kappa(Y^0) = 1, 2$.

(4.1). We first consider the case $\kappa(Y^0) = 1$. The following is a consequence of [Ka1, Theorem 2.3] or [Mi, Ch II, Theorem 6.1.4]. Indeed, in our case, the boundary divisor D^* is fractional and contains no effective integral divisor **(1.3)**.

Theorem. *Let Y be a projective surface with at worst quotient singularities. Suppose that Y is relatively minimal and $\kappa(Y^0) = 1$.*

Then there is a positive integer m such that mK_Y is Cartier and the linear system $|mK_Y|$ is composed with an irreducible pencil Λ without base points. Each general member of Λ is a smooth elliptic curve. So there is an elliptic fibration $Y \to B$.

(4.2). Let Y be a projective surface with at worst quotient singularities. Suppose that Y is relatively minimal and $\kappa(Y^0) = 2$. Then K_Y is nef and big. By [KMM, Theorem 3-1-1], $|mK_Y|$ is base point free for m sufficiently divisible, and hence defines a birational morphism $\varphi : Y \to Z$. This φ is nothing but the contraction of all curves on Y having zero intersection with K_Y. Then Z has at worst log canonical singularities ([Ka1, Theorem 2.9], [Mi, Ch II, Theorem 4.12]). Denote by LC the set of points on Z which is log canonical but not log terminal (i.e., not of quotient singularity). Then we have the following Miyaoka-Yau type inequality proved by [Kb1, 2] and [KNS].

Theorem 4.3. *Let Y be a projective surface with at worst quotient singularities. Suppose that Y is relatively minimal and $\kappa(Y^0) = 2$. Then we have the following, where P runs over all quotient singularities of Z and G_P is the local fundamental* *group at P*

$$K_Y^2 \le 3\{e(Z - LC) - \sum_P (1 - \frac{1}{|G_P|})\}.$$

(4.4). For smooth and minimal surfaces X of general type, we have Noether inequalities

$$p_g(X) \le (1/2)c_1(X)^2 + 2, \quad c_1(X)^2 \ge \frac{1}{5}c_2(X) - \frac{36}{5}.$$

For singular surfaces, we have:

Theorem [TZ, Theorems 1.3 and 2.10].
Let Y be a projective surface with at worst quotient singularities and $\kappa(Y^0) = 2$. Then the logarithmic geometric genus $p_g(Y^0)$ satisfies the optimum upper bound:

$$p_g(Y^0) < K_Y^2 + 3.$$

(4.5). For smooth projective surface X of general type, the famous Miyaoka-Yau inequality asserts that $c_1(X)^2 \le 3c_2(X)$.

Consider log surface (V, D) with V a smooth projective surface and D a reduced divisor with simply normal crossings. Set $\overline{c}_1^2 = (K_V + D)^2$ and $\overline{c}_2 = c_2(V) - e(D)$. Sakai [Sa] proved that

$$\overline{c}_1^2 \le 3\overline{c}_2$$

provided that D is semi-stable and $\kappa(V \setminus D) = 2$. The following is a lower bound of $\overline{c}_1^2$ in terms of $\overline{c}_2$. These two inequalities together give effective restrictions on the region for non-complete algebraic surfaces $V \setminus D$ of general type to exist. In the following, (V, D) is *minimal* if $K_V + D$ is nef and there is no (-1)-curve E with $E.(K_V + D) = 0$.

Theorem [Z5, Cor. to Theorem C, Theorem D]. *Let (V, D) be a log surface with $D \ne 0$. Assume that (V, D) is minimal and $\kappa(V \setminus D) = 2$. Assume further that $\kappa(V) \ge 0$. Then we have*
(1)

$$\overline{c}_1^2 \ge \frac{1}{15}\overline{c}_2 - \frac{8}{5}.$$

(2) *Suppose that $p_g(V \setminus D) \ge 3$ and $|K_V + D|$ is not composed with a pencil. Then*

$$\overline{c}_1^2 \ge \frac{1}{9}\overline{c}_2 - 2.$$

Section 5. Automorphisms of algebraic surfaces - smooth surface case

(5.1). We mention some background of $\mathrm{Aut}(X)$ where X is a smooth projective rational surface. $\mathrm{Aut}(X)$ had been studied by S. Kantor more than one hundred years ago [Kt]. It was continued by Segre, Manin, Iskovskih, Gizatullin and many others [Se], [Ma1, 2], [Is], [Gi]. See also [Ho1], [Ho2]. In [DO], the group of automorphisms of any general del Pezzo surface is described and it turns out that its discrete part is equal to the kernel of the Cremona representation on the moduli space of n points in $\mathbf{P}^2$. Very recently, de Fernex [dF] constructed all the Cremona transformations of $\mathbf{P}^2$ of prime order, where he employed the methods different from those used by Dolgachev and Zhang in [ZD].
In [ZD], minimal pairs (X, G) with prime order $p = |G|$ was considered. In particular, using the recent Mori theory, it was shown there that if the G-invariant sublattice of Pic X has rank 1 then $p \leq 5$ unless $X = \mathbf{P}^2$; the short and precise classification of these pairs, modulo equivariant isomorphism, was also given there.

Generic Enriques surfaces have infinitely many automorphisms. Those Enriques surfaces with finite automorphisms have been classified by S. Kondō ([Kon4], see also [Ni2]); there are seven families of such Enriques surfaces. It follows that K3 covers of Enriques surfaces all have infinite automorphism groups.

For K3 surfaces, much progress on their automorphism groups has been done by Kondō, Keum, Dolgachev, Oguiso, and Zhang ([Kon1, 2, 3], [Ke1, 3], [KK], [DK], [OZ1, 3, 4, 5]).

It is known that minimal surfaces of general type has only finite automorphism groups. It was Xiao [Xi1] who gave a proof of the existence of a bound for the order of the automophism group, which is linear in the Euler number of the surface. For curves C of genus ≥ 2, a classical theorem of Hurwitz gave a sharp bound $|\mathrm{Aut}(C)| \leq 84(g(C) - 1) = -42e(C)$.

Theorem [Xi2, Theorem 2]. *Let X be a minimal surface of general type. Then $|\mathrm{Aut}(X)| \leq (42K_X)^2$, with equality if and only if $X \cong (C \times C)/N$, where C is a curve with $|\mathrm{Aut}(C)| = 84(g(C) - 1)$, N a normal subgroup of* $\mathrm{Aut}(C \times C)$ *acting freely on $C \times C$ and preserving the two projections of $C \times C$.*

(5.2). In [ZD], pairs (X, G) of a smooth *rational* projective surface X and a finite group of automorphisms are considered. A pair is *minimal* if every G-equivariant birational morphism to another pair (Z, G) is an isomorphism.

Theorem [ZD, Therorem 1]. *Let (X, μ_p) be a minimal pair with p a prime number.* (1) *If the invariant sub-lattice* $(\operatorname{Pic} X)^{\mu_p}$ *has rank at least* 2, *then* X *is a Hirzebruch surface and the pair* (X, μ_p) *is birationally equivalent to a pair* $(\mathbf{P}^2, \mu_p)$.
(2) *If* $(\operatorname{Pic} X)^{\mu_p}$ *has rank* 1, *then the pairs are classified in* $[ZD, Theorem1]$; *in particular, we have* $p \leq 5$ *unless* $X = \mathbf{P}^2$.

(5.3). Let X be a K3 surface. The following are well known (cf. [BPV]).
(1) $H^{2,0}(X) = \mathbf{C}\omega_X$,
where ω_X is a nowhere vanishing global holomorphic 2-form on X.
(2) $H^2(X, \mathbf{Z})$ is an even unimodular lattice of signature (3,19) with the cup product, so we have an isomorphism

$$H^2(X, \mathbf{Z}) \cong U \oplus U \oplus U \oplus E_8(-1) \oplus E_8(-1),$$

where U (resp. E_8) is the even unimodular lattice of signature (1,1) (resp. (8,0)).
(3) $\operatorname{Pic}(X)$ is isomorphic to the Néron-Severi group $NS(X)$, and hence can be viewed as a sublattice of $H^2(X, \mathbf{Z})$. The rank of $\operatorname{Pic}(X)$, called the Picard number of X, is denoted by $\rho(X)$. This number can take the value 0, 1, ..., 20. The lattice $\operatorname{Pic}(X)$ is hyperbolic(=Lorentzian) if X is projective, and is semi-negative definite or negative definite if X is not projective.
(4) Recall that all K3 surfaces are Kähler [Siu], so Hodge decomposition holds for them.
(5) Let

$$C(X) \subset H^{1,1}(X, \mathbf{R}) := H^{1,1}(X) \cap H^2(X, \mathbf{R})$$

denote the Kähler cone of X, the set of all classes of symplectic forms of Kähler-Einstein metrics on X. In K3 surface case, $C(X)$ can be numerically characterized as follows:

$C(X) = \{\omega \in H^{1,1}(X, \mathbf{R}) : \langle \omega, \omega \rangle > 0, \langle \omega, R \rangle > 0$ for all smooth rational curves $R\}$

For a compact Kähler manifold, Nakai-Moishezon type criterion, i.e. the characterization of the Kähler cone, is highly non-trivial (see [DP]).

(6) Let r be an element of $\operatorname{Pic}(X)$ with $\langle r, r \rangle = -2d$ $(d > 0)$, and $\langle r, \operatorname{Pic}(X) \rangle \subset d\mathbf{Z}$. Then

$$x \to x + \langle x, r \rangle r/d$$

defines an isometry of $\operatorname{Pic}(X)$, called a $(-2d)$-reflection.

Let $W(\mathrm{Pic}(X))$ (resp.$W(\mathrm{Pic}(X))^{(2)}$) be the subgroup of the orthogonal group $O(\mathrm{Pic}(X))$ generated by all reflections (resp. all (-2)-reflections). These are normal subgroups of $O(\mathrm{Pic}\,(X))$ and, by linearity, acts naturally on $H^{1,1}(X,\mathbf{R})$. The set

$$\{\omega \in H^{1,1}(X,\mathbf{R}) : \langle \omega, \omega \rangle > 0\}$$

has two components, each a cone over a 19-dimensional hyperbolic manifold with constant curvature. $C(X)$ is contained in one of the two components, and the action of $W(\mathrm{Pic}\,(X))^{(2)}$ on this component has $C(X)$ as its fundamental domain.

(7) If X is projective, the ample cone

$$D(X) := C(X) \cap \mathrm{Pic}\,(X) \otimes \mathbf{R}$$

is non-empty and can be numerically characterized as

$D(X) = \{\omega \in \mathrm{Pic}\,(X) \otimes \mathbf{R} : \langle \omega, \omega \rangle > 0, \langle \omega, R \rangle > 0$ for all smooth rational curves $R\}$.

The group $W(\mathrm{Pic}\,(X))^{(2)}$ acts on the component $P_+(X)$ of

$$\{\omega \in \mathrm{Pic}\,(X) \otimes \mathbf{R} : \langle \omega, \omega \rangle > 0\}$$

containing $D(X)$, and has $D(X)$ as its fundamental domain.

Note that $O(\mathrm{Pic}\,(X))$ acts on $P_+(X)$ and is a semi-direct product of the normal subgroup $W(\mathrm{Pic}\,(X))^{(2)}$ and the symmetry group $SymD(X)$ of the cone $D(X)$, i.e.

$$O(\mathrm{Pic}\,(X))/W(\mathrm{Pic}\,(X))^{(2)} \cong SymD(X).$$

(5.4). The Torelli theorem asserts that a K3 surface is determined up to isomorphism by its Hodge structure. More precisely we have:

Theorem ([PSS],[BR]). *Let X and Y be K3 surfaces, and let*

$$\phi : H^2(X,\mathbf{Z}) \to H^2(Y,\mathbf{Z})$$

be an isometry. Extend ϕ to $H^2(X,\mathbf{C})$ or to $H^2(X,\mathbf{R})$ by tensoring with $\mathbf{C}$ or $\mathbf{R}$. Then :

(1) If ϕ sends $H^{2,0}(X)$ to $H^{2,0}(Y)$, then X and Y are isomorphic.

(2) If ϕ also sends $C(X)$ to $C(Y)$, then $\phi = f^*$ for a unique isomorphism $f : Y \to X$.

(5.5). Let X be a projective K3 surface. Torelli theorem shows that there is a map

$$\mathrm{Aut}(X) \to O(\mathrm{Pic}\,(X))/W(\mathrm{Pic}\,(X))^{(2)} \cong SymD(X)$$

which has finite kernel and cofinite image. So in practice, if we want to describe $\mathrm{Aut}(X)$, the main step is to calculate $O(\mathrm{Pic}\,(X))/W(\mathrm{Pic}\,(X))^{(2)}$. This is in general a highly nontrivial arithmetic problem, if the group is infinite. There are 3 cases:

(1) $W(\mathrm{Pic}\,(X))^{(2)}$ is of finite index in $O(\mathrm{Pic}\,(X))$.

(2) $W(\mathrm{Pic}\,(X))^{(2)}$ is of infinite index in $O(\mathrm{Pic}\,(X))$, but $W(\mathrm{Pic}\,(X))$ is of finite index in $O(\mathrm{Pic}\,(X))$. (In this case, we call $\mathrm{Pic}\,(X)$ reflective.)

(3) $W(\mathrm{Pic}\,(X))$ is of infinite index in $O(\mathrm{Pic}\,(X))$, i.e. $\mathrm{Pic}\,(X)$ is not reflective.

Remark. The case (1) occurs if and only if $\mathrm{Aut}(X)$ is finite. If $\rho(X) \geq 3$, this occurs if and only if X contains at least one but finitely many smooth rational curves. If $\rho(X) = 2$, this occurs if and only if X contains a smooth rational curve or an irreducible curve of arithmetic genus 1, if and only if $\mathrm{Pic}\,(X)$ represents -2 or 0 [PSS]. Nikulin [Ni1, Ni4] and Vinberg classified all such lattices of rank ≥ 3 belonging to the case (1). It follows from the classification that every algebraic Kummer surface has an infinite automorphism group (cf. [Ke2]).

The classification of the Néron-Severi lattice is also utilized in [Og1 - Og3], where he has proved the density of the jumping loci of the Picard number of a hyperkähler manifold under small 1-dimensional deformation, where he reveals the structure of hierarchy among all the narrow Mordell-Weil lattices of Jacobian K3 surfaces.

(5.6). For finite groups which can act on a K3 surface, the following results are given by S. Mukai and S. Kondo.

Theorem [Mu],[Kon2]. *Let X be a K3 surface and let G be a finite symplectic subgroup of* $\mathrm{Aut}(X)$, *i.e. G acts trivially on $H^{2,0}(X)$. Then G is isomorphic to a subgroup of the Mathieu group M_{23}, which has at least five orbits on a set Ω of 24 elements. In particular, $|G| \leq 960$.*

Theorem [Kon3]. *The maximum order among all finite groups which can act on a K3 surface is 3840, and is uniquely realized by the group $(\mathbf{Z}/2\mathbf{Z})^4 \cdot A_5 \cdot \mathbf{Z}/4\mathbf{Z}$ acting on the Kummer surface $Km(E_{\sqrt{-1}} \times E_{\sqrt{-1}})$, where $E_{\sqrt{-1}}$ is the elliptic curve with $\sqrt{-1}$ as its fundamental period.*

Some projective K3 surfaces, including all algebraic Kummer surfaces and K3 covers of Enriques surfaces, have infinite automorphism groups. Given a projective K3 surface X with $\mathrm{Aut}(X)$ infinite, it is an interesting problem to determine a set of geometric generators of $\mathrm{Aut}(X)$. This problem has been settled for certain classes of K3 surfaces. These results are given in **(5.7) - (5.10)** below.

(5.7). *Two most algebraic K3 surfaces*

Vinberg[Vin] calculated $\mathrm{Aut}(X)$ for two K3 surface with transcendental lattice

$$T(X) = \begin{pmatrix} 2 & 1 \\ 1 & 2 \end{pmatrix}, \begin{pmatrix} 2 & 0 \\ 0 & 2 \end{pmatrix}$$

respectively. In both cases, the full reflection group $W(\mathrm{Pic}\,(X))$ is of finite index in $O(\mathrm{Pic}\,(X))$.

(5.8). *generic Jacobian Kummer surfaces*

Let C be a smooth curve of genus 2. The Jacobian variety $J(C)$ of C is an abelian surface with a natural involution τ and the quotient variety $J(C)/\tau$ has 16 singularities of type A_1. This surface can be embedded as a quartic surface F in $\mathbf{P}^3$ with 16 nodes. The minimal resolution X of $J(C)/\tau$ is called the *Jacobian Kummer surface* associated with C. We call X *generic* if the Néron-Severi group of $J(C)$ is generated by the class of C. For X generic, the transcendental lattice $T(X)$ can be computed as follows:

$$T(X) = U(2) \oplus U(2) \oplus < -4 > .$$

Note that $\mathrm{Aut}(X)$ is isomorphic to the birational automorphism group $Bir(F)$ of the singular quartic surface F. At the last century it was known that X has many involutions, that is, *sixteen translations* induced by those of $J(C)$ by a 2-torsion point, *sixteen projections* of F from a node, *sixteen correlations* by means of the tangent plane collinear to a trope, and a *switch* defined by the dual map of F. In 1900, Hutchinson found another 60 involutions associated with Göpel tetrads. Since Hutchinson, for generic X no other automorphism had been provided until new 192 automorphisms were given in [Ke1].

Theorem [Ke1]. *For a generic Jacobian Kummer surface, there are* 192 *new automorphisms of infinite order which are not generated by classical involutions.*

Theorem [Kon1]. *The automorphism group of a generic Jacobian Kummer surface is generated by the classical involutions and the* 192 *new automorphisms.*

Theorem [Ke3]. *For F generic, all birational automorphisms of F are induced by Cremona transformations of $\mathbf{P}^3$.*

(5.9). *Kummer surfaces associated with the product of two elliptic curves*
The following four cases were considered. In each case, a set of generators of $\mathrm{Aut}(X)$ is given in [KK].

Case I. $X = Km(E \times F)$ where E and F are non-isogenus generic elliptic curves.

Case II. $X = Km(E \times E)$ where E is an elliptic curve without complex multiplications.

Case III. $X = Km(E_\omega \times E_\omega)$ where ω is a 3rd root of unity and E_τ is the elliptic curve with τ as its fundamental period.

Case IV. $X = Km(E_{\sqrt{-1}} \times E_{\sqrt{-1}})$.

The transcendental lattice $T(X)$ can be computed as follows:

$$T(X) = \begin{cases} U(2) \oplus U(2), & \text{(Case I);} \\ U(2) \oplus <4>, & \text{(Case II);} \\ \begin{pmatrix} 4 & 2 \\ 2 & 4 \end{pmatrix}, & \text{(Case III);} \\ \begin{pmatrix} 4 & 0 \\ 0 & 4 \end{pmatrix}, & \text{(Case IV).} \end{cases}$$

Remark. In Case I the group $W(\mathrm{Pic}\,(X))$ is of finite index in $O(\mathrm{Pic}\,(X))$ and in other cases not.

(5.10). *Quartic Hessian surfaces*

Let $S : F(x_0, x_1, x_2, x_3) = 0$ be a nonsingular cubic surface in $\mathbf{P}^3$. Its Hessian surface is a quartic surface defined by the determinant of the matrix of second order partial derivatives of the polynomial F. When F is general enough, the quartic H is irreducible and has 10 nodes. It contains also 10 lines which are the intersection lines of five planes in general linear position. The union of these five planes is classically known as the Sylvester pentahedron of S. The equation of S can be written as the sum of cubes of some linear forms defining the five planes. A nonsingular model of H is a K3 surface $\tilde{H}$. Its Picard number ρ satisfies the inequality $\rho \geq 16$. Note that $\mathrm{Aut}(\tilde{H}) \cong Bir(H)$. In [DK] an explicit description of the group $Bir(H)$ is given when S is general enough so that $\rho = 16$. In this general case, the transcendental lattice $T(\tilde{H})$ can be computed as follows:

$$T(\tilde{H}) = U \oplus U(2) \oplus A_2(-2).$$

Although H, in general, does not have any non-trivial automorphisms (because S does not), the group $Bir(H) \cong \mathrm{Aut}(\tilde{H})$ is infinite. It is generated by the automorphisms defined by projections from the nodes of H, a birational involution which interchanges the nodes and the lines, and the inversion automorphisms of some elliptic pencils on $\tilde{H}$.

This can be compared with the known structure of the group of automorphisms of the Jacobian Kummer surface **(5.8)**. Indeed, the latter surface is birationally isomorphic to the Hessian H of a cubic surface [Hu] but the Picard number of $\tilde{H}$ is equal to 17 instead of 16.

(5.11). Let us explain the method for computing the automorphism group of an algebraic K3 surface, which was first employed by S. Kondō for generic Jacobian Kummer surface case **(5.8)**[Kon1]. The two cases **(5.9)** and **(5.10)** use this method, and even the first case **(5.7)** can also be calculated by the same method (See [Bor2]).

Let X be an algebraic K3 surface with large Picard number, say $\rho(X) \geq 3$. Suppose that $\mathrm{Aut}(X)$ is infinite. Then the ample cone $D(X)$ is not a (finite) polyhedral cone, i.e. has infinitely many faces. Hence, it is difficult to describe $D(X)$ explicitly. Assume that one can find

- a polyhedral cone D' in $D(X)$,

- a set of automorphisms $\{g_\alpha\}$ of X whose action on $D(X)$ has D' as a fundamental domain.

Then, by **(5.5)**, one can conclude that the automorphisms $\{g_\alpha\}$ generate the whole group $\mathrm{Aut}(X)$, up to finite groups. In addition to $\{g_\alpha\}$, some symmetries of D' (not all elements of $SymD'$ in general) may realize as automorphisms of X and some projectively linear automorphisms, if any, generate the kernel of the map in **(5.5)**.

Remark. In the above, g_α corresponds to a face of D' orthogonal to a vector α, and acts on $D(X)$ like a reflection, i.e. sends one of the half-spaces defined by α to the other half-space defined by α or to one of the two half-spaces corresponding to g_α^{-1}. The second case actually occurs in generic Jacobian Kummer surface case **(5.8)**.

(5.12). To find such a polyhedral cone $D' \subset D(X)$, Kondō used the known structure of the orthogonal group of the even unimodular lattice $II_{1,25}$ of signature (1,25). (Such a lattice is unique up to isomorphism and is isomorphic to $\Lambda \oplus U$, where Λ is the Leech lattice, i.e. the even unimodular negative definite lattice of rank 24 which contains no vectors of norm -2.) To be more precise, the following steps lead to the calculation of $\mathrm{Aut}(X)$.

Step 1. Compute $\mathrm{Pic}\,(X)$.

Step 2. Embed $\mathrm{Pic}\,(X)$ primitively into $II_{1,25} = \Lambda \oplus U$ such that the projection of the Weyl vector $w = (0, (0, 1)) \in \Lambda \oplus U$ onto $\mathrm{Pic}\,(X) \otimes \mathbf{R}$ must be an ample class.

Conway [Co] described a fundamental domain D of the reflection group $W(II_{1,25})^{(2)}$ in terms of the Leech roots (=roots with intersection number 1 with the Weyl vector w). More precisely, he showed that $W(II_{1,25})^{(2)}$ is generated by (-2)-reflections corresponding to Leech roots.

Step 3. The fundamental domain D of the reflection group $W(II_{1,25})^{(2)}$ cuts out a finite polyhedral cone D' inside the ample cone $D(X)$. In other words,

$$D' = D \cap P_+(X),$$

where $P_+(X)$ is the positive component of $\{\omega \in \mathrm{Pic}\,(X) \otimes \mathbf{R} : \langle \omega, \omega \rangle > 0\}$. Indeed, D contains the Weyl vector w and, by Step 2, the projection of w is contained in D'.

Determine the hyperplanes α which bound D'. The reason why D' is polyhedral comes from Borcherds [Bor1]; among infinitely many faces of D, those intersecting $P_+(X)$ bound D', and these faces correspond to Leech roots having a non-zero projection onto $\mathrm{Pic}\,(X) \otimes \mathbf{R}$.

Step 4. Match the faces α of D' with automorphisms g_α such that g_α sends one of the half-spaces defined by α to the other half-space defined by α or to one of the two half-spaces corresponding to g_α^{-1}.

This allows one to prove that the automorphisms g_α generate a group of symmetries of $D(X)$, having D' as its fundamental domain.

Step 5. Take care of $SymD'$. See if which symmetries of D' realize as automorphisms of X. Finally see if there are any projectively linear automorphisms, (which generate the kernel of the map in **(5.5)**).

Remark. In the known cases **(5.7) - (5.10)**, the embedding $\mathrm{Pic}\,(X) \subset \Lambda \oplus U$ is given in such a way that $\mathrm{Pic}\,(X)$ is the orthogonal complement of a root sublattice of $\Lambda \oplus U$. For example, in case **(5.7)** the orthogonal complement of $\mathrm{Pic}\,(X)$ in

$\Lambda \oplus U$ is a primitive sublattice of rank 10 which contains a negative definite root lattice of type $A_5 + A_1^5$, which is of index 2.

In practice, Step 4 seems most complicated. The automorphism g_α works like a reflection, but is not necessarily an involution. It may be of infinite order. At any rate, g_α may be geometrically evident, or can be picked up from a list of already known automorphisms, or may be found by looking at extra structures of X, e.g. elliptic fibrations, double plane structures,..., etc. In worst cases, one has to find an (abstract) effective Hodge isometry of $H^2(X, \mathbf{Z})$ and then realize it geometrically.

The reason why the beautiful combinatorics of the Leech lattice plays a role in the description of the automorphism groups of K3 surfaces is still unclear to us. We hope that the classification of all K3 surfaces whose Picard lattice is isomorphic to the orthogonal complement of a root sublattice of $II_{1,25}$ will shed more light to this question.

Remark (5.13). The method **(5.12)** also works for some supersingular K3 surfaces. I. Dolgachev and S. Kondō [DoKo] have computed the automorphism group of a supersingular K3 surface in characteristic 2 whose Picard lattice is $U \oplus D_{20}$. (For supersingular K3 surfaces Torelli type theorem holds, i.e. an automorphism of a supersingular K3 surface is determined by its action on the Picard lattice.)

An even lattice L is reflective if its reflection group $W(L)$ is of finite index in $O(L)$ (see **(5.5)**). The lattice $U \oplus D_{20}$ is reflective as it is pointed out by Borcherds [Bor1] and it is the only known example, up to scaling, of an even reflective hyperbolic lattice of rank 22. The range of possible rank of an even reflective lattice of signature $(1, r-1), r \geq 1$, is given by Esselmann [Es]: it takes the same range 1, 2,..., 20, 22 as the Picard number of a K3 surface in positive characteristic.

Section 6. Automorphisms of algebraic surfaces - singular surface case

(6.1). In [MM] and [MZ3], one considers pairs (V, G) of surface V and group G of automorphisms, where V may be singular and even non-complete.

(6.2). Let Y be a Gorenstein del Pezzo singular surface of Picard number 1. In [Z9], one classifies all actions on Y by cyclic groups $\mathbf{Z}/(p)$ of prime order $p \geq 5$.

Theorem [Z9, Theorems A and C].
Let Y be a Gorenstein del Pezzo singular surface of Picard number 1. *Then we have:*
(1) *Either* $|\mathrm{Aut}(Y)| = 2^a 3^b$ *for some* $1 \leq a+b \leq 7$, *or* $\mathrm{Aut}(Y) \supseteq \mathbf{Z}/(p)$ *for every prime* $p \geq 5$ *and hence* $|\mathrm{Aut}(Y)| = \infty$.

(2) $|\mathrm{Aut}(Y)|$ *is finite if and only if either* $\mathrm{Sing}\, Y = A_7$ *or* $K_Y^2 = 1$ *and* $|-K_Y|$ *has at least three singular members.*
(3) *Let* $p \geq 5$ *be a prime. Suppose that* Y *is not isomorphic to the quadric cone in* $\mathbf{P}^3$. *Then modulo equivariant isomorphism, there is either none, or only one, or exactly* $p+1$ *action(s) of* $\mathbf{Z}/(p)$ *on* Y. *All actions are given in* [Z9].

Remark 6.3. We like to compare **(6.2)** with known results for smooth del Pezzo surfaces.
(1) If X is a generic rational surface with $K_X^2 \leq 5$, then $|\mathrm{Aut}(X)|$ divides 5! (see [DO], [Ki]) .
(2) If X is a del Pezzo surface of degree 3, 4, then $\mathrm{Aut}(X)$ does not contains $\mathbf{Z}/(p)$ for any prime $p \geq 7$, and modulo equivariant isomorphism there is at most one non-trivial $\mathbf{Z}/(5)$ action on X [Ho1, 2].
(3) Let X be a rational projective surface with a non-trivial $\mathbf{Z}/(p)$-action for some prime p such that the $\mathbf{Z}/(p)$-invariant sublattice of $\mathrm{Pic}\, X$ is of rank 1 (this condition is automatic if the Picard number $\rho(X) = 1$). If X is smooth, then $p \leq 5$ unless $X = \mathbf{P}^2$ [ZD, Theorem 1].

Acknowledgement

We would like to thank the referee for careful reading and suggestions which improve the paper. This article was initiated when both of us were attending the conference – Algebraic Geometry in East Asia – in August 2001, and we would like to thank the organizers for providing the wonderful environment and the encouragement for us to write a survey article. This work was finalized when the second-named author visited Korea Institute for Advanced Study in December 2001, and he likes to thank the institute for the hospitality. This work was partially supported by Korea Science and Engineering Foundation (R01-1999-00004) and an Academic Research Fund of National University of Singapore.

References

[AN] V. A. Alekseev and V. V. Nikulin, Classification of del Pezzo surfaces with log-terminal singularities of index ≤ 2 and involutions on $K3$ surfaces, Soviet Math. Dokl. **39** (1989), 525–528.

[Ar1] M. Artin, Some numerical criteria for contractability of curves on algebraic surfaces. Amer. J. Math. **84** (1962), 485–496.

[Ar2] M. Artin, On isolated rational singularities of surfaces, Amer. J. Math. **88** (1966), 129–136.

[BPV] W. Barth, C. Peters and A. Van de Ven, Compact Complex Surfaces, Springer-Verlag, 1984.

[Bl] R. Blache, The structure of l.c. surfaces of Kodaira dimension zero, I, J. Algebraic Geom. **4** (1995), 137–179.

[Bor1] R. Borcherds, Automorphism groups of Lorentzian lattices, J. Algebra **111** (1987), 133-153.

[Bor2] R. Borcherds, Coxter groups, Lorentzian lattices, and K3 surfaces, International Mathematics Research Notices **19** (1998), 1011-1031.

[Br] E. Brieskorn, Rationale Singularitäten komplexer Flächen, Invent. Math. **4** (1967 / 1968), 336–358.

[BR] D. Burns, Jr. and M. Rapoport, On the Torelli theorem for Kählerian $K3$ surfaces, Ann. Sci. Ecole Norm. Sup. (4) **8** (1975), 235-273.

[Ca] F. Campana, Connexiti rationnelle des variitis de Fano, Ann. Sci. Ecole Norm. Sup. **25** (1992), 539–545.

[CKO] F. Catanese, J. Keum and K. Oguiso, Some remarks on the universal cover of an open K3 surface, Math. Ann. to appear, math.AG/**0104015**.

[Co] J.H. Conway, The automorphism group of the 26 dimensional even Lorentzian lattice, J. Algebra **80** (1983), 159-163.

[CD] F. Cossec and I. V. Dolgachev, Enriques surfaces, I, Progress in Mathematics, **76** (1989), Birkhäuser Boston, Inc.

[dF] Tommaso de Fernex, Birational trasformations of prime order of the projective plane, preprint 2001.

[DP] J. -P. Demailly and M. Paun, Numerical characterization of the Kähler cone of a compact Kähler manifold, math.AG / **0105176**.

[DK] I. V. Dolgachev and J. Keum, Birational automorphisms of quartic Hessian surfaces, Trans. Amer. Math. Soc. 354 (2002), 3031-3057.

[DoKo] I. V. Dolgachev and S. Kondō, A supersingular K3 surface in characteristic 2 and the Leech lattice, math.AG/0112283 v2.

[DO] I. Dolgachev and D. Ortland, Point sets in projective spaces and theta functions, Astérisque, Vol. **165** (1988).

[DZ] I. V. Dolgachev and D. -Q. Zhang, Coble rational surfaces, Amer. J. Math. **123** (2001), 79–114.

[Du] A. H. Durfee, Fifteen characterizations of rational double points and simple critical points, Enseign. Math. **25** (1979), 131–163.

[Es] F. Esselmann, Ueber die maximale Dimension von Lorenz-Gittern mit coendlicher Spiegelungsgruppe, J. Number Theory, **61** (1996), 103-144.

[FKL] A. Fujiki, R. Kobayashi and S. Lu, On the fundamental group of certain open normal surfaces. Saitama Math. J. **11** (1993), 15–20.

[Fu] T. Fujita, On the topology of noncomplete algebraic surfaces, J. Fac. Sci. Univ. Tokyo Sect. IA Math. **29** (1982), 503–566.

[Gi] M. H. Gizatullin, Rational G-surfaces, Math. USSR Izv. **16** (1981), 103–134.

[GHS] T. Graber, J. Harris and J. Starr, Families of rationally connected varieties, math.AG / **0109220**.

[GPa] R. V. Gurjar and A.J. Parameswaran, Affine lines on $\mathbf{Q}$-homology planes, J. of Math. Kyoto Univ. **35** (1995), 63-77.

[GPr] R. V. Gurjar and C. R. Pradeep, Q-homology planes are rational, III, Osaka J. Math. **36** (1999), 259–335.

[GPS] R. V. Gurjar, C. R. Pradeep and A. R. Shastri, On rationality of logarithmic **Q**-homology planes, II. Osaka J. Math. **34** (1997), 725–743.

[GS] R. V. Gurjar and A. R. Shastri, On the rationality of complex homology 2-cells, I; II, J. Math. Soc. Japan, **41** (1989), 37–56; **41** (1989), 175–212.

[GZ1] R. V. Gurjar and D. -Q. Zhang, π_1 of smooth points of a log del Pezzo surface is finite, I, J. Math. Sci. Univ. Tokyo **1** (1994), 137–180.

[GZ2] R. V. Gurjar and D. -Q. Zhang, π_1 of smooth points of a log del Pezzo surface is finite, II, J. Math. Sci. Univ. Tokyo **2** (1995), 165–196.

[GZ3] R. V. Gurjar, and D. -Q. Zhang, Normal algebraic surfaces with trivial bicanonical divisor, J. Algebra **186** (1996), 970–989.

[Ho1] T. Hosoh, Automorphism groups of cubic surfaces, J. Algebra **192** (1997), 651–677.

[Ho2] T. Hosoh, Automorphism groups of quartic del Pezzo surfaces, J. Algebra **185** (1996), 374–389.

[Hu] J. Hutchinson, The Hessian of the cubic surface, Bul. Amer. Math. Soc. **5** (1889), 282-292.

[I1] S. Iitaka, On logarithmic $K3$ surfaces, Osaka J. Math. **16** (1979), 675–705.

[I2] S. Iitaka, A numerical criterion of quasi-abelian surfaces, Nagoya Math. J. **73** (1979), 99–115.

[I3] S. Iitaka, Algebraic geometry – An introduction to birational geometry of algebraic varieties, Graduate Texts in Mathematics, **76**, Springer-Verlag, New York-Berlin, 1982.

[Is] V. A. Iskovskih, Minimal models of rational surfaces over arbitrary fields, Math. USSR Izv. **14** (1981), 17–39.

[Kt] S. Kantor, Theorie der endlichen Gruppen von eindeutigen Transformationen in der Ebene, Berlin: Mayer & Muller, 1895.

[KN] M. Kato and I. Naruki, Depth of rational double points on quartic surfaces, Proc. Japan Acad. Ser. A Math. Sci. 58 (1982), 72–75.

[Ka1] Y. Kawamata, On the classification of noncomplete algebraic surfaces, Lecture Notes in Math. **732** (1979), 215–232. Springer, Berlin.

[Ka2] Y. Kawamata, The cone of curves of algebraic varieties, Ann. of Math. **119** (1984), 603–633.

[KMM] Y. Kawamata, K. Matsuda and K. Matsuki, Introduction to the minimal model problem, Adv. Stud. Pure Math. **10** (1987), 283–360.

[KMc] S. Keel and J. McKernan, Rational curves on quasi-projective surfaces, Mem. Amer. Math. Soc. **140** (1999), no. 669.

[KMo] J. Kollar and S. Mori, Birational geometry of algebraic varieties, Cambridge Tracts in Mathematics, **134** (1998).

[Ke1] J. Keum, Automorphisms of Jacobian Kummer surfaces, Compositio Math. **107** (1997), 269-288.

[Ke2] J. Keum, Every algebraic Kummer surface has infinitely many automorphisms, unpublished manuscript (1996).

[Ke3] J. Keum, Automorphisms of a generic Jacobian Kummer surface, Geom. Ded. **76** (1999), 177-181.

[KK] J. Keum and S. Kondo, The automorphism groups of Kummer surfaces associated with the product of two elliptic curves, Trans. Amer. Math. Soc. **353** (2001), 1469–1487.

[KZ] J. Keum and D. -Q. Zhang, Fundamental groups of open K3 surfaces, Enriques surfaces and Fano 3-folds, J. Pure App. Alg. **170** (2002), 67–91.

[Kb1] R. Kobayashi, Uniformization of complex surfaces. Kähler metric and moduli spaces, Adv. Stud. Pure Math. **18-II** (1990), 313–394.

[Kb2] R. Kobayashi, Einstein-Kähler V-metrics on open Satake V-surfaces with isolated quotient singularities, Math. Ann. **272** (1985), 385–398.

[KNS] R. Kobayashi, S. Nakamura and F. Sakai, A numerical characterization of ball quotients for normal surfaces with branch loci, Proc. Japan Acad. Ser. A Math. Sci. **65** (1989), 238–241.

[Ki] M. Koitabashi, Automorphism groups of generic rational surfaces, J. Algebra **116** (1988), 130 –142.

[Kj1] H. Kojima, Open rational surfaces with logarithmic Kodaira dimension zero, Internat. J. Math. **10** (1999), 619–642.

[Kj2] H. Kojima, Logarithmic del Pezzo surfaces of rank one with unique singular points, Japan. J. Math. (N.S.) **25** (1999), 343–375.

[Kj3] H. Kojima, Complements of plane curves with logarithmic Kodaira dimension zero, J. Math. Soc. Japan **52** (2000), 793–806.

[Kj4] H. Kojima, Rank one log del Pezzo surfaces of index two, submitted Oct 2000.

[KoMiMo] J. Kollar, Y. Miyaoka and S. Mori, Rationally connected varieties, J. Alg. Geom. **1** (1992), 429–448.

[Kon1] S. Kondō, The automorphism group of a generic Jacobian Kummer surface, J. Alg. Geom. **7** (1998), 589-609.

[Kon2] S. Kondō, Niemeier lattices, Mathieu groups, and finite groups of symplectic automorphisms of K3 surfaces (with an appendix by S. Mukai), Duke Math. J. **92** (1998), 593-603.

[Kon3] S. Kondō, The maximum order of finite groups of automorphisms of $K3$ surfaces, Amer. J. Math. **121** (1999), 1245-1252.

[Kon4] S. Kondō, Enriques surfaces with finite automorphism groups, Japan. J. Math. **12** (1986), 191-282.

[Ma1] Yu. I. Manin, Rational surfaces over perfect fields, II, Math. USSR Sb. **1** (1967), 141–168.

[Ma2] Yu. I. Manin, Cubic forms : Algebra, geometry, arithmetic, 2nd ed, North-Holland Math. Library, **4** (1986), North-Holland Publ. Co., Amsterdam-New York.

[Mi] M. Miyanishi, Open algebraic surfaces, CRM Monograph Series, **12**, American Mathematical Society, 2001.

[MM] M. Miyanishi and K. Masuda, Open algebraic surfaces with finite group actions, Transform. Group, to appear.

[MT1] M. Miyanishi and S. Tsunoda, Noncomplete algebraic surfaces with logarithmic Kodaira dimension $-\infty$ and with nonconnected boundaries at infinity, Japan. J. Math. (N.S.) **10** (1984), 195–242.

[MT2] M. Miyanishi and S. Tsunoda, Logarithmic del Pezzo surfaces of rank one with noncontractible boundaries, Japan. J. Math. (N.S.) **10** (1984), 271–319.

[MZ1] M. Miyanishi and D. -Q. Zhang, Gorenstein log del Pezzo surfaces of rank one, J. Algebra **118** (1988), 63–84.

[MZ2] M. Miyanishi and D. -Q. Zhang, Gorenstein log del Pezzo surfaces, II, J. Algebra **156** (1993), 183–193.

[MZ3] M. Miyanishi and D. -Q. Zhang, Equivariant classification of Gorenstein open log del Pezzo surfaces with finite group actions, Preprint 2001.

[Mu] S. Mukai, Finite groups of automorphisms of $K3$ surfaces and the Mathieu group, Invent. Math. **94** (1988), 183-221.

[Ni1] V.V. Nikulin, On the quotient groups of the automorphism group of hyperbolic forms by the subgroups generated by 2-reflections, J. Soviet Math. **22** (1983), 1401-1476.

[Ni2] V.V. Nikulin, On the description of groups of automorphisms of Enriques surfaces, Soviet Math. Dokl. **277** (1984), 1324-1327.

[Ni3] V.V. Nikulin, Del Pezzo surfaces with log-terminal singularities, I; II; III, Math. USSR-Sb. **66** (1990), 231–248; Math. USSR-Izv. **33** (1989), 355–372; Math. USSR-Izv. **35** (1990), 657–675.

[Ni4] V. V. Nikulin, Discrete reflection groups in Lobachevsky spaces and algebraic surfaces, Proc. Intern. Cong. Math. Berkeley, Calif. 1986, pp.654–671.

[Og1] K. Oguiso, Picard numbers in a family of hyperkähler manifolds and applications, J. Alg. Geom. to appear

[Og2] K. Oguiso, Picard numbers in a family of hyperkähler manifolds - A supplement to the article of R. Borcherds, L. Katzarkov, T. Pantev, N. I. Shepherd-Barron, math.AG / **0011258**.

[Og3] K. Oguiso, Automorphism groups in a family of K3 surfaces, math.AG / **0104049**.

[OZ1] K. Oguiso and D. -Q. Zhang, On the most algebraic $K3$ surfaces and the most extremal log Enriques surfaces, Amer. J. Math. **118** (1996), 1277–1297.

[OZ2] K. Oguiso and D. -Q. Zhang, On the complete classification of extremal log Enriques surfaces, Math. Z. **231** (1999), 23–50.

[OZ3] K. Oguiso and D. -Q. Zhang, K3 surfaces with order 11 automorphisms, math.AG / **9907020**.

[OZ4] K. Oguiso and D. -Q. Zhang, On Vorontsov's theorem on $K3$ surfaces with non-symplectic group actions, Proc. Amer. Math. Soc. **128** (2000), 1571–1580.

[OZ5] K. Oguiso and D. -Q. Zhang, The Simple Group of Order 168 and K3 Surfaces, in : Complex Geometry, Collection of papers dedicated to Hans Grauert; Bauer, Catanese, Kawamata, Peternell and Siu (ed), Springer, 2002; math.AG / **0011259**.

[Oh] K. Ohno, Toward determination of the singular fibers of minimal degeneration of surfaces with $\kappa = 0$, Osaka J. Math. **33** (1996), 235–305.

[PSS] I. Piatetski-Shapiro, I.R. Shafarevich, A Torelli theorem for algebraic surfaces of type K3, Math. USSR Izv. **5** (1971), 547-587.

[PS] C. R. Pradeep and A. R. Shastri, On rationality of logarithmic $\mathbf{Q}$-homology planes, I, Osaka J. Math. **34** (1997), 429–456.

[Re1] M. Reid, Young person's guide to canonical singularities, Proc. Sympos. Pure Math. Part 1 **46** (1987), 345–414.

[Re2] M. Reid, Campedelli versus Godeaux – Problems in the theory of surfaces and their classification, Sympos. Math., XXXII (1991); 309–365.

[Sa] F. Sakai, Semistable curves on algebraic surfaces and logarithmic pluricanonical maps, Math. Ann. **254** (1980), 89–120.

[Se] B. Segre, The Non-singular cubic surfaces, Oxford University Press, Oxford, 1942.
[SZ] I. Shimada and D. -Q. Zhang, Classification of extremal elliptic $K3$ surfaces and fundamental groups of open $K3$ surfaces, Nagoya Math. J. **161** (2001), 23–54.
[Su] K. Suzuki, On Morrison's cone conjecture for Klt surfaces with $K_X \equiv 0$, Comment. Math. Univ. St. Paul **50** (2001), 173–180.
[Siu] Y. T. Siu, Every $K3$ surface is Kähler, Invent. Math. **73** (1983), 139-150.
[Ta] S. Takayama, Simple connectedness of weak Fano varieties, J. Algebraic Geom. **9** (2000), 403–407.
[Ts] S. Tsunoda, Structure of open algebraic surfaces, I, J. Math. Kyoto Univ. **23** (1983), 95–125.
[TZ] S. Tsunoda and D. -Q. Zhang, Noether's inequality for noncomplete algebraic surfaces of general type, Publ. Res. Inst. Math. Sci. **28** (1992), 21–38.
[Vin] E.B. Vinberg, The two most algebraic K3 surfaces, Math. Ann. **265** (1985), 1-21.
[Xi1] G. Xiao, Bound of automorphisms of surfaces of general type. I, Ann. of Math. (2) **139** (1994), no. 1, 51-77.
[Xi2] G. Xiao, Bound of automorphisms of surfaces of general type. II, J. Algebraic Geom. **4** (1995), no. 4, 701-793.
[Z1] D. -Q. Zhang, On Iitaka surfaces, Osaka J. Math. **24** (1987), 417–460.
[Z2] D. -Q. Zhang, Logarithmic del Pezzo surfaces of rank one with contractible boundaries, Osaka J. Math. **25** (1988), 461–497.
[Z3] D. -Q. Zhang, Logarithmic del Pezzo surfaces with rational double and triple singular points, Tohoku Math. J. **41** (1989), 399–452.
[Z4] D. -Q. Zhang, Logarithmic Enriques surfaces, I; II, J. Math. Kyoto Univ. **31** (1991), 419–466; **33** (1993), 357-397.
[Z5] D. -Q. Zhang, Noether's inequality for noncomplete algebraic surfaces of general type, II, Publ. Res. Inst. Math. Sci. **28** (1992), 679–707.
[Z6] D. -Q. Zhang, Algebraic surfaces with nef and big anti-canonical divisor, Math. Proc. Cambridge Philos. Soc. **117** (1995), 161–163.
[Z7] D. -Q. Zhang, The fundamental group of the smooth part of a log Fano variety, Osaka J. Math. **32** (1995), 637–644.
[Z8] D. -Q. Zhang, Algebraic surfaces with log canonical singularities and the fundamental groups of their smooth parts, Trans. Amer. Math. Soc. **348** (1996), 4175–4184.
[Z9] D. -Q. Zhang, Automorphisms of finite order on Gorenstein del Pezzo surfaces, Trans. Amer. Math. Soc. **354** (2002), 4831–4845.
[ZD] D. -Q. Zhang, Automorphisms of finite order on rational surfaces, with an appendix by I. Dolgachev, J. Algebra **238** (2001),560–589.

J. Keum
Korea Institute for Advanced Study
207-43 Cheongryangri-dong, Dongdaemun-gu

Seoul 130-012, Korea
E-mail : jhkeum@kias.re.kr

D. -Q. Zhang
Department of Mathematics
National University of Singapore
2 Science Drive 2, Singapore 117543
Republic of Singapore
E-mail: matzdq@math.nus.edu.sg

LINEAR SERIES OF IRREGULAR VARIETIES

JUNGKAI A. CHEN (JOINT WITH CHRISTOPHER D. HACON)

1. Introduction

Let X be a nonsingular projective variety over complex number $\mathbb{C}$ of dimension n and D be a divisor on X. The linear series $|D|$ is defined to be the set of effective divisors which are linear equivalent to D. It's natural to ask under what circumstance the linear series $|D|$ is non-empty or base point free. Or one might even want to know if $|D|$ defines a barional map or an embedding.

Regarding the base point freeness of linear series, Fujita conjectured the following:

Conjecture 1.1. *If L is an ample divisor on X, then $|K_X + mL|$ is free for $m \geq n+1$.*

This conjecture can be easily proved in dimension 1. For $\dim X = 2, 3, 4$, the statement is verified by Reider [Re], Ein and Lazarsfeld [EL1], and Kawamata [Ka] respectively. For higher dimensional varieties, a remarkable breakthrough is made by Angehrn and Siu [AS], which assert that $|K_X + mL|$ is free for $m \geq \frac{1}{2}(n^2+n)+1$. Their proofs use various form of vanishing theorem essentially.

The purpose of this note is to introduce some application of Fourier-Mukai transform in the study of linear series. To setup, let A be an abelian variety and $\mathcal{F}$ be a coherent sheaf on A. $\mathcal{F}$ is asaid to be IT^0 if $H^i(A, \mathcal{F} \otimes P) = 0$ for all $P \in \operatorname{Pic}^0(A)$ and $i > 0$. By using Fourier-Mukai transform, it's easy to see the following results.

Lemma 1.2. *If $\mathcal{F} \neq 0$ is IT^0, then $h^0(A, \mathcal{F}) \neq 0$.*

Lemma 1.3. *If $\mathcal{F}$ is IT^0 and there is a surjective map $\mathcal{F} \to k(y)$, then the induced map $H^0(A, \mathcal{F} \otimes P) \to H^0(A, k(y) \otimes P)$ is surjective for general P.*

Consider now a variety with $q(X) > 0$. Let $\mathrm{alb} : X \to \mathrm{Alb}(X)$ be the Albanese map. And let $a(X)$ be the dimension of image of alb. We say X is of maximal Albanese dimension if $a(X) = \dim X$. In general, our idea here can realized as following: In order to study the linear series $|D|$, we consider the push-forward $\mathcal{F} := \mathrm{alb}_* \mathcal{O}(D)$. If $\mathcal{F}$ satisfied IT^0, then one can obtain some information on base points by Lemma 1.3. For example, we are able to show the following result.

Theorem 1.4. *If $\dim X - a(X) \leq 2$, then $|7K_X|$ defines a birational map.*

2000 MSC. 14C20, 14K12

In particular, for a 3-fold with $q > 0$, the $|7K_X|$ defines a birational map.

2. Fourier-Mukai Transforms

In this section, we are going to describe the Fourier-Mukai transform (cf. [Mu]) briefly. And then we are going to demonstrate some easy application of Fourier-Mukai transform. In particular, we prove a generalization of Lefschetz base point freeness theorem to coherent sheaves. This generalization turns out to be useful in the study of irregular varieties.

Let A be an abelian variety, and denote the corresponding dual abelian variety by $\hat{A}$. Let $\mathcal{P}$ be the normalized Poincaré bundle on $A \times \hat{A}$. For any point $y \in \hat{A}$, let $P_y = \mathcal{P}|_{A\times\{y\}}$ denote the associated topological trivial line bundle. Define the functor $\hat{\mathcal{S}}$ of $\mathcal{O}_A$-modules into the category of $\mathcal{O}_{\hat{A}}$-modules by

$$\hat{\mathcal{S}}(M) = \pi_{\hat{A},*}(\mathcal{P}\otimes\pi_A^* M).$$

The derived functor $R\hat{\mathcal{S}}$ of $\hat{\mathcal{S}}$ then induces an equivalence of categories between the two derived categories $D(A)$ and $D(\hat{A})$. In fact, by [Mu]: *There are isomorphisms of functors:*

$$R\mathcal{S} \circ R\hat{\mathcal{S}} \cong (-1_A)^*[-g]$$

and

$$R\hat{\mathcal{S}} \circ R\mathcal{S} \cong (-1_{\hat{A}})^*[-g],$$

where $[-g]$ denotes "shift the complex g places to the right".

A coherent sheaf $\mathcal{F}$ on A is said to be IT^i if there exists an integer $i(\mathcal{F})$ such that for all $j \neq i(\mathcal{F})$, $H^j(A, \mathcal{F}\otimes P) = 0$ for all $P \in \mathrm{Pic}^0(A)$. It is easily seen that if $\mathcal{F}$ satisfies the IT^i, then $R^j\hat{\mathcal{S}}(\mathcal{F}) = 0$ for all $j \neq i(\mathcal{F})$. We will denote the coherent sheaf $R^{i(\mathcal{F})}\hat{\mathcal{S}}(\mathcal{F})$ on $\hat{A}$ by $\hat{\mathcal{F}}$ and call it the *Fourier-Mukai transform of $\mathcal{F}$*. We remark that that is a map

$$R^j\hat{\mathcal{S}}(\mathcal{F})\otimes k(y) \to H^j(A, \mathcal{F}\otimes P_y).$$

Lemma 2.1. *If $\mathcal{F} \neq 0$ is IT^i, then $H^i(A, \mathcal{F}) \neq 0$.*

Proof. $\hat{\mathcal{F}}$ is a locally free sheaf of rank $h^i(A, \mathcal{F})$. If $h^i(A, \mathcal{F}) = 0$, then $\hat{\mathcal{F}} = 0$. It follows that $\mathcal{F} = (-1)^*\hat{0} = 0$. Which is a contradiction. □

Definition 2.2. *Let $\mathcal{F}$ be a coherent sheaf on an irregular variety X. $\mathcal{F}$ is said to have an essential base point at y if there is a surjective map $\mathcal{F} \to k(y)$ such that for all $P \in \mathrm{Pic}^0(X)$, the induced map $H^0(X, \mathcal{F}\otimes P) \to H^0(X, k(y))$ is zero.*

Proposition 2.3. *If $\mathcal{F}$ is a non-zero sheaf on an abelian variety satisfying IT^0, then $\mathcal{F}$ has no essential base points.*

Proof. Suppose that $\mathcal{F}$ has an essential base point at y. Let $\phi \neq 0$ be a surjective map as above, $\mathcal{F}_y := ker(\mathcal{F} \xrightarrow{\phi\neq 0} k(y))$. Then the exact sequence

$$0 \to \mathcal{F}_y \otimes P \to \mathcal{F}\otimes P \to k(y) \to 0$$

gives

$$H^0(A, \mathcal{F}_y\otimes P) \xrightarrow{\cong} H^0(A, \mathcal{F}\otimes P),$$
$$H^0(A, k(y)) \xrightarrow{\cong} H^1(A, \mathcal{F}_y\otimes P),$$

for all $P \in \mathrm{Pic}^0(A)$. Computing the Fourier-Mukai transform, one has

$$0 \to R^0\hat{\mathcal{S}}(\mathcal{F}_y) \to R^0\hat{\mathcal{S}}(\mathcal{F}) \to P_y \to R^1\hat{\mathcal{S}}(\mathcal{F}_y) \to 0.$$

By Grauert's theorem, $V := R^0\hat{\mathcal{S}}(\mathcal{F})$ and $R^0\hat{\mathcal{S}}(\mathcal{F}_y)$ are vector bundles of the same rank, and $R^1\hat{\mathcal{S}}(\mathcal{F}_y)$ is a line bundle. If follows that $P_y \to R^1\hat{\mathcal{S}}(\mathcal{F}_y)$ is an isomorphism and hence $V \to P_y$ is the zero map. Therefore, one has a triangle

$$R^{\bullet}\hat{\mathcal{S}}(\mathcal{F}_y) \longrightarrow R^{\bullet}\hat{\mathcal{S}}(\mathcal{F}) \xrightarrow{0} R^{\bullet}\hat{\mathcal{S}}(k(y))$$

in $D(\hat{A})$. There is a second triangle in $D(\hat{A})$ given by (cf. [Ha] TR2)

$$R^{\bullet}\hat{\mathcal{S}}(\mathcal{F}) \xrightarrow{0} R^{\bullet}\hat{\mathcal{S}}(k(y)) \longrightarrow R^{\bullet}\hat{\mathcal{S}}(\mathcal{F}_y)[1]$$

which is easily seen to be isomorphic to (cf. [Ha] TR3)

$$V \cong R^{\bullet}\hat{\mathcal{S}}(\mathcal{F}) \xrightarrow{0} P_y \cong R^{\bullet}\hat{\mathcal{S}}(k(y)) \to P_y \oplus V[1].$$

Taking the Fourier-Mukai transform again, one has $\mathcal{F}_y \cong \mathcal{F} \oplus k(y)[1]$ in $D(A)$. This is the required contradiction. □

Corollary 2.4. *Let L be an ample line bundle on an abelian variety A. Then $|2L|$ is base point free.*

Proof. L is IT^0 and there is a surjective map $L \to k(y)$ for any given point $y \in A$. Thus for general $P \in \mathrm{Pic}^0(A)$, $H^0(A, L\otimes P) \to H^0(A, k(y))$ is surjective. Which means that y is not a base point of $|L + P|$ for general P. One can pick P so that y is not a base point of $|L + P|$ and $|L - P|$. Hence y is not a base point of $|2L|$. □

3. Adjoint linear series of irregular varieties

In order to apply the criterion for essential base point freeness to linear series of irregular varieties, we will need the following result:

Theorem 3.1. *Let X be a smooth projective varieties with $q(X) > 0$. Let $a : X \to A$ be a non-constant morphism to an abelian variety A. Let L be a line bundle on X and fix a point $x \in X$. Suppose that*

i) $h^i(A, a_*L\otimes P) = 0$ for all $i > 0$ and $P \in \mathrm{Pic}^0(A)$.

ii) $a_*(L\otimes\mathcal{I}_x) \neq a_*(L)$.

Then x is not an essential base point of $|L|$. In particular, x is not a base point of $|2L|$.

Proof. For a fixed point x in X, we start by considering the sequence

$$0 \to L \otimes \mathcal{I}_x \to L \to k(x) \to 0,$$

and its push-forward via $a : X \to A$

$$0 \to a_*(L \otimes \mathcal{I}_x) \to a_*(L) \to a_*(k(x)) \to R^1 a_*(L \otimes \mathcal{I}_x) \to \dots.$$

By assumption $a_*(L \otimes \mathcal{I}_x) \to a_*(L)$ is not an isomorphism. It is clear that $a_*(k(x)) = k(y)$ where $y = a(x)$. Let $\mathcal{Q}$ be the kernel of $k(y) \to R^1 a_*((L) \otimes \mathcal{I}_x)$. It follows that $\mathcal{Q} \neq 0$. However, $k(y)$ has no non-zero sub-sheaf other than itself. Hence $\mathcal{Q} = k(y)$. We therefore have an exact sequence:

$$0 \to a_*(L \otimes \mathcal{I}_x) \to a_*(L) \to k(y) \to 0.$$

$a_*(L)$ satisfies the I.T. of index 0. By Proposition 2.3, y is not an essential base point of $a_*(L)$. Therefore, x is not a base point of $L \otimes a^* Q$ for some $Q \in \mathrm{Pic}^0(A)$. By the semi-continuity of $h^0(X, L \otimes \mathcal{I}_x \otimes a^* Q)$ as Q varies in $\mathrm{Pic}^0(A)$, it follows that x is not a base point of $|L + a^* Q|$ for general $Q \in \mathrm{Pic}^0(X)$, and hence the theorem. □

As an application, we have the following result.

Corollary 3.2. *Let X be a variety of maximal Albanese dimension, and L a nef and big line bundle on X. Then $Bs|2K_X + 2L|$ is contained in the exceptional locus of $a : X \longrightarrow \mathrm{Alb}(X)$.*

Proof. Let $a : X \to A$ be the Albanese map. We first note that $a_*(\omega_X \otimes L)$ is a sheaf satisfying IT^0 (cf. [Ko] Corollary 10.15 and [HP] Lemma 2.7). Clearly, it is non-zero since a is generically finite onto its image.

Let $x \in X$ be a point which is not in the exceptional divisor. There is an open neighborhood U of x such that $a : U \to a(U)$ is a finite morphism. Hence condition *ii*) in Theorem 3.1 is satisfied and we are done. □

One can generalize this result to a more general setting.

Theorem 3.3. *Let $a : X \longrightarrow Z := \mathrm{alb}_X(X) \subset A$ be the Albanese morphism of a nonsingular projective variety X to a subvariety Z contained in an abelian variety A. Let L be an ample line bundle on X, $e := \dim(X) - \dim(Z)$ and $m := \frac{e^2+e+2}{2}$. Let U be an open subset of Z and $V := a^{-1}(U)$ such that $a|_V : V \longrightarrow U$ is a smooth morphism. Then $|K_X + mL|$ has no essential base points contained in U. In particular $Bs|2K_X + 2mL| \subset X - V$.*

Proof. See [CH]. □

4. Non-vanishing of Adjoint Linear Series

For a non-zero sheaf satisfying IT^0, we have seen the non-vanishing of its cohomology. This leads to a non-vanishing of linear series in a natural way.

Lemma 4.1. *Let X be a variety with $q(X) > 0$ and L be a nef and big divisor. And $a : X \to A$ is a non-trivial morphism to an abelian variety A with general fiber F. Then $h^0(X, K_X + L) \neq 0$ if and only if $h^0(F, K_F + L_F) \neq 0$.*

Proof. Let $\mathcal{F} := \mathrm{alb}_*(\mathcal{O}(K_X + L)$. It's easy to check that $\mathcal{F}$ is IT^0. By Lemma 1.1, one sees that $h^0(X, K_X + L) \neq 0$ if and only if $\mathcal{F} \neq 0$. We are done since $\mathcal{F} \neq 0$ if and only $h^0(F, K_F + L_F) \neq 0$. □

Theorem 4.2. *Let X be a variety with $q(X) > 0$. Suppose that* $\mathrm{dim}X - a \leq 2$. *Let L be a nef and big divisor such that $K_X + L$ is nef. Then $H^0(X, K_X + L) \neq 0$.*

Proof. By the previous Lemma, it suffices to show the statement for X_y. The assertion is true for 1-dimensional varieties. If X_y has dimension 2 with $q(X_y) > 0$, then we proceed inductively. If X_y has dimension 2 with $q(X_y) = 0$, then we use Riemann-Roch theorem.

$$h^0(F_y, K_{F_y} + L_{X_y}) = \chi(F_y, K_{F_y} + L_{X_y}) = 1 + p_g + \frac{1}{2}(K_{x_y} + L_{X_y})L_{X_y} \geq 1.$$

This completes the proof. □

5. Pluricanonical Maps

In this section, we are going to show the application to the study of linear series. Let X be a smooth projective variety and $Z = \mathrm{alb}_X(X) \subset \mathrm{Alb}(X)$ its Albanese image.

Lemma 5.1. *Let $g : X \longrightarrow S$ be a surjective morphism of smooth projective varieties, L a nef and big Cartier divisor on S. Let X_s be a general fiber of $X \longrightarrow S$. For any integer $m \geq 2$, assume that $P_m(X_s) > 0$ and write $|mK_{X_s}| = |H^m| + F^m$ where $|H^m|$ has no fixed divisors. Assume furthermore that K_X is S-big and $\kappa(S) \geq 0$. Then, after replacing X by an appropriate birational model, there exists an integer $b > 0$ (any sufficiently big and divisible integer $b > 0$ will do) such that for any sufficiently big and divisible integers $a > 0$, there is a divisor $B \in |ab(m-1)K_X - bg^*L|$, B has normal crossings support and $\lfloor \frac{B}{ab} \rfloor |_{X_s} \prec \lfloor \frac{B}{ab}|_{X_s} \rfloor \prec F^m$.*

Proof. $(m-1)K_X$ is S-big, so for all sufficiently big and divisible integers r_0, the linear series $|r_0(m-1)K_X - 2g^*L|$ is non-empty. Let D be a general member of this linear series.

Replacying X by an appropriate birational model, we may assume that $|mK_{X_s}| = |H^m| + F^m$ where $|H^m|$ is base point free and F^m has normal crossings support. Moreover we may assume that

$$|(r_0(m-1)K_X - 2g^*L)|_{X_s}| = |N| + G$$

where $|N|$ is base point free, $G+F^m$ has normal crossings support. Let $D=\sum d_iD_i$ with D_i reduced irreducible divisors, and let $d:=\sum d_i$. Since $s\in S$ is a general point, also $D_i|_{X_s}$ is reduced and so for $G=\sum g_iG_i$ we also have $g_i\le d$. Let $t>\frac{md}{r_0}$, then

$$\lfloor \frac{m}{m-1}F^m+\frac{G}{tr_0}\rfloor \prec F^m.$$

Since $\kappa(S)\ge 0$, for r sufficiently big and divisible, the linear series $g^*|(t-1)r(m-1)K_S|$ is non-empty.

By [V], for all sufficiently big and divisible integers $r>0$, there is a surjection

$$H^0(X,\omega_{X/S}^{(t-1)r(m-1)}\otimes g^*L^{r/r_0})\otimes\mathbb{C}(X)\longrightarrow H^0(X_s,\omega_{X_s}^{(t-1)r(m-1)}).$$

Assume that r_0m divides r. Let B_1,B_2,B_3 be general divisors in the linear series $|\omega_{X/S}^{(t-1)r(m-1)}\otimes g^*L^{r/r_0}|$, $|r_0(m-1)K_X-2g^*L|$, $g^*|(t-1)r(m-1)K_S|$ respectively. It is easy to see that

$$B_1|_{X_s}=\bar{H}^{(t-1)r(m-1)}+F^{(t-1)r(m-1)}\prec \bar{H}^{(t-1)r(m-1)}+\frac{(t-1)r(m-1)}{m}F^m,$$

where $\bar{H}^{(t-1)r(m-1)}$ is a general member of $|H^{(t-1)r(m-1)}|$.

Consider now the map of linear series

$$|(t-1)r(m-1)K_{X/S}+\frac{r}{r_0}g^*L|\times|r_0(m-1)K_X-2g^*L|^{\times r/r_0}$$

$$\times g^*|(t-1)r(m-1)K_S|\longrightarrow\!\!\!\!\!\rightarrow |tr(m-1)K_X-\frac{r}{r_0}g^*L|.$$

We remark that $B_3|_{X_s}=0$ and $B_2|_{X_s}=\bar{N}+G$ where $\bar{N}$ is a general member of the base point free linear series $|N|$. It is easy to see that

$$\lfloor\frac{B_1+\frac{r}{r_0}B_2+B_3}{tr}|X_s\rfloor\prec\lfloor\frac{\bar{H}^{(t-1)r(m-1)}}{tr}+\frac{(t-1)r(m-1)}{trm}F^m+\frac{\bar{N}}{tr_0}+\frac{G}{tr_0}\rfloor$$

$$\prec\lfloor\frac{\bar{H}^{(t-1)r(m-1)}}{tr}+\frac{\bar{N}}{tr_0}+\frac{m-1}{m}F^m+\frac{G}{tr_0}\rfloor\prec F^m.$$

Define $B:=B_1+\frac{r}{r_0}B_2+B_3$. Let $\nu:X'\longrightarrow X$ be a birational morphism such that $|tr(m-1)K_{X'}-\frac{r}{r_0}\nu^*g^*L|=\Phi+|M|$ where $\Phi+K_{X'/X}$ is a divisor with normal crossings support and $|M|$ is base point free. We have $\nu^*B+tr(m-1)K_{X'/X}\in\Phi+|M|$. X is smooth and $\{\frac{B}{tr}|_{X_s}\}$ is klt and hence

$$\lfloor\nu^*\{\frac{B}{tr}|_{X_s}\}\rfloor\prec K_{X'_s/X_s}.$$

Since

$$|mK_{X'_s}|=|mK_{X_s}|+mK_{X'_s/X_s}=f^*|H^m|+F^m+mK_{X'_s/X_s},$$

$f^*|H^m|$ is free and $F'^m=f^*F^m+mK_{X'_s/X_s}$ is the fixed divisor of $|mK_{X'_s}|$.

We remark that $\frac{B}{tr}|_{X_s} = \{\frac{B}{tr}|_{X_s}\} + \lfloor\frac{B}{tr}|_{X_s}\rfloor$, and for general $s \in S$, $\nu^*(B|_{X_s}) = (\nu^*B)|X'_s$. For B' a general member of $\Phi + |M|$, one has

$$\lfloor\frac{B'}{tr}|_{X'_s}\rfloor = \lfloor\frac{\Phi}{tr}|_{X'_s}\rfloor \prec \lfloor\frac{\nu^*B}{tr}|_{X'_s} + (m-1)K_{X'_s/X_s}\rfloor =$$

$$= \lfloor\nu^*\{\frac{B}{tr}|_{X_s}\} - K_{X'_s/X_s}\rfloor + \nu^*\lfloor\frac{B}{tr}|_{X_s}\rfloor + mK_{X'_s/X_s} \prec \nu^*F^m + mK_{X'_s/X_s}.$$

So for $a = tr_0$, $b = \frac{r}{r_0}$, and for a general divisor $B' \in |ab(m-1)K_{X'} - b\nu^*g^*L|$, one sees that B' has normal crossings support and $\lfloor\frac{B'}{ab}|_{X'_s}\rfloor \prec F'^m$. Replacyng B with B' and X with X' the lemma follows. □

Theorem 5.2. *Let X be a smooth projective variety of general type and* $\mathrm{alb}_X(X) \subset \mathrm{Alb}(X)$ *its Albanese image. Let $F := X_z$ be the fiber over a general $z \in \mathrm{alb}_X(X)$. Let $m_1 \geq m_2 \geq 2$ be any integers such that $|m_iK_F|$ is non empty and $|(m_1+m_2)K_F|$ defines a birational map. Then $|(m_1+2m_2)K_X + P|$ induces a birational map for any $P \in \mathrm{Pic}^0(X)$.*

Proof. We sketch the proof here. For more detail, please see [CH]. Let $a := \mathrm{alb}_X : X \longrightarrow A$ be the Albanese map, $\nu : Z \longrightarrow \mathrm{alb}_X(X)$ an appropriate desingularization. Fix H an ample line bundle on A. Let $m = m_1+m_2$. For $r \gg 0$ sufficiently divisible, pick a divisor

$$B \in |r(m_1-1)(m_2-1)(m-1)K_X - a^*H|.$$

Replacing X by an appropriate birational model, one may assume that alb_X factors through a morphism $f : X \longrightarrow Z$ and that B has normal crossings support and by Lemma 5.1 that for general $z \in Z$ (after replacying H by an appropriate multiple) $\lfloor\frac{B}{r(m-1)(m_2-1)}\rfloor|_{X_z} \prec F^{m_1}$, $\lfloor\frac{B}{r(m-1)(m_1-1)}\rfloor|_{X_z} \prec F^{m_2}$ and $\lfloor\frac{B}{r(m_1-1)(m_2-1)}\rfloor|_{X_z} \prec F^m$.

We define

$$L_1 := (m_1-1)K_X - \lfloor\frac{B}{r(m-1)(m_2-1)}\rfloor$$

$$\equiv \frac{a^*H}{r(m-1)(m_2-1)} + \{\frac{B}{r(m-1)(m_2-1)}\},$$

$$L_2 := (m_2-1)K_X - \lfloor\frac{B}{r(m-1)(m_1-1)}\rfloor$$

$$\equiv \frac{a^*H}{r(m-1)(m_1-1)} + \{\frac{B}{r(m-1)(m_1-1)}\},$$

$$L_3 := (m-1)K_X - \lfloor\frac{B}{r(m_1-1)(m_2-1)}\rfloor$$

$$\equiv \frac{a^*H}{r(m_1-1)(m_2-1)} + \{\frac{B}{r(m_1-1)(m_2-1)}\}.$$

Therefore, since $a^*H = f^*(\nu^*H)$, L_1, L_2 and L_3 are numerically equivalent to the pull back of a nef and big $\mathbb{Q}$-divisor on Z plus a klt divisor. In particular

$$h^i(Z, f_*(\omega_X \otimes L_j) \otimes \nu^* P) = h^i(A, a_*(\omega_X \otimes L_j) \otimes P) = 0$$

for all $i > 0$, $j = 1, 2$, and all $P \in \mathrm{Pic}^0(A)$. Therefore, $a_*(\omega_X \otimes L_j)$ is IT^0 for $j = 1, 2, 3$.

By a general fiber, we mean a smooth fiber over $z \in Z$ such that X_z satifies the assumption and

$$R^i a_*(\omega_X \otimes L_j) \otimes k(z) \cong H^i(X_z, \omega_{X_z} \otimes (L_j)),$$

for all i, j. By a general point $y \in X_z$, we mean a point in X_z which is not in the base loci of $|K_{X_z} + L_j|$ for all j.

The idea is to seperate general points y_1, y_2. For any such y_i, y_i is not an essential base point of $|K_X + L_j|$ thanks to the local isomorphism and the choice that y_i is not in the base loci of $|K_{X_z} + L_j|$. It follows that y_i is not a base point of $|2K_X + L_1 + L_2 + P|$ for all P. By comparing the base loci, y_i is not a base point of $|K_X + L_3 + P|$ for all P. We then shows that $a_*(\omega_X \otimes L_3 \otimes \mathcal{I}_{y_i})$ is IT^0. Hence for general $P \in \mathrm{Pic}^0(A)$, one can find a divisor in $|K_X + L_3 + P|$ passing through y_1 but not y_2. We pick a divisor in $|K_X + L_1 - P|$ not passing through y_1, y_2. Thus one sees that $|2K_X + L_3 + L_1|$ seperates y_1 and y_2.

Turning into a more detail discussion, we proceed the comparison of base loci. We consider the divisor

$$\beta := \llcorner \frac{B}{r(m_1 - 1)(m_2 - 1)} \lrcorner - \llcorner \frac{B}{r(m - 1)(m_1 - 1)} \lrcorner - \llcorner \frac{B}{r(m - 1)(m_2 - 1)} \lrcorner.$$

Write $B = \sum b_i B_i$ where the B_i are the distinct reduced irreducible components of B. Then from the inequalities

$$\llcorner \frac{b_i}{r(m_1 - 1)(m_2 - 1)} \lrcorner \geq \llcorner \frac{b_i}{r(m - 1)(m_1 - 1)} + \frac{b_i}{r(m - 1)(m_2 - 1)} \lrcorner \geq$$

$$\geq \llcorner \frac{b_i}{r(m - 1)(m_1 - 1)} \lrcorner + \llcorner \frac{b_i}{r(m - 1)(m_2 - 1)} \lrcorner$$

it follows that β is effective. Therefore, we have a map of linear series

$$|K_X + L_3| \xrightarrow{+\beta} |2K_X + L_1 + L_2|.$$

It is possible to arrange that $\llcorner \frac{B}{r(m_1-1)(m_2-1)} \lrcorner \prec Bs|mK_X|$. Comparing base loci, one sees that

$$h^0(X, \mathcal{O}_X(K_X + L_3)) = h^0(X, \mathcal{O}_X(2K_X + 2L)) = h^0(X, \mathcal{O}_X(mK_X)).$$

Moreover, it is easy to see that

$$h^0(X, \mathcal{O}_X(K_X + L_3) \otimes P) = h^0(X, \mathcal{O}_X(K_X + L_3))$$

for all $P \in \mathrm{Pic}^0(A)$. By [HP] Proposition 2.15, one also sees that $h^0(X, \mathcal{O}_X(mK_X)) = h^0(X, \mathcal{O}_X(mK_X) \otimes P)$ for all $P \in \mathrm{Pic}^0(A)$. In particular, for all $P \in \mathrm{Pic}^0(A)$ one has an isomorphism

$$|K_X + M + P| \xrightarrow{+\beta} |2K_X + L_1 + L_2 + P|$$

and therefore if $y \notin Bs|m_1 K_{X_z}| \cup |m_2 K_{X_z}|$, one has that $y \notin Bs|K_X + L_3 + P|$.

We claim that if $y_1, y_2 \in X$ are general points, then for general $P \in \mathrm{Pic}^0(A)$ one has $\varphi_{|K_X+L_3+P|}(y_1) \neq \varphi_{|K_X+L_3+P|}(y_2)$.

To see this, Let $z_i = a(y_i)$. Assume that $z = z_1 = z_2$ (if $z_1 \neq z_2$ the assertion will follow in a similar but easier fashion). Since $y_i \in X$ are general, we may assume that $z \in Z$ is also general. Let $F = X_z$, since $\varphi_{|mK_F|} = \varphi_{|(K_X+L_3)|_F|}$ is birational, we may assume that the homomorphism $H^0(F, (\omega_X \otimes L_3)|_F) \longrightarrow \mathbb{C}(y_1) \oplus \mathbb{C}(y_2)$ is surjective.

Since $z \in Z$ is general, we have

$$a_*(\omega_X \otimes L_3) \otimes \mathbb{C}(\nu(z)) \cong f_*(\omega_X \otimes L_3) \otimes \mathbb{C}(z) \cong H^0(F, (\omega_X \otimes L_3)|_F).$$

One sees that

$$a_*(\omega_X \otimes L_3 \otimes \mathcal{I}_{y_1}) \otimes \mathbb{C}(\nu(z)) =$$
$$= ker\left(a_*(\omega_X \otimes L_3) \otimes \mathbb{C}(\nu(z)) \longrightarrow H^0(F, (\omega_X \otimes L_3)|_F) \longrightarrow \mathbb{C}(y_1)\right),$$
$$a_*(\omega_X \otimes L_3 \otimes \mathcal{I}_{y_1,y_2}) \otimes \mathbb{C}(\nu(z)) =$$
$$= ker\left(a_*(\omega_X \otimes L_3) \otimes \mathbb{C}(\nu(z)) \longrightarrow H^0(F, (\omega_X \otimes L_3)|_F) \longrightarrow \mathbb{C}(y_1) \oplus \mathbb{C}(y_2)\right).$$

It follows that

$$a_*(\omega_X \otimes L_3 \otimes \mathcal{I}_{y_1,y_2}) \neq a_*(\omega_X \otimes L_3 \otimes \mathcal{I}_{y_1}).$$

Consider the exact sequence

$$0 \longrightarrow a_*(\omega_X \otimes L_3 \otimes \mathcal{I}_{y_1,y_2}) \longrightarrow a_*(\omega_X \otimes L_3 \otimes \mathcal{I}_{y_1}) \longrightarrow \mathcal{O}z \longrightarrow 0.$$

Since $h^0(a_*(\omega_X \otimes L_3 \otimes \mathcal{I}_{y_1} \otimes P)) = h^0(a_*(\omega_X \otimes L_3 \otimes P)) - 1$ for any $P \in \mathrm{Pic}^0(A)$. Since $h^i(a_*(\omega_X \otimes L_3 \otimes P)) = 0$ for all $i > 0$, it follows that $a_*(\omega_X \otimes L_3 \otimes \mathcal{I}_{y_1})$ satisfies the I.T. and hence by Proposition 2.3

$$h^0(a_*(\omega_X \otimes L_3 \otimes \mathcal{I}_{y_1,y_2}) \otimes Q) = h^0(a_*(\omega_X \otimes L_3 \otimes \mathcal{I}_{y_1}) \otimes Q) - 1$$

for general $Q \in \mathrm{Pic}^0(A)$. Therefore for general $Q \in \mathrm{Pic}^0(A)$

$$h^0(\omega_X \otimes L_3 \otimes \mathcal{I}_{y_1,y_2} \otimes Q) = h^0(\omega_X \otimes L_3 \otimes \mathcal{I}_{y_1} \otimes Q) - 1 = h^0(\omega_X \otimes L_3 \otimes Q) - 2$$

It's then easy to see that the linear series $|(m_1 + 2m_2)K_X + P|$ also induces birational morphisms. □

Corollary 5.3. *Let X be of general type, then i) if $\dim(X) - a(X) \leq 1$ then $|6K_X|$ induces a birational morphism, ii) $\dim(X) - a(X) = 2$ then $|7K_X|$ induces a birational morphism.*

Proof. For curves of general type, $|2K_X| \neq \emptyset$ and $|3K_X|$ defines a birational map. Thus it suffices to take $m_1 = m_2 = 2$ as in the previous Theorem. And for surface of general type, $|2K_X| \neq \emptyset$ and $|5K_X|$ defines a birational map by [Bo]. Thus it suffices to take $m_1 = 3, m_2 = 2$. □

When X is of maximal Albanese dimension, it is possible to improve the above result.

Theorem 5.4. *Let X be of general type and maximal Albanese dimension. If $\chi(\omega_X) > 0$, then $|3K_X|$ is birational.*

Proof. By [EL2], since $\chi(\omega_X) > 0$, it follows that for all $P \in \mathrm{Pic}^0(X)$, $h^0(X, \omega_X \otimes P) > 0$. There exists an open subset $U \subset X$ such that for any $x \in U$, and general $Q \in \mathrm{Pic}^0(X)$ (depending on x), there is a section of $H^0(X, \omega_X \otimes Q)$ not vanishing at x. Therefore, for any $x \in U$ and for any $P \in \mathrm{Pic}^0(X)$, there is a section of $H^0(X, \omega_X^2 \otimes P)$ not vanishing at x.

Let x_1, x_2 be two general points in U. Proceeding as in the proof of the previous theorem, we see that for general $Q \in \mathrm{Pic}^0(X)$, there is a section of $\omega_X^2 \otimes Q$ vanishing on x_1 but not on x_2. For a fixed $P \in \mathrm{Pic}^0(X)$, we choose general $Q \in \mathrm{Pic}^0(X)$ and then consider the map of linear series

$$|K_X - Q + P| \times |2K_X + Q| \longrightarrow |3K_X + P|.$$

We conclude that for any fixed $P \in \mathrm{Pic}^0(X)$, there is a section of $\omega_X^3 \otimes P$ vanishing on x_1 but not on x_2. □

REFERENCES

[AS] U. Angehrn, Y. T. Siu, *Effective freeness and point seperation for adjoint bundles*, Invent. Math. **122** (1995) 2, 291-308.

[Bo] E. Bombieri, *Canonical models of surfaces of general type*, IHES Publ. Math. No. **42** (1973), 171-219.

[CH] J. A. Chen, C. D. Hacon, *Effective generation of adjoint linear series of irregular varieties*, Preprint.

[EL1] L. Ein, R. Lazarsfeld, *Global generation of pluricanonical and adjoint linear series on smooth projective threefolds*, Jour. AMS **6**, No. 4 (1993), 875-903.

[EL2] L. Ein, R. Lazarsfeld, *Singularities of theta divisors, and birational geometry of irregular varieties*, Jour. AMS **10**, No. 1 (1997), 243-258.

[Ha] R. Hartshorne, *Residue and Duality*, Lecture Notes in Mathematics **20**. Springer-Verlag, 1966.

[HP] C. D. Hacon and R. Pardini, *On the birational geometry of varieties of maximal Albanese dimension* To appear in Jour. für die Reine Angew. Math.

[Ka] Y. Kawamata, *On Fujita's freeness conjecture for* 3*-folds and* 4*-folds*, Math. Ann. **308**, (1997) 491-505.

[Ko] J. Kollár, *Shafarevich Maps and Automorphic Forms*, Princeton University Press, 1995.

[Mu] S. Mukai, *Duality between $D(X)$ and $D(\hat{X})$ with its application to Picard sheaves*, Nagoya Math. Jour. **81**, (1981) 153-175.

[Re] I. Reider, *Vector bundles of rank* 2 *and linear systems on algebraic surfaces*, Ann. Math. **127**, (1988) 221-233.

[V] E. Viehweg, *Weak positivity and the additivity of the Kodaira dimension for certain fiber spaces* Algebraic Varieties and analytic Varieties (Tokyo 1981), 329-353, Adv. Stud. Pure Math. 1. North-Holland, Amsterdam, 1983.

Hecke curves on the moduli space of vector bundles over an algebraic curve

Jun-Muk Hwang [1]

Throughout the paper, we will work over the complex numbers.

In [Hw], we studied the geometry of the moduli space M_i of vector bundles of rank 2 with a fixed determinant of degree i over a smooth projective algebraic curve of genus $g \geq 2$ via Hecke curves on it. We specifically obtained the following two results:

(**[Hw,Theorem 1]**) *The tangent bundle of M_i is stable if i is odd.*

(**[Hw,Theorem 2]**) *For $g \geq 5$, the set of generically finite surjective morphisms from a fixed complete variety to M_i is countable for any i.*

In this paper, we give partial generalizations of these to the cases of moduli spaces of vector bundles of higher rank. More precisely, let M be the moduli space of semi-stable bundles of rank r with a fixed determinant and M^s be the smooth locus of M. Then we have the following.

Corollary 1 *The tangent bundle $T(M^s)$ is simple if $g \geq 4$.*

Corollary 3 *If X is a general curve of genus $g > 2r + 1$, the set of generically finite surjective morphisms from a fixed complete variety to M is countable.*

These corollaries follow from our main results, Theorem 3 and Theorem 4 stated in Section 3, concerning the variety of standard rational tangents on M. Each of the two theorems is of independent interest in its own right and perhaps more important than the corollaries above.

It seems hard to strengthen Corollary 1 to the stability of $T(M^s)$. In the rank 2 case of [Hw], we were able to exploit the simple structure of ruled surfaces. In order to generalize the same idea to higher rank cases, we need a good knowledge of (double) projective bundles of higher rank over curves. On the other hand, a refinement of our technique may improve Corollary 3 to include some cases with $g \leq 2r + 1$. For $r = 2$, the proof of Corollary 3 in Section 3 gives another proof of [Hw,Theorem 2] for general curves of genus $g > 5$. Our proof of Corollary 3 below is rather elementary, while the proof in [Hw] uses deeper results about M_i.

M.S. Narasimhan ([Na]) informed us that he had a different proof of the simplicity of $T(M)$, which applies to any $g \geq 2$ (unpublished).

[1]Supported by Grant No. 98-0701-01-5-L from the KOSEF.

As in [Hw], our main tools are the theory of standard rational tangents developed in our joint work with N. Mok ([HM]) and the geometry of Hecke curves on M. Hecke curves were introduced by Narasimhan and Ramanan and most of their properties needed here are found in [NR2,Section 5]. However, since the main concern in [NR2] was Hecke cycles, which are not curves, the properties of the Hecke curves were only implicitly studied there. For the readers' convenience, we will rewrite some of the details in Sections 2 and 3 below with references to the relevant part of [NR2].

There are a number of interesting open questions about Hecke curves. We mention two of them, Question 1 and Question 2 below, which are most interesting from the viewpoint of the theory of standard rational tangents on Fano varieties.

Acknowledgment We are grateful to Professor M.S. Narasimhan for valuable discussions and the referee for suggestions which improved the clarity of the paper.

1 Generalities on the variety of standard rational tangents

In this section, we recall some results from the theory of the variety of standard rational tangents ([HM]). Let M be a Fano variety of Picard number 1. We will assume that there exists a component H of the Hilbert scheme of complete curves on M such that

(1) generic members of H lie in the smooth locus M^s of M and sweep out an open subset of M^s and that

(2) the subscheme $H_y \subset H$ consisting of members of H passing through a generic point $y \in M$ is an irreducible smooth *complete* variety of which every member is an irreducible smooth rational curve contained in M^s.

For a smooth point $y \in M$, let $T_y(M)$ be the tangent space to M at y. Define the **tangent morphism**

$$\tau_y : H_y \;\rightarrow\; \mathbf{P}T_y(M)$$

by sending $[C] \in H_y$ corresponding to a smooth rational curve $C \subset M^s$ to

$$\tau_y([C]) \;:=\; \mathbf{P}T_y(C).$$

From The assumption (2) above, it follows that a generic member of H_y has normal bundle of the form $\mathcal{O}(1)^p \oplus \mathcal{O}^q, p+q = \dim M$ ([Ko,IV.2.9]), implying that τ_y is generically finite over its image. The image of τ_y will be denoted by $\mathcal{C}_y$ and called the **variety of standard rational tangents** at the generic point y associated to the family H. The dimension of $\mathcal{C}_y$ is p, the number of $\mathcal{O}(1)$-factors in the decomposition of the normal bundle of a generic member of H_y. The following is a special case of [HM, Theorem 1.4] (Note the condition in [HM,Theorem 1.2] is satisfied when $\mathcal{C}_y$ is smooth and irreducible by the remark after [HM,Theorem 1.2].). In [HM], it was assumed that M is smooth. But the proof works verbatim even when M has singularity under the assumptions (1) and (2) above. Let Δ denote a unit disc in $\mathbf{C}$.

Theorem 1 *In the above situation, assume that τ_y is an embedding. Then for any complete variety Z and a family of generically finite surjective morphisms $\{f_t : Z \to M, t \in \Delta\}$, there exists a family of biregular automorphisms $\{g_t \in Aut(M), g_0 = id_M, |t| < \epsilon\}$ such that $f_t = g_t \circ f_0$ for all $|t| < \epsilon$ for some $\epsilon > 0$.*

In particular, if M has no vector fields, we have the following rigidity:

Theorem 1' *In Theorem 1, assume furthermore that $Aut(M)$ is 0-dimensional. Then any family of generically finite surjective morphisms $\{f_t : Z \to M, t \in \Delta\}$ from a fixed complete variety Z is trivial, namely, $f_t = f_0$ for all $t \in \Delta$.*

Let $T(M^s)$ be the tangent bundle of the smooth locus of M. Recall that a vector bundle is simple if the scalar multiplications are the only endomorphisms of the vector bundle.

Theorem 2 *Let M be a Fano variety which has a component H of the Hilbert scheme satisfying the conditions (1) and (2) above. Assume in addition that the variety of standard rational tangents $\mathcal{C}_y$ at a generic point $y \in M$ is non-degenerate in $\mathbf{P}T_y(M)$. Then the tangent bundle $T(M^s)$ is simple.*

Proof. Let χ be an endomorphism of $T(M^s)$. Let $y \in M$ be a generic point and $v \in T_y(M)$ be a vector tangent to a generic member $C \subset M^s$ of H_y. Let $\tilde{v}$ be a vector field on C extending v having two distinct zeroes. Then $\chi(\tilde{v})$ is a section of $T(M^s)|_C$ vanishing at two distinct points where $\tilde{v}$ vanishes. From the decomposition type of the normal bundle of C, $\chi(\tilde{v})$ must be tangent to C. Since $\tilde{v}$ and $\chi(\tilde{v})$ have two distinct zeroes at the same points, they must be proportional. It follows that v is an eigenvector of χ in $T_y(M^s)$. Since this is true for any generic choice of v in $\mathcal{C}_y$ which is non-degenerate in $\mathbf{P}T_y(M^s)$, we see that every vector of $T_y(M^s)$ is an eigenvector of χ. Thus χ acts as a scalar multiplication on $T_y(M)$ for generic $y \in M$. The eigenvalues must be constant along members of H because $\tilde{v}$ and $\chi(\tilde{v})$ are just a constant multiple of each other. Since M has Picard number one, any point of M^s is joined to x by a connected chain of curves $C_i \in H$ ([HM, Section 3]). Noting that the eigenvalue of the scalar transformation χ is constant on C_i, we see that χ is globally constant scalar multiplication on M^s, meaning the simplicity of $T(M^s)$. □

2 Hecke curves

Let X be a smooth projective curve of genus g and let M denote the moduli space of semi-stable bundles over X of rank r and with a fixed determinant of degree d. It is well-known that M is a Fano variety of Picard number 1. The smooth locus M^s of M corresponds to stable bundles on X. Let $\mathcal{U}(r,d)$ be the moduli space of semi-stable bundles of rank r and degree d. There is a natural surjective morphism $\mathcal{U}(r,d) \to Jac(X)$ whose fiber is isomorphic to M. The dimension of $\mathcal{U}(r,d)$ is $r^2(g-1)+1$ and the dimension of M is $(r^2-1)(g-1)$.

Given two nonnegative integers k and l, a vector bundle W of rank r and degree d is **(k,l)-stable**, if, for every proper subbundle V of W, we have

$$\frac{deg(V)+k}{rk(V)} < \frac{deg(W)+k-\ell}{rk(W)}.$$

The usual stability is equivalent to $(0,0)$-stability. A (k,l)-stable bundle is $(k,l-1)$-stable for $l>0$. The dual bundle of a (k,l)-stable bundle is (l,k)-stable. We need the next two Propositions proved in [NR2] to explain the definition of Hecke curves.

Proposition 1 ([NR2,5.4]) *If $g \geq 4$, a generic point $[W] \in M$ corresponds to a $(1,1)$-stable bundle W.*

Proposition 2 ([NR2,5.5]) *Given an exact sequence $0 \to V \to W \to \mathcal{O}_x \to 0$, where V, W are vector bundles on X and $\mathcal{O}_x$ is the skyscraper sheaf at a point $x \in X$, if W is (k,l)-stable, then V is $(k,l-1)$-stable.*

Throughout this section, we assume that $g \geq 4$ so that a generic point $[W] \in M$ corresponds to a $(1,1)$-stable bundle W over X by Proposition 1. Denote by W^* the dual bundle and $\mathbf{P}W$ the projectivization consisting of lines through the origin on each fiber. For $x \in X$ and $\zeta \in \mathbf{P}W^*_x$, consider a new vector bundle W^ζ defined by

$$0 \longrightarrow W^\zeta \longrightarrow W \longrightarrow (W_x/\zeta^\perp)\otimes \mathcal{O}_x \longrightarrow 0$$

where $\zeta^\perp$ denotes the hyperplane in W_x annihilated by ζ. Let $\iota : W^\zeta_x \to W_x$ be the homomorphism between the fibers at x induced by the sheaf map $W^\zeta \to W$. The kernel of ι, $Ker(\iota)$, is a 1-dimensional subspace of the fiber W^ζ_x and its annihilator $(Ker(\iota))^\perp$ is a hyperplane in $(W^\zeta)^*_x$. Let $\mathbf{l}$ be a line in $\mathbf{P}W^\zeta_x$ containing the point $[Ker(\iota)]$. For each point $[l] \in \mathbf{l}$ corresponding to a 1-dimensional subspace $l \subset W^\zeta_x$, consider the vector bundle $\widetilde{W}^l$ defined by

$$0 \longrightarrow \widetilde{W}^l \longrightarrow (W^\zeta)^* \longrightarrow [(W^\zeta)^*_x/l^\perp]\otimes \mathcal{O}_x \longrightarrow 0$$

where $l^{\perp} \subset (W^{\varsigma})^*_x$ is the hyperplane annihilating l. This vector bundle $\widetilde{W}^l$ is stable for each $[l] \in \mathbf{l}$ by Proposition 2. It is easy to check that for $l = Ker(\iota)$,

$$\widetilde{W}^{Ker(\iota)} \cong W^*.$$

It follows that $\{(\widetilde{W}^l)^*; l \in \mathbf{l}\}$ defines a rational curve passing through $[W]$ in M^s. A rational curve on M^s constructed this way is called a **Hecke curve**. Using [NR2,5.9], one can show that a Hecke curve is smooth. In view of [NR2,5.16], it is easy to check that a Hecke curve has degree $2r$ with respect to K_M^{-1}. The following question seems to be of great interest.

Question 1 Is a Hecke curve a rational curve of minimal degree passing through a generic point of M? Conversely, is a rational curve of minimal degree passing through a generic point of M necessarily a Hecke curve?

By [DN], $K_M^{-1} = 2n\theta$ where θ is the generator of $Pic(M)$ and $n = (r, d)$. Thus for $d = 0$, a Hecke curve has degree 1 with respect to θ and the first part of Question 1 is affirmative in this case, as well as in case $r = 2$ ([Hw,Proposition 8]).

Let $\pi : \mathbf{P}W^* \to X$ be the natural projection. On $\mathbf{P}W^*$, consider the relative cotangent bundle Ω^{π} of the fibration π. The projective bundle $\mathbf{P}\Omega^{\pi}$ over $\mathbf{P}W^*$ is a smooth projective variety of dimension $2r-2$. The set of all lines in $\mathbf{P}W^{\varsigma}_x$ containing the point $[Ker(\iota)]$ is naturally isomorphic to $\mathbf{P}(W^{\varsigma}_x/Ker(\iota)) \cong \mathbf{P}\Omega^{\pi}_{\varsigma}$. The argument of [NR2,5.13] shows that Hecke curves associated to two distinct points of $\mathbf{P}\Omega^{\pi}$ are distinct rational curves on M. Thus $\mathbf{P}\Omega^{\pi}$ is naturally isomorphic to the variety of all Hecke curves through $[W]$.

Proposition 3 *Hecke curves are dense in an irreducible component of the Hilbert scheme of curves on M.*

Proof. Since $\dim \mathbf{P}\Omega^{\pi} = 2r - 2$, the set of all Hecke curves have dimension $\dim(M)+2r-3$. A generic Hecke curve must have semi-positive normal bundle. Thus the space of all deformations of a Hecke curve C has dimension $\dim(M)+C\cdot K_M^{-1}-3$. Since $C \cdot K_M^{-1} = 2r$, all small deformations of a Hecke curve are again Hecke curves. □

It follows that the component H of the Hilbert scheme of M corresponding to Hecke curves satisfies the conditions (1) and (2) for H in Section 1.

3 Tangent morphism for Hecke curves

The tangent morphism $\tau_{[W]} : H_{[W]} = \mathbf{P}\Omega^{\pi} \to \mathbf{P}T_{[W]}(M)$ for a (1,1)-stable bundle W on X is defined by associating to $\eta \in \mathbf{P}\Omega^{\pi}$ the tangent vector of the corresponding

Hecke curve at $[W]$, namely,

$$\tau_{[W]}(\eta) \quad := \quad \mathbf{P}T_{[W]}(C_\eta)$$

where $C_\eta \subset M$ is the Hecke curve defined by η.

Theorem 3 *The tangent morphism $\tau_{[W]}$ for a $(1,1)$-stable bundle W is the morphism defined by the complete linear system of the line bundle*

$$L_W \quad := \quad \xi \otimes \psi^* K_X$$

on $\mathbf{P}\Omega^\pi$ where ψ is the composition of the two projections

$$\mathbf{P}\Omega^\pi \xrightarrow{\varphi} \mathbf{P}W^* \xrightarrow{\pi} X,$$

and ξ is the dual tautological line bundle of the projectivization satisfying $\varphi_*\xi = T^\pi$, T^π denoting the relative tangent bundle of π.

Proof. The proof is essentially the same as that of [Hw,Proposition 11], once we note that $H^0(H_{[W]}, L_W)^*$ is canonically isomorphic to $T_{[W]}(M)$. Let us just check this fact.

$$\begin{aligned} H^0(\mathbf{P}\Omega^\pi, L_W) &= H^0(\mathbf{P}W^*, \varphi_*\xi \otimes \pi^* K_X) \\ &= H^0(\mathbf{P}W^*, T^\pi \otimes \pi^* K_X) \\ &= H^0(X, \pi_* T^\pi \otimes K_X) \\ &= H^0(X, ad_W \otimes K_X) \end{aligned}$$

where ad_W is the bundle of traceless endomorphisms of W on X. By Serre duality, the last line is canonically dual to $H^1(X, ad(W))$ which is canonically isomorphic $T_{[W]}(M)$ by standard deformation theory of bundles. □

From Theorem 3, the variety of standard rational tangents $\mathcal{C}_{[W]} = \tau_{[W]}(H_{[W]})$ is non-degenerate in $\mathbf{P}T_{[W]}(M)$. By Theorem 2, we get

Corollary 1 *The tangent bundle of M^s is simple if $g \geq 4$.*

By the result of Kebekus ([Ke,Theorem 3.4]), the tangent morphism is finite over its image, whence follows

Corollary 2 *The line bundle $\xi \otimes \psi^* K_X$ on $\mathbf{P}\Omega^\pi$ is ample for any $(1,1)$-stable vector bundle W on X.*

When $r = 2$, Corollary 2 is well-known in terms of geometry of ruled surfaces (cf. [Hw]). It is probably possible to give a more direct proof also for higher ranks.

In regard to Theorem 1, one of the most intriguing questions about the tangent morphism is the following:

Question 2 *Given a sufficiently generic stable bundle W on X, is the line bundle L_W very ample on $\mathbf{P}\Omega^{\pi}$?*

We have the following partial answer.

Theorem 4 *If X is a general curve of genus $g > 2r+1$, the tangent morphism $\tau_{[W]} : H_{[W]} \to \mathbf{P}T_{[W]}(M)$ is an embedding for a generic point $[W] \in M$.*

For the proof, we need two lemmas. The first one is an honest generalization of [NR2,Proposition 5.4].

Lemma 1 *If $g > \ell + 1$, then a generic point of M is $(0,\ell)$-stable.*

Proof. As in [NR2,Proposition 5.4], we count the dimension of the space of non-$(0,\ell)$-stable bundles. Recall that a vector bundle W of rank r and degree d is $(0,\ell)$-stable, if the rank q and the degree δ of any proper subbundle satisfies $q\ell < dq - \delta r$. Suppose a stable bundle W is not $(0,\ell)$-stable and $V \subset W$ be a subbundle of rank q and degree δ with $q\ell \geq dq - \delta r$. Since non-stable bundles can be deformed to stable bundles ([NR1,Proposition 2.6]), we may assume that V and W/V are stable in dimension counting. The dimension of deformations of V is equal to

$$\dim \mathcal{U}_X(q,\delta) = q^2(g-1)+1$$

while W/V has deformation of dimension

$$\dim \mathcal{U}_X(r-q, d-\delta) = (r-q)^2(g-1)+1.$$

The dimension of extensions of W/V by V is

$$\begin{aligned} & h^1(X, Hom(W/V,V)) \\ &= -deg(Hom(W/V,V)) + (g-1)\cdot rk(Hom(W/V,V)) + h^0(X, Hom(W/V,V)) \\ &= -q(\delta-d) - (r-q)\delta + (g-1)q(r-q) \end{aligned}$$

because $h^0(X, Hom(W/V,V)) = 0$ from the stability of W. Thus the dimension of non-trivial extensions is $qd - r\delta + (g-1)q(r-q) - 1$. It follows that the dimension of the space of non-$(0,\ell)$-stable bundles is at most

$$\begin{aligned} &[q^2(g-1)+1] + [(r-q)^2(g-1)+1] + [qd - r\delta + (g-1)q(r-q) - 1] \\ &= (g-1)(q^2+r^2-rq) + qd - r\delta + 1. \end{aligned}$$

The right hand side is strictly smaller than $\dim \mathcal{U}_X(r,d) = r^2(g-1)+1$ if

$$(g-1)(rq-q^2) \quad > \quad qd - r\delta.$$

But from the assumption $g-1 > \ell$ and the hypothesis $q\ell \geq qd - r\delta$ on q and δ,

$$(g-1)q(r-q) \; > \; \ell q \; \geq \; qd - r\delta. \; \Box$$

The next lemma is a generalization of [NR2,Lemma 5.6].

Lemma 2 *Let r be a positive integer. Assume that $h^0(X, \mathcal{O}(rx+ry)) = 1$ for any $x, y \in X$. Then for any $(0,2r)$-stable bundle W of rank r and any stable bundle W' or rank r with $det(W') = det(W) \otimes \mathcal{O}(-rx-ry)$, we have*
(i) if $f : W' \to W$ is a non-zero homomorphism, f is of maximal rank,
(ii) $h^0(X, Hom(W', W)) \leq 1$.

Proof. If f is not of maximal rank, let

$$\begin{array}{ccccc} W' & \longrightarrow & V' & \longrightarrow & 0 \\ & & \downarrow & & \\ W & \longleftarrow & V & \longleftarrow & 0 \end{array}$$

be the factorization of f with $V \cong Im(f)$ and $V' \cong W'/Ker(f)$ so that $q := rk(V) = rk(V') < r$ and $V' \to V$ is of maximal rank, implying $deg(V) \geq deg(V')$. Since W' is stable,

$$\frac{deg(V)}{q} \geq \frac{deg(V')}{q} > \frac{deg(W')}{r} = \frac{deg(W)-2r}{r}$$

which contradicts the $(0,2r)$-stability of W. This proves (i).

Suppose f and g are two linearly independent homomorphisms $W' \to W$. Choose $z \neq x, y$. Then $h = af + bg \neq 0$ for some $a, b \in \mathbf{C}$ must induce a singular element h_z of $Hom(W'_z, W_z)$. By (i) h has maximal rank. Let

$$\Lambda^r h : \Lambda^r W' \cong det(W) \otimes \mathcal{O}(-rx-ry) \quad \longrightarrow \quad \Lambda^r W = det(W)$$

be the determinant of h. Since $h^0(X, \mathcal{O}(rx+ry)) = 1$, $\Lambda^r h$ has zero precisely at $rx+ry$. This is a contradiction to the fact that h_z is singular. This finishes the proof of (ii). $\Box$

Proof of Theorem 4. Write L for L_W for simplicity. For any $x \in X$, the line bundle L restricted to the fiber $\psi^{-1}(x) = \mathbf{P}(\Omega^1(\mathbf{P}W^*_x))$ is very ample. Thus L is very ample on $\mathbf{P}\Omega^\pi$ if for any $x, y \in X$, the case $x = y$ included, the restriction map

$$H^0(\mathbf{P}\Omega^\pi, L) \quad \longrightarrow \quad H^0(\psi^{-1}(x+y), L|_{\psi^{-1}(x+y)})$$

is surjective. From the exact sequence

$$0 \longrightarrow L \otimes \psi^* \mathcal{O}(-x-y) \longrightarrow L \longrightarrow L|_{\psi^{-1}(x+y)} \longrightarrow 0,$$

the surjectivity is guaranteed if

$$H^1(\mathbf{P}\Omega^\pi, L \otimes \psi^* \mathcal{O}(-x-y)) \quad = \quad H^1(X, ad_W \otimes K_X(-x-y))$$

or its dual $H^0(X, ad_W \otimes \mathcal{O}(x+y))$ vanishes. By Lemma 1, we may assume that W is $(0, 2r)$-stable from $g > 2r+1$.

If X is a general curve of genus $g > 2r+1$, we see that the assumption of Lemma 2 is satisfied, in view of the dimension count of curves which have meromorphic functions with poles of the type $rx + ry$ as in [ACGH,p.213]. By Riemann-Hurwitz formula, there will be at most $(2g-2) + 4r - 2(r-1) = 2g + 2r$ branch points of the meromorphic function $X \to \mathbf{P}_1$ in addition to the pole. Thus up to the automorphism of $\mathbf{P}_1$ preserving ∞, there are at most $2g + 2r - 2$ branch points. It follows that if $3g - 3 > 2g + 2r - 2$, namely, $g > 2r+1$, then a generic genus g curve cannot have such a meromorphic function.

Setting $W' = W \otimes \mathcal{O}(-x-y)$ in Lemma 2, we get

$$\begin{aligned} 1 &\geq h^0(X, Hom(W', W)) \\ &= h^0(X, End(W) \otimes \mathcal{O}(x+y)) \\ &= h^0(X, \mathcal{O}(x+y)) + h^0(X, ad_W \otimes \mathcal{O}(x+y)) \end{aligned}$$

meaning that $H^0(X, ad_W \otimes \mathcal{O}(x+y)) = 0$, as desired. □

Noting that $Aut(M)$ is 0-dimensional by [NR1], we apply Theorem 1' to get

Corollary 3 *For a general curve X of genus $g > 2r+1$ and any complete variety Z of dimension equal to $\dim M$, the set of dominant morphisms from Z to M is countable.*

REFERENCES

[ACGH] Arbarello, E., Cornalba, M., Griffiths, P. and Harris, J.: *Geometry of algebraic curves, Volume I.* Grundlehren der math. Wissenschaften 267, Springer Verlag, 1985

[DN] Drezet, J.-M. and Narasimhan, M. S.: Groupe de Picard des variétés de modules de fibrés semi-stables sur les courbes algébriques. Invent. math. **97** (1989) 53-94

[HM] Hwang, J.-M. and Mok, N.: Cartan-Fubini type extension of holomorphic maps for Fano manifolds of Picard number 1. J. Math. Pures Appl. **80** (2001) 563-575

[Hw] Hwang, J.-M.: Tangent vectors to Hecke curves on the moduli space of rank 2 bundles over an algebraic curve. Duke Math. Journal **101** (2000) 179-187

[Ke] Kebekus, S.: Families of singular rational curves. J. Alg. Geom. **11** (2002) 245-256

[Ko] Kollár, J.: *Rational curves on algebraic varieties.* Ergebnisse der Mathematik und ihrer Grenzgebiete, 3 Folge, Band 32, Springer Verlag, 1996

[Na] Narasimhan, M. S.: private communication.

[NR1] Narasimhan, M.S. and Ramanan, S.: Deformations of the moduli space of vector bundles over an algebraic curve. Ann. Math. **101** (1975) 391-417

[NR2] Narasimhan, M.S. and Ramanan, S.: Geometry of Hecke cycles I. in *C. P. Ramanujam-a tribute.* Springer Verlag, 1978, p.291-345

Korea Institute for Advanced Study
207-43 Cheongryangri-dong
Seoul, 130-012, Korea
jmhwang@ns.kias.re.kr

MINIMAL RESOLUTION VIA GRÖBNER BASIS

YUKARI ITO

In memory of Meeyoung Kim

ABSTRACT. The minimal resolution of cyclic quotient singularity is well known and the McKay correspondence can be defined as a correspondence between the exceptional curves of the minimal resolution and the special representations. In this paper, we give a construction of the minimal resolution and explain the generalized McKay correspondence with the special representations in terms of the Gröbner basis.

CONTENTS

1. INTRODUCTION

Let G be a finite small group in $GL(2,\mathbb{C})$ and X the quotient $\mathbb{C}^2/G$. In this case, it is known that there exists the minimal resolution of the quotient singularity, where a group is small if it acts freely on $\mathbb{C}^2 \setminus \{0\}$.

The history of the minimal resolution is very long and many algebraic geometers have contributed to the construction of the resolution. To reach the resolution, several tools have been established such as blowing-up, Newton diagram, continued fraction and toric resolution.

1991 *Mathematics Subject Classification.* 14C05, 14E15.

Key words and phrases. quotient singularity, minimal resolution, special representation, Gröbner basis.

The author is partially supported by JSPS, the Grant-in-aid for Scientific Research (No.13740019).

In this paper, we would like to introduce a new construction of the minimal resolution via the Gröbner basis. It is deeply related with the theory of the G-Hilbert schemes and based on the author's recent work [5]. The G-Hilbert scheme was introduced in a paper ([6]) by Nakamura and the author and is isomorphic to the minimal resolution of the quotient singularity. We will obtain the G-Hilbert scheme in terms of the Gröbner basis as follows:

Theorem 1.1. *Let G be a finite small cyclic group G in $GL(2,\mathbb{C})$, and I_G be an ideal of $\mathcal{O}_{\mathbb{C}^2}$ defined by the free G-orbit. Then the Gröbner fan for the G-homogenous ideal I_G determines a toric variety which is isomorphic to the minimal resolution of $\mathbb{C}^2/G$.*

As an application of this theorem, we can explain the generalized McKay correspondence via the Gröbner basis. The generalized McKay correspondence gives a bijection between the set of the exceptional curves of the minimal resolution and the set of the special irreducible representations of G. The special representation was defined by Wunram and Riemenschneider in terms of vector bundles on the minimal resolution, and the difference between special representation and non-special representation can be understood by the following theorem:

Theorem 1.2. *There is a bijection between an irreducible special representation ρ_k of G and a binomial generator in the initial ideal $in_w(I)$.*

By this theorem, we can calculate all the irreducible special representations in terms of the Gröbner basis without any knowledge of vector bundles. We will show these two theorems in the following sections. This paper is organized as follows: First, we recall the results on the G-Hilbert scheme in the following section, and give some basic definitions in the theory of the the Gröbner basis and the proof of Theorem 1.1 in the next section. Then we apply the result to the generalized McKay correspondence with the special representations in section 4. In the the final section, we give an example with a picture of the Gröbner fan.

2. G-Hilbert scheme

In this section, we discuss the G-Hilbert scheme in dimension 2.

Let G be a finite small subgroup of $GL(2,\mathbb{C})$ and X the quotient $\mathbb{C}^2/G$.

Definition 2.1. A *G-Cluster Z* is a G-invariant subscheme $Z \subset \mathbb{C}^2$ for which $H^0(Z,\mathcal{O}_Z)$ is the regular representation of G, and the *G-Hilbert scheme* is the moduli space of G-clusters.

Remark 2.2. The G-Hilbert scheme $\mathrm{Hilb}^G(\mathbb{C}^2)$ is a roughly 2-dimensional component of G-fixed part of $\mathrm{Hilb}^n(\mathbb{C}^2)$, where $n = |G|$, and can be written as

$$\mathrm{Hilb}^G(\mathbb{C}^2) = \{I \in \mathcal{O}_{\mathbb{C}^n} | I : \ G\text{-invariant ideal}, \mathcal{O}_{\mathbb{C}^n}/I \cong \mathbb{C}[G]\}.$$

We have the following 2-dimensional results on the G-Hilbert scheme.

Theorem 2.3. *For any finite small subgroup G of $GL(2,\mathbb{C})$, the G-Hilbert scheme is the minimal resolution of $\mathbb{C}^2/G$.*

It was proved by

$$\begin{cases} \text{Ito-Nakamura([6]):} & G \subset SL(2,\mathbb{C}), \\ \text{Kidoh([7]):} & G \text{ is a cyclic subgroup of } GL(2,\mathbb{C}), \\ \text{Ishii([4]):} & G \text{ is a small subgroup of } GL(2,\mathbb{C}). \end{cases}$$

In the proof of the first case, they used the fact that the Hilbert scheme of n points on $\mathbb{C}^2$ $\mathrm{Hilb}^n(\mathbb{C}^2)$ is a minimal resolution of the n-th symmetric products of $\mathbb{C}^2$. Under the action of $G \subset SL(2,\mathbb{C})$, the smoothness of the fixed part and symplectic form are preserved, and then the G-Hilbert scheme, the 2-dimensional irreducible component of the G-fixed part of $\mathrm{Hilb}^n(\mathbb{C}^2)$ is the minimal resolution of the quotient singularity $\mathbb{C}^2/G$.

In the second case, we can show that the G-Hilbert scheme is a toric variety and discuss the stratification in terms of Hirzebruch-Jung continued fraction.

The third case contains the above two cases, and the connectedness of the G-Hilbert schemes was proved in terms of derived category as an analogy of the result in dimension three by Bridgeland-King-Reid ([1]).

Let G be the cyclic group $C_{r,a}$, generated by the matrix $\begin{pmatrix} \epsilon & 0 \\ 0 & \epsilon^a \end{pmatrix}$ where $\epsilon^r = 1$ and $\gcd(r,a) = 1$, and consider the character map $\mathbb{C}[x,y] \longrightarrow \mathbb{C}[t]/t^r$ given by $x \mapsto t$ and $y \mapsto t^a$. Then we have a corresponding character for each monomial in $\mathbb{C}[x,y]$.

Let us recall the following theorem for the cyclic singularities. Here we can write down the G-Hilbert scheme as a set of ideals. This result gives the correspondence in Theorem 1.2.

Theorem 2.4 ([5]Theorem 3.9). *For every finite cyclic groups in $GL(2,\mathbb{C})$ and every G-cluster Z, the generator of the ideal I_Z can be chosen as the system of 3 equations:*

$$\begin{cases} x^a = \alpha y^c, \\ y^b = \beta x^d, \\ x^{a-d} y^{b-c} = \alpha\beta, \end{cases}$$

where α and β are complex numbers and both x^a and y^c (resp. y^b and x^d) correspond to the same representation (or character).

The pair (α, β) is a local affine coordinate near the point p and it is also obtained from the calculation with toric geometry. Moreover, each axis of the affine chart is just a exceptional curve or the original axis of $\mathbb{C}^2$. The exceptional curve is isomorphic to a $\mathbb{P}^1$ and the points on it are written in a ratio sych as $(x^a : y^b)$ (resp.

$(x^d : y^c))$ which corresponds to a special representation ρ_a (resp. ρ_d). The fixed point p is the intersection point of two exceptional curves E_a and E_d.

Thus we can obtain the whole space of exceptional locus by deforming the point p and patching the affine pieces.

Remark 2.5. We will define the special representations in section 4: it is enough to undestand that there exists a set of the irreducible representations of G for which each corresponds to an exceptional divisor. The relation between them in terms of the G-Hilbert scheme has already been discussed in the author's previous paper [5].

In the generalization to the higher dimensional cases, we have the following result, which was shown by Nakamura for abelian groups and Bridgeland-King-Reid for general cases:

Theorem 2.6 (Nakamura[8], Bridgeland-King-Reid[1]). *For any finite subgroup G of $SL(3,\mathbb{C})$, the G-Hilbert scheme is a crepant resolution of $\mathbb{C}^3/G$.*

Here we use a similar defintion for 3-dimensional G-Hilbert schemes and a crepant resolution is a minimal resolution with trivial canonical bundle. Moreover, few examples of the G-Hilbert schemes are known in the higher dimension, and there are many problems yet to be proved, for instance, when the G-Hilbert scheme will be a smooth or crepant resolution.

3. Proof of Theorem 1.1

Let us recall the basic notations for the Gröbner basis and define the Gröbner fan before the proof of Theorem 1.1. The author's notations are based on a book by Sturmfels [12].

Let $k[x_1, \cdots, x_n]$ be a polynomial ring, and the monomials be denoted $\mathbf{x}^{\mathbf{a}} = x_1^{a_1} x_2^{a_2} \cdots x_n^{a_n}$.

Fix a weight vector $w = (w_1, \cdots, w_n) \in \mathbb{R}^n$.

Definition 3.1. Let $w \geq 0$ and $\prec$ an arbitrary term order. Then the *weight term order* $<_w$ for $\forall \mathbf{a}, \mathbf{b} \in \mathbb{N}^n$ is defined as follows:

$$\mathbf{a} <_w \mathbf{b} \Longleftrightarrow w \cdot \mathbf{a} < w \cdot \mathbf{b} \text{ or } (w \cdot \mathbf{a} = w \cdot \mathbf{b} \text{ and } \mathbf{a} \prec \mathbf{b}).$$

Definition 3.2. For any non-zero polynomial $f = \sum c_i \mathbf{x}^{\mathbf{a_i}}$, the *initial form* $in_w(f)$ is the sum of all terms $c_i \mathbf{x}^{\mathbf{a_i}}$ such that the inner product $w \cdot \mathbf{a_i}$ is maximal.

Definition 3.3. For an ideal I, the *initial ideal* $in_w(I)$ is the ideal generated by all initial forms:

$$in_w(I) := \langle in_w(f) \mid f \in I \rangle.$$

Definition 3.4. A finite subset $\mathcal{G} \subset I$ is a *Gröbner basis* for I w.r.t. $<_w$ if $in_w(I)$ is generated by $\{in_w(g) | g \in \mathcal{G}\}$.

Definition 3.5. The Gröbner basis $\mathcal{G}$ is *reduced* if for any two distinct elements g, $g' \in \mathcal{G}$, no term of g' is divisible by $in_w(g)$.

Definition 3.6. The *universal Gröbner basis* $\mathcal{U}$ for I is the union of distinct reduced Gröbner basis for I.

Definition 3.7. 2 weight vectors w, $w' \in \mathbb{R}^n$ are called *equivalent* w.r.t. I if and only if $in_w(I) = in_{w'}(I)$.

Then we can consider the equivalence class of weight vectors:

Proposition 3.8 ([12]Proposition 2.3). *Each equivalence class of weight vectors,*

$$c[w] = \{w' \in \mathbb{R}^n_{\geq 0} \mid in_w(g) = in_{w'}(g) \text{ for } \forall g \in \mathcal{G}\}$$

where $\mathcal{G}$ is the reduced Gröbner basis of I w.r.t. $<_w$, is a relatively open convex polyhedral cone.

Definition 3.9. The *Gröbner fan* for I is the set of closed cones $\overline{c[w]}$ for $\forall w \in \mathbb{R}^n$.

Proposition 3.10 ([12]Proposition 2.4). *The Gröbner fan is a fan.*

Let us consider our case. Let G be a finite cyclic group in $GL(2, \mathbb{C})$ and $\mathbb{C}[x, y]$ be a coordinate ring for the original space $\mathbb{C}^2$. Then we will discuss the minimal resolution of the quotient singularity $\mathbb{C}^2/G$.

Let us see the proof of Theorem 1.1.

Proof. First, we recall the toric resolution to obtain the minimal resolution of a cyclic quotient singularity.

Let $\mathbb{R}^2$ be the 2-dimensional real vector space, $\{e^i | i = 1, 2\}$ its standard base, L the lattice generated by e^1 and e^2, $N := L + \sum \mathbb{Z}v$, where $v = 1/r(1, a) \in G = C_{r,a}$, and

$$\sigma := \left\{ \sum_{i=1}^{2} x_i e^i \in \mathbb{R}^2, \quad x_i \geq 0, \forall i, 1 \leq i \leq 2 \right\}$$

a rational convex polyhedral cone in $N_{\mathbb{R}} = N \otimes_{\mathbb{Z}} \mathbb{R}$. The corresponding affine torus embedding Y_σ is defined as $\mathrm{Spec}(\mathbb{C}[\check{\sigma} \cap M])$, where M is the dual lattice of N and $\check{\sigma}$ the dual cone of σ in $M_{\mathbb{R}}$ defined as $\check{\sigma} := \{\xi \in M_{\mathbb{R}} | \xi(x) \geq 0, \forall x \in \sigma\}$.

Then $X = \mathbb{C}^2/G$ corresponds to the toric variety which is induced by the cone σ within the lattice N.

Lemma 1 We can construct a simplicial decomposition S with the vertices on the Newton Boundary, that is, the boundary of the convex hull of the lattice points in σ except origin.

Lemma 2 If $\widetilde{X} := X_S$ is the corresponding torus embedding, then X_S is non-singular. Thus, we obtain the minimal resolution $\pi = \pi_S : \widetilde{X} = X_S \longrightarrow \mathbb{C}^2/G = Y$. Moreover, each lattice point of the Newton boundary corresponds to an exceptional divisor.

By Lemma 1, we obtained r lattice points p_i in N on the Newton boudary, where r is the number of the exceptional curves and $1 \leq i \leq r$. If we assume that the coordinates of the point p_i are $1/r(a_i, b_i)$, then the binomial $x^{b_i} - y^{a_i}$ is a G-semi-invariant by the definition of the toric variety.

Moreover, the vectors $\mathbf{p}_i = 1/r(a_i, b_i)$ are the primitive vectors in N and they are the basis of all lattice points in N.

On the other hand, the Gröbner fan for the G-homogenous ideal I_G is a union of the cones in the area $\{x \geq 0, y \geq 0\}$, and each cone corresponds to an equivalece class of weight vectors. The weight vectors can be regarded as the multiplicity of the variables x, y, and the monomials in x and y are weighted monomials under the group action, where the weight of the variables depends on the group action.

The Gröbner fan is a fan and we have 1-dimensional cones which give the boudary of the 2-dimensional cones in this case. These 1-dimensional cones must pass through the points p_i in N if the Gröbner fan gives the minimal resolution.

Let us consider a line l_i through p_i. For each i, there exsists a G-semi-invariant binomial $x^{b_i} - y^{a_i}$ as we saw above. Note that these binomials are contained in the reduced Gröbner basis, because all of them are given by the lattice point on the Newton boundary. Therefore none of them are diveded by any other binomial generators. If you take the weight vector $(w_x, w_y) = (a_i, b_i)$, then it is on the line and the corresponding initial ideal has a binomial generator $x^{b_i} - y^{a_i}$. However if you take it $(w_x, w_y) = (a_i + \epsilon, b_i)$ (resp. $= (a_i, b_i + \epsilon)$), where ϵ is a small positive number, then we have the corresponding initial ideal has a monomial generator x^{b_i} (resp. (y^{a_i})). Here we have two different equivalence classes and there exists a line l_i between these corresponding 2-dimensional cones, that is, the boudary line is the same as the line l_i. Each 1-dimensional cone can be appear only as an intersection of the 2-dimensional cones in this situation and all these lines pass through the lattice points p_is on the Newton boundary. Therefore there are no other 1-dimensional cones except l_is.

Now let us consider the relation between (a_i, b_i) in Theorem 1.1 and (a, b, c, d) in Theorem 2.4 to identify two fans: one was given as a toric resolution and the other one was Gröbner fan. The eqation in Theorem 2.4 gives an affine piece of the minimal resolution and right hand sides of the first two equation, x^a and y^b, are the same as the monomial generators of the initial ideals in a 2-dimensional cone of the Gröbner fan. The exceptional curves is the intersection of two affine pieces. The coordinate is given by binomials such as $x^a - y^c$ and $y^b - x^d$ in terms of Theorem 2.4. They can be the binomial generators in the initial ideals for the 1-dimensional cones in the Gröbner fan such as $x^{b_i} - y^{a_i}$ in terms of Theorem 1.1. Thus we have the assertion. □

4. Application to the generalized McKay correspondence

In this section, we give an explanation on the generalized McKay correspondence using Theorem 1.1. To describe the correspondence in terms of the Gröbner basis, we discuss the special representations.

Let G be a finite small subgroup of $GL(2,\mathbb{C})$, that is, the action of the group G is free outside the origin, and ρ be a representation of G on V. G acts on $\mathbb{C}^2 \times V$ and the quotient is a vector bundle on $(\mathbb{C}^2 \setminus \{0\})/G$ which can be extended to a reflexive sheaf $\mathcal{F}$ on $X :\, = \mathbb{C}^2/G$.

For any reflexive sheaf $\mathcal{F}$ on a rational surface singularity X and the minimal resolution $\pi\colon \widetilde{X} \to X$, we define a sheaf $\widetilde{\mathcal{F}} :\, = \pi^*\mathcal{F}/\text{torsion}$.

Definition 4.1. ([2]) The sheaf $\widetilde{\mathcal{F}}$ is called a *full sheaf* on $\widetilde{X}$.

Theorem 4.2. ([2]) *A sheaf $\widetilde{\mathcal{F}}$ on $\widetilde{X}$ is a full sheaf if the following conditions are fulfilled:*

1. *$\widetilde{\mathcal{F}}$ is locally free,*
2. *$\widetilde{\mathcal{F}}$ is generated by global sections,*
3. *$H^1(\widetilde{X}, \widetilde{\mathcal{F}}^\vee \otimes \omega_{\widetilde{X}}) = 0$, where $\vee$ means the dual.*

Note that a sheaf $\widetilde{\mathcal{F}}$ is indecomposable if and only if the corresponding representation ρ is irreducible. Therefore we obtain an indecomposable full sheaf $\widetilde{\mathcal{F}_i}$ on $\widetilde{X}$ for each irreducible representation ρ_i, but in general, the number of the irreducible representations is larger than that of irreducible exceptional components. Therefore Wunram and Riemenschneider introduced the notion of speciality for full sheaves:

Definition 4.3. ([11]) A full sheaf is called *special* if and only if

$$H^1(\widetilde{X}, \widetilde{\mathcal{F}}^\vee) = 0.$$

A reflexive sheaf $\mathcal{F}$ on X is *special* if $\widetilde{\mathcal{F}}$ is so.

A representation ρ is *special* if the associated reflexive sheaf $\mathcal{F}$ on X is special.

With these definitions, the following equivalent conditions for the speciality hold:

Theorem 4.4. ([11], [13])

1. *$\widetilde{\mathcal{F}}$ is special $\Longleftrightarrow$ $\widetilde{\mathcal{F}} \otimes \omega_{\widetilde{X}} \to [(\mathcal{F} \otimes \omega_{\widetilde{X}})^{\vee\vee}]^\sim$ is an isomorphism,*
2. *$\mathcal{F}$ is special $\Longleftrightarrow$ $\mathcal{F} \otimes \omega_{\widetilde{X}}/\text{torsion}$ is reflexive,*
3. *ρ is a special representation $\Longleftrightarrow$ the map $(\Omega^2_{\mathbb{C}^2})^G \otimes (\mathcal{O}_{\mathbb{C}^2} \otimes V)^G \to (\Omega^2_{\mathbb{C}^2} \otimes V)^G$ is surjective.*

Then we have the following nice generalized McKay correspondence for quotient surface singularities:

Theorem 4.5. ([13]) *There is a bijection between the set of special non-trivial indecomposable reflexive modules $\mathcal{F}_i$ and the set of irreducible components E_i via*

$c_1(\widetilde{\mathcal{F}}_i)E_j = \delta_{ij}$ *where* c_1 *is the first Chern class, and also a one-to-one correspondence with the set of special non-trivial irreducible representations.*

As a corollary of this theorem, we can obtain the original McKay correspondence for finite subgroups of $SL(2,\mathbb{C})$ because in this case all irreducible representations are special. You can find the first observation of the McKay correspondence in McKay's paper [10] and the first geometric proof similar to the above was given by Gonzalez-Sprinberg and Verdier in [3].

Let us show Theorem 1.2:

Proof. It can be proved as a collorary of Theorem 1.1 with the results above. As we saw in this section, each special irreducible representation corresponds to an exceptional curve. On the other hand, the exceptional curves can be written as lattice points p_i on the Newton boudary in terms of toric resolution. In terms of the Gröbner basis we have the Gröbner fan which determines the minimal resolution by Theorem 1.1, and each cone for the equivalence class of the weight vectors defines an affine piece of the resolution. Next we can write down the exceptional curves as an intersection of these affine pieces, i.e., lines through the origin and the lattice points p_i. One can then find a binomial generator corresponding to the points p_i as in the proof of Theorem 1.1 in section 3. Thus we have a non-monomial generator in the corresponding initial ideals.

Moreover, we can find the corresponding representations for each monomial via the G-Hilbert scheme as we discussed in section 2. By Theorem 2.4 and the discussion in [5], it is known that each special representation corresponde to each exceptional curve. The exceptional curveis given by a 1-dimensional cone in the Gröbner fan for I_G.

Therefore we have a bijection between an irreducible special representation and non-monomial generator in the initial ideals for the G-homogenous ideal I_G with respect to the weight $\mathbf{w}$, where $\mathbf{w}$ is on the boundary of the equivalece classes of the weight vectors. Thus we have proved Theorem 1.2 □

5. Example

We will see an example of the minimal resolution via the Gröbner basis in this section.

Let G be a cyclic group of type $C_{7,3}$ which is generated by the matrix $\begin{pmatrix} \epsilon & 0 \\ 0 & \epsilon^3 \end{pmatrix}$ where $\epsilon^7 = 1$. The toric resolution of this quotient singularity is given by the triangulation of the lattice $N: = \mathbb{Z}^2 + \frac{1}{7}(1,3)\mathbb{Z}$ with the lattice points: See Figure 5.1.

It is isomorphic to the G-Hilbert scheme and the defining equations of the G-clusters are:

$G_1 : x = a_1y^5, y^7 = b_1, xy^2 = a_1b_1,$

$G_2 : x^2 = a_2y^3, y^5 = b_2x, xy^2 = a_2b_2,$

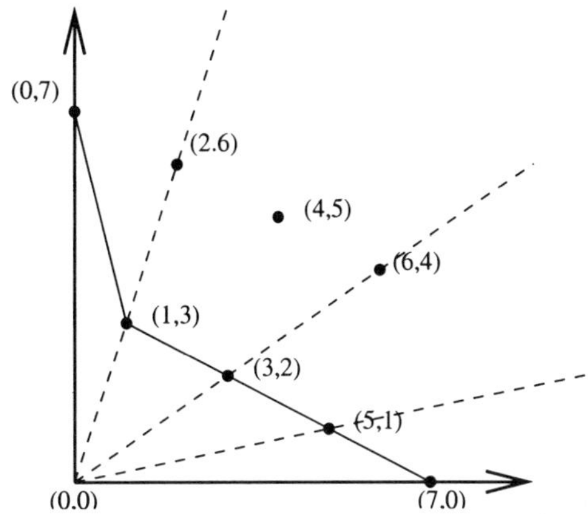

FIGURE 5.1. toric resolution of $\mathbb{C}^2/G$

$G_3 : x^3 = a_3y, y^3 = b_3x^2, xy^2 = a_3b_3,$
$G_4 : x^7 = a_4, y = b_4x^3, x^4y = a_4b_4,$
where a_i, $b_i \in \mathbb{C}$.

On the other hand, the universal Gröbner basis $\mathcal{U}$ for I_G can be written as follows:

$$\mathcal{U} = \{x^7 - 1, y^7 - 1, xy^2 - 1, x^4y - 1, x - y^5, x^2 - y^3, x^3 - y\}.$$

There are four 2-dimensional cones in the Gröbner fan. Therefore we have four equivalence classes of weight vectors $(w_x, w_y) \in \mathbb{R}^2_{\geq 0}$ for the Gröbner basis $\mathcal{U}$ and the corresponding initial ideals:

$c_1[w]: \ w_x > 5w_y \geq 0, \ in(I) = \langle x, y^7 \rangle,$
$c_2[w]: \ 5w_y > w_x > \frac{3}{2}w_y \geq 0, \ in(I) = \langle x^2, y^5 \rangle,$
$c_3[w]: \ \frac{3}{2}w_y > w_x > \frac{1}{3}w_y \geq 0, \ in(I) = \langle x^3, y^3 \rangle,$
$c_4[w]: \ w_x > \frac{1}{3}w_y \geq 0, \ in(I) = \langle x^7, y \rangle.$

Thus we can obtain the minimal resolution via the Gröbner basis. It can be oveserved that the same monomials appear in the the defining equations of the G-clusters and the initial ideals above. Moreover, we have the corresponding irreducible representations for each exceptional curve which is defined as an intersection of 2 affine pieces.

There is a line which is defined as the intersection of the closure of $c_i[w]$s. Let U_i be an affine piece deterined by the cone $\overline{c_i[w]}$. For an exceptional curve E_{12} defined in $U_1 \cap U_2$, we have a non-monomial generator $x - y^5$ in the initial ideal corresponding to a character ϵ or an irreducible representation ρ_1.

Similarly, we obtain the other two non-monomial generators of the initial ideal and the corresponding representations:

$E_{23}: \ x^2 - y^3 \ \leftrightarrow \ \rho_2,$
$E_{34}: \ x^3 - y \ \leftrightarrow \ \rho_3.$

Thus we can obtain the special representations by Theorem 1.2.

After writing up this paper, the author met D. Maclagan and she mentioned her similar result for toric Hilbert schemes.[9]

REFERENCES

[1] T. Bridgeland, A. King and M. Reid, *The McKay correspondence as an equivalence of derived categories*, J. AMS, **14**, (2001), 535–554.
[2] H. Esnault, *Reflexive modules on quotient surface singularities*, J. Reine Angew. Math. **362** (1985), 63–71.
[3] G. Gonzalez-Sprinberg and J.L. Verdier, *Construction géometrique de la correspondance de McKay*, Ann. Sci. École Norm. Sup. **16** (1983), 409–449.
[4] A. Ishii, *On the McKay correspondence for a finite small subgroup of* $GL(2, \mathbb{C})$, to appear in J. Reine Angew. Math.
[5] Y. Ito, *Special McKay correspondence*, Actes de l'école d'été "Géométrie des variétés toriques" (Grenoble, 2000), Séminaires et Congrès, **6** (2002), (Société Mathématique de France), 213–225.
[6] Y. Ito and I. Nakamura, *Hilbert schemes and simple singularities*, in: New trends in Algebraic Geometry (Warwick, June 1996), K. Hulek and others Eds., CUP (1999), 151–233.
[7] R. Kidoh, *Hilbert schemes and cyclic quotient singularities*, Hokkaido Mathematical Journal, **30** (2001), 91–103.
[8] I. Nakamura, *Hilbert schemes of abelian group orbits*, J. Algebraic Geom., **10**, (2001), 757–779.
[9] D. Maclagan and R. R. Thomas, *The toric Hilbert scheme of a rank two lattice is smooth and irreducible*, preprint, math.AG/0208031.
[10] J. McKay, *Graphs, singularities and finite groups*, Proc. Symp. Pure Math., **37** (1980), Amer. Math. Soc. 183–186.
[11] O. Riemenschneider, *Characterization and application of special reflexive modules on rational surface singularities*, Institut Mittag-Leffler Report No.3 (1987).
[12] B. Sturmfels, *Gröbner Bases and Convex Polytopes* Univ. Lect. Series, **8**, AMS (1995).
[13] J. Wunram, *Reflexive modules on quotient surface singularities*, Math Ann. **279** (1988), 583–598.

DEPARTMENT OF MATHEMATICS, TOKYO METROPOLITAN UNIVERSITY, HACHIOJI, TOKYO 192-0397, JAPAN
E-mail address: yukari@comp.metro-u.ac.jp

Deformation theory of smoothable semi log canonical surfaces

Yongnam Lee [1]

abstract. The aim of the paper is to study the $\mathbf{Q}$-Gorenstein deformation theory of log surfaces under suitable conditions. As its application, we consider compactifications of the space of double cover of $\mathbf{P}^2$ branched over curves of degree 4 or 6.

Introduction

Via the geometric invariant theory, Gieseker [Gi] proved the existence of a quasiprojective coarse moduli space $M_{K^2,\chi}$ for smooth projective surfaces of general type with fixed numerical invariants K^2 and χ. With the proof of bounds for log surfaces with given K^2, Alexeev [A] clarified the construction of projective coarse moduli space of surfaces of general type with fixed K^2 that was started in [K-SB]. The compactified moduli space should include (possibly reducible) surfaces with ordinary double curves and certain other mild singularities. These surfaces are called smoothable stable surfaces. This notion of a smoothable stable surface can be generalized to a smoothable stable log surface similar as the generalization of a stable curve to a stable pointed curve.

As applications of stable log surfaces, Hassett [Has1] [Has2] considered a compactification $\mathcal{P}_d^1$ of plane curves of degree d ($d \geq 4$) by using the stable log surfaces and the $\mathbf{Q}$-Gorenstein deformation theory of stable log surfaces. Then he succeeded to prove that $\mathcal{P}_4^1$ is isomorphic to Deligne-Mumford compactification of moduli space of curves of genus 3. He considered all possible plane curve singularities on the boundary of Deligne-Mumford compactification, and then he constructed corresponding stable log surfaces by using local stable reduction theorem. But it is already too complicated to manage if $d \geq 5$.

Hacking [Hac] considered instead the family of compactifications given by moduli space $\mathcal{P}_d^\alpha$ of log surfaces (X, B) such that $K_X + \alpha B$ is semi log canonical singularities and ample where $\frac{3}{d} < \alpha \leq 1$. The compactification is simpler for lower α because it is allowed worse singularities for lower α. He gave a compactification $\mathcal{P}_d$ of plane curves of degree d by allowable family of stable pairs of degree d. Then he proved that $\mathcal{P}_d$ is a separated proper Deligne-Mumford stack and $\mathcal{P}_d$ is smooth if $3 \nmid d$.

[1] The research was supported by the Scientist Exchange Program by JSPS and KOSEF, and by the grant 1999-2-102-002-3 from the interdisciplinary Research program of the KOSEF.

2000 *AMS Subject Classification.* 14J10, 14J17.

The paper goes as follows:
Section 1 provides several notions of singularities as preliminaries. Section 2 presents the **Q**-Gorenstein deformation theory of pairs of log surfaces (X, B) under the following assumptions:
Assume that (X, B) is a log surface such that
1. X is a semi log canonical surface,
2. B is a reduced Weil divisor on X,
3. B is Cartier on the index one cover of X.

The third condition is necessary to compare between the **Q**-Gorenstein deformation of X and the **Q**-Gorenstein deformation of pairs (X, B). In paricular, the notion of stable pairs of degree d satisfies our assumtion. Our intention is to generalize the results in [L].

Theorem. *Assume that X is a locally smoothable semi log terminal surface. Then*

$$H^0(\tilde{\mathcal{T}}^1_X) = \sum_{p \in \mathrm{Sing}(X,D)} d_p \mathbf{C}_p + \sum_{D_i} H^0(\mathcal{O}_{D_i}(\sum_q q))$$

where $\mathrm{Sing}(X, D)$ *is the set of singular points outside double curves D, singular point p is the type $\frac{1}{d_p r^2}(1, d_p ar - 1)$ with $(a, r) = 1$* [K-SB], *D_i is a connected component of double curves D, and a singular point q in D_i is not locally **Q**-factorial point.*

Therefore we obtain the infinitesimal smoothable deformations explicitly. Then the existence of smoothing can be interpreted by the relation between the tangent bundle of its minimal resolution and the normal bundle of the exceptional divisor. It holds the following:

Theorem. *Let X be a locally smoothable semi log canonical surface and let $p \in \mathrm{Sing}(X, D)$. Assume that p is permissible ($d_p = 1$ and it admits a semistable resolution without changing a base* [Ka]*). Let (S, E) be the pair of the minimal resolution of the germ (X, p) and the exceptional divisor in S. If $H^2(\mathcal{T}_S(-E)) = 0$ then $H^0(\tilde{\mathcal{T}}^1_{X,p}) = \mathbf{C}_p \xrightarrow{0} H^2(\tilde{\mathcal{T}}^0_X)$.*

In Section 3, as applications we consider compactifications of the space of double cover of $\mathbf{P}^2$ branched over curves of degree 4 or 6. If (X, B) be a stable pair of degree 4 or 6, then $(X, \frac{1}{2}B)$ has semi log canonical singularities. Let Y be a double cover of X branched over B. Then Y has also semi log canonical singularities. By using the **Q**-Gorenstein deformation theory presented in Section 2, we construct compactifications $\mathcal{R}_4$ and $\mathcal{R}_6$ of the space of double covering surfaces directly over $\mathcal{P}_4$ and $\mathcal{P}_6$.

Theorem. *There are compactifications of the space of double cover of $\mathbf{P}^2$ branched over curves of degree 4 or 6.*

In [KL], we proved that if D is a plane curve of degree d and $(\mathbf{P}^2, D)$ is a stable pair of degree d then D is a stable plane curve. The converse is also true under a suitable condition. These results make us possible to compare between the compactifications $\mathcal{P}_d$ and the compactifications of plane curves by Geometric Invariant Theory. In particular, we obtain the same classification of $\mathcal{P}_4$ as in [Hac] by construction of the stable pairs of degree 4 corresponding the semistable plane curves of degree 4. In a future work, we intend to classify the cases in $\mathcal{P}_6$ and intend to compare to this compactification $\mathcal{R}_6$ with Shah's compactification [S] by using Geometric Invariant Theory of plane curves of degree 6.

We work throughout over the complex number field $\mathbf{C}$. The notation here follows Hartshorne's Algebraic Geometry.

1. Preliminaries

The notion of discrepancy is the fundamental measure of the singularities of (X, D) (cf. [K et] or [KM]).

Definition. Let X be a normal variety and $D = \sum d_i D_i$ an effective $\mathbf{Q}$-divisor such that $K_X + D$ is $\mathbf{Q}$-Cartier. Let $f\colon Y \to X$ be a proper birational morphism. Then we can write

$$K_Y + f_*^{-1}(D) \equiv f^*(K_X + D) + \sum a(E, D)E$$

where $f_*^{-1}(D)$ is the proper transform of D, the sum runs over distinct prime divisors $E \subset Y$, and $a(E, D) \in \mathbf{Q}$. This $a(E, D)$ is called the *discrepancy* of E with respect to (X, D); it only depends on the divisor E, and not on the partial resolution Y.

We define $\operatorname{discrep}(X, D) = \inf_E\{a(E, D) | E \text{ is an exceptional divisor of } X\}$. We say that (X, D), or $K_X + D$ is

$$\left.\begin{array}{l}\text{terminal}\\ \text{canonical}\\ \text{purely log terminal}\\ \text{log canonical}\end{array}\right\} \quad \text{if} \quad \operatorname{discrep}(X, D) \quad \left\{\begin{array}{l} > 0,\\ \geq 0,\\ > -1,\\ \geq -1.\end{array}\right.$$

Moreover, (X, D) is *Kawamata log terminal* (klt) if (X, D) is purely log terminal and $d_i < 1$ for every i; and (X, D) is *divisorial log terminal* (dlt) if there exists a log resolution such that the exceptional locus consists of divisors with all $a(E, D) > -1$.

And we say that a log surface (X, D) has semi log canonical (resp. semi log terminal) singularities if

1. X satisfies Serre's condition S_2,
2. X has normal crossing singularities in codimension one,
3. $K_X + D$ is $\mathbf{Q}$-Cartier, and for any birational morphism $f : Y \to X$ from a semi-smooth surface Y we have

$$K_Y \equiv f^*(K_X + D) + \sum a_i E_i$$

where all $a_i \geq -1$ (resp. $a_i > -1$). A surface Y is called semi-smooth if every closed point of Y is either smooth or normal crossing or a pinch point.

Definition. A stable log surface is the pair (X, B) where
1. X is a connected projective surface and B a reduced Weil divisor on X,
2. (condition on singularities) the pair (X, B) has semi log canonical singularities,
3. (numerical condition) $K_X + B$ is ample.

One can replace a general relatively log minimal model of a deformation of a smoothable stable log surface $(\mathcal{X}, \mathcal{B}) \to \Delta$ with another relatively log minimal model, which has better singularities via base change and simultaneous resolution [Ka]. The notion of permissible singularities is due to Kawamata [Ka].

Definition. Let $\pi : (\mathcal{X}, \mathcal{B}) \to \Delta$ be a deformation of a smoothable stable log surface (X, B). We say that π is a *permissible degeneration* if the following conditions are satisfied: The analytic germ of each point $p \in (X, B)$ is isomorphic to one of the following germs, where $\pi = t$:
1. $(xyz = t)$,
2. $(xy = t) \subset A$, $A = \frac{1}{r}(a, r-a, 1, 0)$ for $(a, r) = 1$,
3. $(xy = z^r + t^n) \subset A$ for some n.

In a permissible degeneration of (X, B), the central fiber has smooth double curves, and each component of (X, B) has cyclic quotient singularities of the following two types: $\frac{1}{r}(1, a)$ on the double curves and B, and $\frac{1}{r^2}(a, r-a)$ outside the double curves and B. And $(\mathcal{X}, \mathcal{B})$ admits a semistable model by the succession of weighted blowing up of singular points. This construction is explicitly given in [C]. We say that a pair (X, B) has a *permissible singularity at* p if local germ at p has a permissible degeneration.

2. Q-Gorenstein deformation theory of pairs

Let $\mathcal{X}/\mathcal{A}$ be a flat family of semi log canonical surfaces over a noetherian scheme. $\mathcal{X}/\mathcal{A}$ is called $\mathbf{Q}$-Gorenstein if $\omega_{\mathcal{X}/\mathcal{A}}^{[i]}$ commutes with base change for all $i \in \mathbf{Z}$. And $\mathcal{X}/\mathcal{A}$ is called weakly $\mathbf{Q}$-Gorenstein if $\omega_{\mathcal{X}/\mathcal{A}}$ is $\mathbf{Q}$-Cartier. In general, $\mathbf{Q}$-Gorenstein implies weakly **Q**-Gorenstein. Hacking [Hac] showed that weakly **Q**-Gorenstein is the same as **Q**-Gorenstein if $\mathcal{A}$ is a discrete valuation ring. Let X be a semi log canonical surface. Consider the sheaves of deformation $\mathcal{T}_X^i = \mathcal{E}xt_X^i(\Omega_X, \mathcal{O}_X)$ for $i = 0,1,2$. Define the subsheaves $\tilde{\mathcal{T}}_X^i$ of $\mathcal{T}_X^i$ that represent the infinitesimal **Q**-Gorenstein deformations. These subsheaves $\tilde{\mathcal{T}}_X^i$ can be constructed via local index one cover. Let $U \subset X$ be an analytic neighborhood with an index one cover U'. For the case of the field **C**, this index one cover is unique up to isomorphism. Define $\tilde{\mathcal{T}}_X^i(U)$ as the invariant part of $\mathcal{T}_{U'}^i$. And we define $\tilde{\mathcal{T}}_X^i$ by hypercohomology. It is obvious from the definition that $\tilde{\mathcal{T}}_X^i \subset \mathcal{T}_X^i$ and $\tilde{\mathcal{T}}_X^0 = \mathcal{T}_X^0$. Then $\tilde{\mathrm{T}}_X^0 = \mathrm{T}_X^0$, and $\tilde{\mathrm{T}}_X^1$ as the elements of T_X^1 mapped to $H^0(\tilde{\mathcal{T}}_X^1)$. And the obstruction map can be computed by $\tilde{\mathrm{T}}_X^1 \subset \mathrm{T}_X^1 \overset{ob}{\to} \mathrm{T}_X^2$. We consider the **Q**-Gorenstein deformation theory under the following assumption :

Assumption. Let (X, B) be a log surface such that
1. X is a semi log canonical surface,
2. B is a reduced Weil divisor on X,
3. B is Cartier on the index one cover of X.

Let X be a semi log canonical surface with B a reduced Weil divisor on X. By the formalism of deformation of pairs [R] we have the long exact sequence $0 \to \mathrm{T}_{X,B}^0 \to \mathrm{T}_B^0 \oplus \mathrm{T}_X^0 \to \mathrm{Hom}_X(\Omega_X, \mathcal{O}_B) \to \mathrm{T}_{X,B}^1 \to \mathrm{T}_B^1 \oplus \mathrm{T}_X^1 \to \mathrm{Ext}_X^1(\Omega_X, \mathcal{O}_B) \to \mathrm{T}_{X,B}^2 \to \mathrm{T}_X^2$. The short exact sequence $0 \to \mathcal{O}_X(-B) \to \mathcal{O}_X \to \mathcal{O}_B \to 0$, where $\mathcal{O}_X(-B)$ denotes the ideal sheaf of B, induces the long exact sequence: $0 \to \mathrm{T}_X^0(-B) \to \mathrm{T}_X^0 \to \mathrm{Hom}_X(\Omega_X, \mathcal{O}_B) \to \mathrm{T}_X^1(-B) \to \mathrm{T}_X^1 \to \mathrm{Ext}_X^1(\Omega_X, \mathcal{O}_B) \to \mathrm{T}_X^2(-B) \to \mathrm{T}_X^2$ where $\mathrm{T}_X^i = \mathrm{Ext}_X^i(\Omega_X^1, \mathcal{O}_X)$, $\mathrm{T}_X^i(-B) = \mathrm{Ext}_X^i(\Omega_X^1, \mathcal{O}_X(-B))$ for $i = 0,1,2$. By these two long exact sequences, we obtain a long exact sequence for deformation of pairs, which is a generalization of the long exact sequence in a smooth case (cf. [Has1]):

$$\begin{aligned} 0 &\to \mathrm{T}_X^0(-B) \to \mathrm{T}_{X,B}^0 \to \mathrm{T}_B^0 \\ &\to \mathrm{T}_X^1(-B) \to \mathrm{T}_{X,B}^1 \to \mathrm{T}_B^1 \\ &\to \mathrm{T}_X^2(-B) \to \mathrm{T}_{X,B}^2 . \end{aligned}$$

Similarily, there are subsheaves $\tilde{\mathcal{T}}_X^i(-B)$ of $\mathcal{T}_X^i(-B)$ and $\tilde{\mathcal{T}}_{X,B}^i$ of $\mathcal{T}_{X,B}^i$ that represent **Q**-Gorenstein deformations ($\omega_{\mathcal{X}/\mathcal{A}}^{[i]}$ and $\mathcal{O}_{\mathcal{X}/\mathcal{A}}(\mathcal{B})^{[i]}$ commute with base change for all $i \in \mathbf{Z}$). By the Assumption it is well defined on the index one cover of X. Then there is a long exact sequence for **Q**-Gorenstein deformation of pairs,

$$0 \to \mathrm{T}_X^0(-B) \to \mathrm{T}_{X,B}^0 \to \mathrm{T}_B^0 \to \tilde{\mathrm{T}}_X^1(-B) \to \tilde{\mathrm{T}}_{X,B}^1 \to \mathrm{T}_B^1 .$$

Furthermore, if (X, B) is a stable log surface ($K_X + B$ is ample) then it holds the long exact sequence $0 \to \tilde{\mathrm{T}}^1_X(-B) \to \tilde{\mathrm{T}}^1_{X,B} \to \mathrm{T}^1_B$ by the stability condition. Also $\tilde{\mathrm{T}}^1_X$ and $\tilde{\mathrm{T}}^1_X(-B)$ can be computed by the local-global spectral sequence ([Go], §7.3):
$0 \to H^1(\mathcal{T}^0_X) \to \tilde{\mathrm{T}}^1_X \to \ker[H^0(\tilde{\mathcal{T}}^1_X) \to H^2(\mathcal{T}^0_X)] \to 0$
$0 \to H^1(\mathcal{T}^0_X(-B)) \to \tilde{\mathrm{T}}^1_X(-B) \to \ker[H^0(\tilde{\mathcal{T}}^1_X(-B)) \to H^2(\mathcal{T}^0_X(-B))] \to 0$. Let B' be the inverse image of B on the index one cover. Also by Assumption, $0 \to \mathcal{O}_{U'}(-B') \to \mathcal{O}_{U'} \to \mathcal{O}_{B'} \to 0$ gives a resolution of $\mathcal{O}_{B'}$ by invertible sheaves. It produces the long exact sequence

$$0 \to \mathrm{T}^0_{X,B} \to \mathrm{T}^0_X \to H^0(B, \mathcal{O}_B(B)) \to \tilde{\mathrm{T}}^1_{X,B} \to \tilde{\mathrm{T}}^1_X \to H^1(B, \mathcal{O}_B(B)).$$

If $H^1(B, \mathcal{O}_B(B)) = 0$ then the forgetful functor $\tilde{\mathrm{T}}^1_{X,B} \to \tilde{\mathrm{T}}^1_X$ is smooth from the following commutative digram:

$$\begin{array}{ccccc} H^0(B, \mathcal{O}_B(B)) & \to & \tilde{\mathrm{T}}^1_{X,B} & \to & \tilde{\mathrm{T}}^1_X \\ \downarrow ob & & \downarrow ob & & \downarrow ob \\ H^1(B, \mathcal{O}_B(B)) & \to & \tilde{\mathrm{T}}^2_{X,B} & \to & \tilde{\mathrm{T}}^2_X. \end{array}$$

Since $H^1(\mathcal{T}^0_X)$ represents deformations of X for which the singularities remain locally a product, our concern is to understand the kernel of the map $H^0(\tilde{\mathcal{T}}^1_X) \to H^2(\mathcal{T}^0_X)$. Assume that X is a locally smoothable semi log terminal surface. Locally smoothability means that for each point p of X there is a threefold $\mathcal{X}$ over a discrete valuation ring T with the central fiber (X, p) such that $\omega_{\mathcal{X}/T}$ is $\mathbf{Q}$-Cartier and X_t is smooth for $t \in T^*$. Then $H^0(\tilde{\mathcal{T}}^1_X)$ can be computed explicitly.

Theorem 2.1. *Assume that X is a locally smoothable semi log terminal surface. Then*

$$H^0(\tilde{\mathcal{T}}^1_X) = \sum_{p \in \mathrm{Sing}(X,D)} d_p \mathbf{C}_p + \sum_{D_i} H^0(\mathcal{O}_{D_i}(\sum_q q))$$

where $\mathrm{Sing}(X, D)$ *is the set of singular points outside double curves D, a singular point $p \in \mathrm{Sing}(X, D)$ is the type of $\frac{1}{d_p r^2}(1, d_p a r - 1)$ with $(a, r) = 1$ [K-SB], D_i is a connected component of double curves D, and a singular point q in D_i is not locally $\mathbf{Q}$-factorial point.*

Proof. First recall that the sheaf $\tilde{\mathcal{T}}^1_X$ is supported only on singular points and $H^0(\tilde{\mathcal{T}}^1_X) = \sum_{p \in \mathrm{Sing}(X,D)} \mathrm{Def}(X, p) + \sum_{D_i} H^0(\mathcal{O}_{D_i}(\sum_{q'} q'))$ where q' is a singular point in D_i. The type of singular point p in $\mathrm{Sing}(X, D)$ is $\frac{1}{d_p r^2}(1, d_p a r - 1)$ with $(a, r) = 1$, and the dimension of $\mathrm{Def}(X, p)$ is d_p ([K-SB], Lemma 3.21). Let q' be a singular point in D_i and let $\mathcal{X}$ be a $\mathbf{Q}$-Gorenstein smoothing of the germ (X, q'). If q' is a $\mathbf{Q}$-factorial point in $\mathcal{X}$ then the index one cover of $(\mathcal{X}, q')$ is smooth with d-semistable central fiber as the notion of Friedman ([F], Definition 1.13). Hence the $\mathbf{Q}$-factorial point q' does not affect $H^0(\tilde{\mathcal{T}}^1_X)$. If q' is not a $\mathbf{Q}$-factorial point in $\mathcal{X}$ then the preimage q'' of q' in index one cover is a $\mathbf{Q}$-factorial point. Consider a semistable resolution of index one cover by the succession of weighted blowing of $\mathbf{Q}$-factorial point q''. Then we obtain the proof by the argument in ([F], Corollary 1.12). □

In some cases, it is possible to compute $H^2(\mathcal{T}^0_X)$. Lemma 2.2 is in [Ma].

Lemma 2.2. *Let X be a normal surface with log terminal singulairties. Assume that $h^0(-K_X) \neq 0$ and $q(X) = 0$. Then $H^2(\mathcal{T}^0_X) = 0$.*

Proof. Let S be the minimal resolution of X. Then $H^2(\mathcal{T}^0_X) = H^0(\Omega^{**}_X \otimes K_X) \hookrightarrow H^0(\Omega^{**}_X) = H^0(\Omega_S) = 0$. □

Lemma 2.3. *Let X be a semi log canonical surface and let $X^\nu \to X$ be the normalization. Let (Y, C) be a pair of component of X^ν and double curves in Y. If $H^2(\mathcal{T}^0_Y(-C)) = 0$ for all (Y, C) then $H^2(\mathcal{T}^0_X) = 0$.*

Proof. The proof is obtained directly from the short exact sequence,

$$0 \to \sum_Y \mathcal{T}^0_Y(-C) \to \mathcal{T}^0_X \to \mathcal{S} \to 0$$

where $\mathcal{S}$ is supported only on double curves. □

Lemma 2.4. *Let X be a smooth surface and E a divisor of X. Assume that $-(K_X + E)$ is ample. Then $H^2(T_X(-E)) = 0$.*

Proof. The proof is obtained directly from Akizuki-Nakano vanishing theorem [AN] i.e. $H^0(\Omega_X(K_X + E)) = 0$. □

Theorem 2.5. *Let X be a locally smoothable semi log canonical surface and let $p \in \mathrm{Sing}(X, D)$. Assume that p is permissible. Let (S, E) be the pair of the minimal resolution of the germ (X, p) and the exceptional divisor E in S. If $H^2(\mathcal{T}_S(-E)) = 0$ then $H^0(\tilde{\mathcal{T}}^1_{X,p}) = \mathbf{C}_p \xrightarrow{0} H^2(\tilde{\mathcal{T}}^0_X)$.*

Proof. Let $p \in \mathrm{Sing}(X, D)$. Let $\mathcal{X}$ be a $\mathbf{Q}$-Gorenstein smoothing of the germ (X, p) over a small disk $\Delta \subset \mathbf{C}$. Since p is permissible, by the succession of weighted blowing up of the singular point p, $\mathcal{X}$ admits a semi-stable model $\mathcal{Y} \to \mathcal{X} \to \Delta$ without base change and a general fiber ([Ka], Theorm 1.3). Let Y be the central fiber of $\mathcal{Y}$. It is obvious that the existence of a $\mathbf{Q}$-Gorenstein smoothing of (X, p)

is the same as the existence of a smoothing of Y. Then $Y = \sum_{k=0}^{n} W_k$ where W_0 is a resolution of the germ (X,p) and W_k is a rational surface with $-(K_{W_k} + D_k)$ ample for $k = 1,\dots,n$. Since (W_0, D_0) is obtained by blowing up of (S,E), $H^2(\mathcal{T}_{W_0}(-D_0)) = 0$. The proof is obtained by Lemma 2.3 and 2.4. □

3. Compactfications of the space of double cover of $\mathbf{P}^2$ branched over curves of degree 4 and 6.

Hacking [Hac] considered the family of compactifications given by moduli space $\mathcal{P}_d^{\alpha}$ of log surfaces (X,B) such that $K_X + \alpha B$ is semi log canonical singularties and ample where $\frac{3}{d} < \alpha \le 1$. The compactification is simpler for lower α. He gave a compactification $\mathcal{P}_d$ of plane curves of degree d by allowable family of stable pairs of degree d.

Definition. A log surface (X,B) is called a stable pair of degree d if X is a proper connected surface, B effective Weil divisor with the following properties:
1. There is a $\epsilon > 0$ such that $K_X + (\frac{3}{d} + \epsilon)B$ is semi log canonical and ample.
2. $dK_X + 3B \sim 0$ ($\frac{d}{3}K_X + B \sim 0$ if $3|d$).
3. There is a $\mathbf{Q}$-Gorenstein smoothing (deformation $\mathcal{X}$ of X over a discrete valuation ring T with smooth general fiber such that $K_{\mathcal{X}/T}$, $\mathcal{B}/T$ are $\mathbf{Q}$-Cartier).

Define $\mathcal{P}_d(S) = \{(\mathcal{X},\mathcal{B})/S \mid \text{allowable family of stable pairs of degree d}\}$. The family $(\mathcal{X},\mathcal{B})/S$ is an allowable family if $\omega_{\mathcal{X}/S}^{[i]}$, $\mathcal{O}_{\mathcal{X}}(\mathcal{B})^{[i]}$ commute with base change for all i.

Theorem 3.1. [Hac] *1. $\mathcal{P}_d$ is a separated proper Deligne-Mumford stack.*
2. $\mathcal{P}_d$ is smooth if $3 \nmid d$.

Also the following results are obtained in [Hac] without difficulty by the definition of stable pair of degree d.

Corollary 3.2. [Hac] *Let (X,B) be a stable pair of degree d. Then*
1. $-K_X$ is ample,
2. $Index_p K_X \le d$,
3. Let $\pi : U' \to U$ be the index one cover. Then $\pi^* B$ is Cartier,
4. $H^1(B, \mathcal{O}_B(B)) = 0$.

In particular, if (X,B) is a stable pair of degree d then (X,B) satisfies Assumption in Section 2. In [KL], we proved that if D is a plane curve of degree d and $(\mathbf{P}^2, D)$ is a stable pair of degree d then D is a stable plane curve. The converse is also true under a suitable condition. These results make us possible to compare between the compactifications $\mathcal{P}_d$ and the compactifications of plane curves by Geometric Invariant Theory. In particular, we obtain the same classification of $\mathcal{P}_4$ as in [Hac] by construction of the stable pairs of degree 4 corresponding the semistable plane curves of degree 4.

Remark 3.3. Consider a plane curve $(f = 0)$ of degree 4 and let (X, B) be a stable pair of degree 4. By the result [KL], $X = \mathbf{P}^2$ if and only if B is a stable plane curve of degree 4. Hence we need only to consider the strictly semistable cases. The strictly semistable case is a double smooth conic or f's with a tacnode ([Mu], 1.12). The stable curve corresponding a double smooth conic is a hyperelliptic curve and its corresponding stable log surface is $\mathbf{P}(1,1,4)$ ([Has1], 4.1). Also the stable log surace corresponding f's with a tacnode is $\mathbf{P}(1,1,2) \cup_D \mathbf{P}(1,1,2)$ where $D = \mathbf{P}^1$ ([Has2], Section 5) and the singular point of $\mathbf{P}(1,1,2)$ is in D.

Lemma 3.4. *Let (X, B) be a stable pair of degree 4. Then there is a double cover Y of X branched over B.*

Proof. By Remark 3.3, we need to consider three case. Assume that $X = \mathbf{P}^2$ and B is a stable plane curve of degree 4. Then $B = 2L$ where L is a Cartier divisor. Assume that $X = \mathbf{P}(1,1,4)$. In this case X is obtained by contraction of -4-section C in the rational ruled surface $\mathbf{F}_4 = \mathbf{P}(\mathcal{O}_{\mathbf{P}^1} \oplus \mathcal{O}_{\mathbf{P}^1}(4))$ and the proper transform B' of B has no intersection with C. Let $\pi : \mathbf{F}_4 \to X = \mathbf{P}(1,1,4)$ be the contraction of C. Then $K_{\mathbf{F}_4} \sim -2C - 6F = \pi^* K_X - \frac{1}{2}C$ where F is a fiber of $\mathbf{F}_4 \to \mathbf{P}^1$, and $\pi^* B = B'$. Since (X, B) is a stable pair of degree 4,

$$\begin{aligned} 0 &\sim \pi^*(4K_X + 3B) \\ &= 4K_{\mathbf{F}_4} + 2C + 3B' \\ &= -6C - 24F + 3B'. \end{aligned}$$

Hence, we have $B' \sim 2C + 8F = 2(C + 4F)$, so $B' = 2L$ where L is a Cartier divisor.

Assume that $X = \mathbf{P}(1,1,2) \cup_D \mathbf{P}(1,1,2)$ where $D = \mathbf{P}^1$. Then each component $Z = \mathbf{P}(1,1,2)$ of X is obtained by the contraction of -2-section C in $\mathbf{F}_2$, and the proper transform B' of B has no intersection with C. Let $\pi : \mathbf{F}_2 \to Z = \mathbf{P}(1,1,2)$ be the contraction of C. Then $K_{\mathbf{F}_2} \sim -2C - 4F = \pi^* K_Z$ where F is a fiber of $\mathbf{F}_2 \to \mathbf{P}^1$, and $\pi^* B = B'$. Since (X, B) is a stable pair of degree 4,

$$\begin{aligned} 0 &\sim \pi^*(4K_Z + 4D + 3B) \\ &= 4K_{\mathbf{F}_2} + 4F + 2C + 3B' \\ &= -6C - 12F + 3B'. \end{aligned}$$

Hence, we have $B' \sim 2C + 4F = 2(C + 2F)$, so $B' = 2L$ where L is a Cartier divisor.

Since $B' = 2L$ for each case, there is a double cover Y' of $\mathbf{P}(1,1,4)$ or $\mathbf{P}(1,1,2) \cup_D \mathbf{P}(1,1,2)$ branched over B'. Then the double cover Y of X is obtained by contraction of curves correspoding negative sections. □

Lemma 3.5. *Let (X, B) be a stable pair of degree 6. Then there is a double cover Y of X branched over B.*

Proof. Since (X, B) is a stable pair of degree 6, $K_X + \frac{1}{2}B \sim 0$. So $\frac{1}{2}B \sim -K_X$, and it is a Cartier divisor on the index one cover. The double cover Y of X branched over B is obtained by consideration of the index one cover. □

If (X, B) be a stable pair of degree 4 or 6, then $(X, \frac{1}{2}B)$ has semi log canonical singularities. By Lemmas 3.4 and 3.5, there is a double cover Y of X branched over B. Then Y has also semi log canonical singularities ([Ko], Proposition 3.16), because $K_Y \equiv f^*(K_X + \frac{1}{2}B)$. Since the double curve cannot be a component of B, it is reduced to the case of log canonical singularities.

Theorem 3.6. *There are compactifications of the space of double cover of $\mathbf{P}^2$ branched over curves of degree 4 or 6.*

In a future work, we intend to classify the cases in $\mathcal{P}_6$ and intend to compare to this compactification $\mathcal{R}_6$ with Shah's compactification [S] by using Geometric Invariant Theory of plane curves of degree 6.

Acknowledgements

The author acknowledges the invitation and the hospitality of Research Institute for Mathematical Sciences, Kyoto University in the summer of 2001, where the work was started. He would like to thank Professors S. Mori and N. Nakayama for helpful discussions during his stay at RIMS. Also he would like to thank Organizing Committee for the invitation of the Conference of Algebraic Geometry in East Asia.

REFERNCES

[A] V. Alexeev, Boundness and K^2 for log surfaces, Internat. J. Math. **5** (1994), 779–810.

[AN] Y. Akizuki and S. Nakano, Note on Kodaira-Spencer's proof of Lefshetz's theorem, Proc. Jap. Acad. Ser. A **30** (1954), 266-272.

[C] A. Corti, Semistable 3-fold flips, Preprint.

[F] R. Friedman, Global smoothings of varieties with normal crossings, Ann. Math. **118** (1983), 75–114.

[Gi] D. Gieseker, Global moduli for surfaces of general type, Invent. Math. **43** (1977), 233–282.

[Go] R. Godement, *Topologie Algébrique et Théorie des Faisceau*, Hermann, Paris, 1958.

[Hac] P. HACKING, *A compactification of the space of plane curves*, Ph.D. Thesis, Cambridge University, 2001.

[Has1] B. HASSETT, Stable log surfaces and limits of quartic plane curves, Manuscripta Math. **100** (1999), 469-497.

[Has2] B. HASSETT, Local stable reduction of plane curve singularities, J. Reine Angew. Math. **520** (2000), 169-194.

[Ka] Y. KAWAMATA, Moderate degenerations of algebraic surfaces, *in Complex algebraic varieties Bayreuth 1990* **1507** of Lecture Notes in Math. Springer-Verlag, 1992.

[KL] H. KIM AND Y. LEE, Log canonical thresholds of semistable plane curves, Preprint.

[Ko] J. KOLLÁR, Singularity of pairs, Proc. Symp. Pure Math. **62.1** (1997), 221-287.

[K et] J. KOLLÁR ET AL, *Flips and abundance for algebraic threefolds*, Astérisque **211** 1992.

[KM] J. KOLLÁR AND S. MORI, *Birational geometry of algebraic varieties*, **134** Cambridge Tracts in Mathematics, 1998.

[K-SB] J. KOLLÁR AND SHEPHERD-BARRON, Threefolds and deformations of surface singularities, Invent. Math. **91** (1988) 299–338.

[L] Y. LEE, Bounds and **Q**-Gorenstein smoothings of smoothable stable log surfaces, To appear in Contemp. Math.

[Ma] M. MANETTI, Normal degenerations of the complex projective plane, J. Reine Angew. Math. **419** (1991), 89-118.

[Mu] D. MUMFORD, Stability of projective varieties, L'Ens. Math. **23** (1977), 39-110.

[R] Z. Ran, Deformation of maps, *in Algebraic curves and projective geometry (Trento 1988)*, **1389** of Lecture Notes in Math. Springer-Verlag, 1989.

[S] J. SHAH, A complete moduli space for K3 surfaces of degree 2, Ann. Math. **112** (1980), 485-510.

Department of Mathematics
Sogang University
Sinsu-dong, Mapo-gu
Seoul 121-742, Korea

Modular Curves and Some Related Issues

by Viet NguyenKhac [1]

Hanoi Institute of Mathematics

1991 Mathematics Subject Classification. Primary 11G30, 11G05, 11G09, 14H10.

Abstract

This is an extended exposition of a talk given by the author at the Symposium "Algebraic Geometry in East Asia" held in Kyoto on August last year. The purpose was to collect a unified account of recent results and problems on classical and Drinfel'd modular curves with regards to the d-gonality problem and various relations concerning (directly, or indirectly) strong uniform boundedness conjecture, torsion points, modular parametrization, Seshadri's constant, model-theoretic aspect, cryptology *etc.* (a large part of results presented here remained some time as folklore). We discuss also some open problems, related conjectures and perspective.

1. Introduction. The departure point of our investigation was a function field analogue of the so-called strong Uniform Boundedness Conjecture (abbreviated as sUBC). Let $\Phi(d)$ be the set of all isomorphism classes of finite abelian groups occurring as the full groups of torsion in the Mordell-Weil groups of elliptic curves over number fields K with absolute degree $[K : \mathbb{Q}] \leq d$. Then sUBC states that $\Phi(d)$ is finite. After long time efforts of many mathematicians (Yu. Manin, B. Mazur, S. Kamienny, M. Kenku, F. Momose, *etc.*) the problem was finally solved recently by L. Merel, and also by P. Parent with an effective estimate (*cf.* [7] and references therein).

About five years ago in collaboration with M.-H. Saito we succeeded in carrying out an algebraic proof of a function field analogue of sUBC that was reduced to study the problem of d-gonality of $X_0(N)$ (see below). As a result it appeared in Preprint Series of Kyoto University ([14]) and also in Duke e-print (alg-geom/9603024). *Post factum* D. Abramovich communicated to us an analytical approach (characteristic

[1] Institute of Mathematics, P.O.Box 631 Bo Ho, 10000 Hanoi, Vietnam
e-mail: socnhim@hn.vnn.vn, nkviet@thevinh.ncst.ac.vn.

zero case) essentially due to S.-T. Yau (*cf.* [24]) strikingly yielding a much sharper bound. Later we became aware of a paper by P. Zograf ([25]), in which the author firstly extends Yang-Yau result ([23]), giving an upper bound for the first non-zero eigenvalue of a compact Riemann surface admitting a meromorphic function of degree d, to the non-compact case. Furthermore P. Zograf proves a precise form of linear bound for the d-gonality problem of modular curves, based on Selberg's eigenvalue theorem. Henceforth we will quote subsequently this case of the theorem as the Yau-Zograf theorem.

Our results were reported on Algebraic Geometry Symposiums at Sendai, Kinosaki, at a meeting of MSJ, circulated among specialists and cited by various authors. Correspondences during this period allowed us to enlarge the paper, in particular this includes an asymptotically good bound (later carried out to the Drinfel'd modular case, see below), a relation with a conjecture proposed in [15], and a complete list of (non-p-)torsions of elliptic surfaces over a hyperelliptic base. It should be noted that the analytical proof of Yau-Zograf is true for congruence subgroup case. To proceed algebraically one has to involve the counting point "trick" often used by coding theorists. This trick also allows us to have a complete list of (non-p-)torsions of elliptic surfaces over trigonal bases. There is another approach using the class number first proposed in [12], and later completed by F. Momose ([11]) which turned out to be enough efficient in practice. There might be also a Seshadri constant approach to the d-gonality problem of modular curves which seems to the author quite closely related to the Yau-Zograf theorem. Most recently the d-gonality problem of $X_0(N)$ found an interesting application in solving the Coleman-Kaskel-Ribet conjecture for modular curves, independently by M. Baker, K. Ribet and A. Tamagawa.

On the other hand at the same time one of the authors of [14] realized that our approach is applicable to the case of Drinfel'd modular curves. It was reported on "International Algebraic Conference Dedicated to the Memory of A. G. Kurosh" ([13]). In this instance it poses several questions and related problems. Generally speaking it seems to the author of this paper that it is worth having a unified picture around the ideas exposed here.

2. Function field analogue of sUBC. Let d be a positive integer, and let C be a smooth projective curve defined over a field k of characteristic $p \neq 2, 3$. Set $K = k(C)$ the function field of C. We say that C is d-gonal, if there exists a finite morphism $f\colon C \longrightarrow \mathbb{P}^1_{\overline{k}}$ of degree d. For example, if a smooth curve C is 1-gonal, C is isomorphic to $\mathbb{P}^1_{\overline{k}}$, and if C is 2-gonal, then either $g(C) \leq 1$, or C is a hyperelliptic curve of genus $g(C) \geq 2$. If C is d-gonal, then there exists an extension of fields

$\overline{k}(\mathbb{P}^1_{\overline{k}}) = \overline{k}(t) \hookrightarrow \overline{k}(C)$ of degree d. Note that such a field extension may not be unique even if we assume that d is minimal.

Now let E be a non-constant elliptic curve defined over K (non-constancy means that its K/k-trace is trivial). We denote by

$$E(K)'_{tors} := \{x \in E(K)_{tors} : \ p \ \text{not dividing the order of} \ x\}$$

the non-p-part of the torsion subgroup of the Mordell-Weil group. Let $\Phi_{fun}(d)$ be the set of all isomorphism classes of finite abelian groups occurring as the full non-p-parts of torsions in the Mordell-Weil groups of non-constant elliptic curves over function field K with an extension $\overline{k}(\mathbb{P}^1) \hookrightarrow \overline{k}(C)$ of degree $\leq d$. Then the function field analogue of sUBC states that $\Phi_{fun}(d)$ *is finite.*

Assume the Mordell-Weil group $E(K)$ has a torsion element of order N. Then from the moduli theory of Deligne-Rapoport it follows that there exists a finite morphism $C \longrightarrow X_0(N)$ onto the modular curve $X_0(N)$, corresponding to the Hecke congruence subgroup, *i.e.* subgroup in the full modular group $PSL_2(\mathbb{Z})$, consisting of matrices $\begin{pmatrix} a & b \\ c & d \end{pmatrix}$ with c, divisible by N. Thus the problem is reduced to the problem of d-gonality of $X_0(N)$ in view of the following lemma.

Descending Lemma (see, e.g. [14], Lemma 1.3 for a proof valid in any characteristic). Let $C_1 \to C_2$ be a finite morphism between two smooth projective curves over an algebraically closed field. If C_1 is d-gonal, then C_2 also is d-gonal.

At present there are at least three approaches to d-gonality problem for modular curves: our algebraic approach generalizing Ogg's method from [17]; class number approach ([12], [11]); and the analytical approach in characteristic zero (Theorem of Yau-Zograf, [24], [25]).

3. Generalization of Ogg's method. It is well known (Igusa) that $X_0(N)$ can be defined over $\mathbb{Q}$ canonically and this model has good reduction outside $p \mid N$. So by taking reduction at $p \nmid N$, one can also consider $X_0(N)$ as the curve defined over $\mathbb{F}_p$. Let $\nu(N)$ denote the number of distinct prime divisors of N. It is known that

$$[PSL_2(\mathbb{Z}) : \Gamma_0(N)] = \phi(N) := N \prod_{p \mid N} (1 + 1/p).$$

We have the following theorem (*cf.* [14]) which gives a purely algebraic proof of function field analogue of sUBC proposed above.

Theorem 3.1. *Let $X_0(N)$ be the Hecke modular curve of level N, and let $k_0 = \mathbb{Q}$, or $\mathbb{F}_p$ with $p \nmid N$. Assume $X_0(N)_{\overline{k}_0}$ is d-gonal, i.e. if it admits a finite morphism $f\colon X_0(N)_{\overline{k}_0} \longrightarrow \mathbb{P}_{\overline{k}_0}$ of degree d. Then one has:*

A) *With $k_0 = \mathbb{Q}$:*

1) *If N is odd, then*

$$\frac{1}{12}\phi(N) + 2^{\nu(N)} \leq \begin{cases} 5d & \text{if } d = 1,2 \\ 4(d-1)^2 + 5 & \text{if } d \geq 3 \end{cases} \tag{3.1.1}$$

2) *If $3 \nmid N$, then*

$$\frac{1}{6}\phi(N) + 2^{\nu(N)} \leq \begin{cases} 10d & \text{if } d = 1,2 \\ 6(d-1)^2 + 10 & \text{if } d \geq 3 \end{cases} \tag{3.1.2}$$

3) *In general, as $d \to \infty$ we have the following asymptotic bound*

$$\phi(N) \leq O(d^2) \tag{3.1.3}$$

B) *If $k_0 = \mathbb{F}_p$, then*

$$\frac{p-1}{12}\phi(N) + 2^{\nu(N)} \leq \max\{p^2 + 1 + 2p\cdot(d-1)^2, (p^2+1)\cdot d\} \tag{3.1.4}$$

For the proof one proceeds as follows. On one hand by counting supersingular points on $X_0(N)_{\mathbb{F}_{p^2}}(\mathbb{F}_{p^2})$ ([17]) for $p \nmid N$ we have the folllowing lower bound for the number $\mathbb{F}_{p^2}$-rational points on $X_0(N)_{\mathbb{F}_{p^2}}(\mathbb{F}_{p^2})$

$$\# X_0(N)_{\mathbb{F}_{p^2}}(\mathbb{F}_{p^2}) \geq 2^{\nu(N)} + \frac{p-1}{12}\phi(N).$$

On the other hand for an upper bound we need the "Tower Theorem" (Theorem 3.2 below) with a subsequent application of Weil's bound.

Theorem 3.2 ([14], Theorem 2.1). *Let C be a smooth projective curve defined over a perfect field k, $f\colon C \to \mathbb{P}^1{}_{\overline{k}}$ a dominant morphism defined over $\overline{k}$ of degree d. Then there exist a smooth projective curve C' and a dominant morphism*

$$f'\colon C \to C'$$

of degree d' dividing d, both are defined over k, such that

$$g(C') \leq (d/d' - 1)^2 \tag{3.2.1}$$

Lemma 3.3. *There exists a tower of (smooth projective) curves over* $\overline{k}$

$$C \longrightarrow C_n \longrightarrow \cdots \longrightarrow C_2 \longrightarrow C_1 \longrightarrow C_0 \simeq \mathbb{P} \tag{3.3.1}$$

satisfying the following conditions:

(i) $h_i\colon C_i \longrightarrow C_{i-1}$ *and* $f_n\colon C \longrightarrow C_n$ *are morphisms defined over* $\overline{k}$, *of degrees* $e_i \geq 2$ *and* d_n *respectively.*

(ii) Putting $f_i\colon C \longrightarrow C_i$ *and* $d_i := \deg f_i$, *we have* $d = d_0 > d_1 > \cdots > d_n \geq 1$ *and for each* $0 \leq i \leq n$

$$g(C_i) \leq (d/d_i - 1)^2.$$

(iii) For every element $\sigma \in G_k = Gal(\overline{k}/k)$ *morphism*

$$f_n \times f_n^\sigma : C \longrightarrow C_n \times C_n^\sigma$$

has degree $d_n = \deg f_n$ *onto its image.*

Theorem 3.2 follows from Lemma 3.3 and a general theory of twisting (or Galois descent). In its turn to get Tower (3.3.1) from Lemma 3.3 (and also bound (3.2.1)) one uses the fundamental inequality of Castelnuovo-Severi. For completeness we formulate this inequality in the following form.

3.4. Inequality of Castelnuovo-Severi ([1], [9], V, and references therein). *Assume we are given two surjective morphisms* $\pi_i\colon C \longrightarrow C_i$, $i = 1, 2$ *between smooth projective curves* C, C_1, C_2 *of genera* g, g_1, g_2 *respectively. Assume further that* π_1 *and* π_2 *are independent in the sense that they do not factor via a morphism* $h\colon C \longrightarrow C'$ *(i.e. there doesn't happen that* $\pi_i = \pi_i' \circ h$, *for certain morphisms* $\pi_i'\colon C' \longrightarrow C_i$, $i = 1, 2$*) with* $\deg h > 1$. *Then*

$$g \leq d_1 g_1 + d_2 g_2 + (d_1 - 1)(d_2 - 1) \tag{3.4.1}$$

where $d_i := \deg \pi_i, i = 1, 2$. *Moreover in the case of equality in* (3.4.1) *all divisors* $\{\pi_{1*} \circ \pi_2^*(Q)\}_{Q \in C_2}$ *on* C_1 *(respectively* $\{\pi_{2*} \circ \pi_1^*(P)\}_{P \in C_2}$ *on* C_2*) are linearly equivalent (but not necessarily form a linear system).*

For the other formulations of this inequality on languages of correspondences, and also endomorphisms between the corresponding Jacobians, see *loc.cit.* and [15]. In this connection note a promising conjecture proposed in [15].

4. Class number approach. Let $X := X_0(N)$ and let $w := w_N$ be the main Atkin-Lehner involution defined by $\begin{pmatrix} 0 & -1 \\ N & 0 \end{pmatrix}$ which normalizes $\Gamma_0(N)$. It is a rational involution on X, so the quotient curve $X^+ := X/\langle w \rangle$ is also defined over $\mathbb{Q}$. The number of fixed points of w is given by Fricke's formula

$$s := s(N) = \delta_n h(-4N),$$

where $h(-4N)$ denotes the class number (of quadratic forms of discriminants $-4N$), and

$$\delta_n = \begin{cases} 2, & \text{if } n \equiv 7 \pmod 8 \\ \frac{4}{3}, & \text{if } n \equiv 3 \pmod 8 \\ 1, & \text{otherwise} \end{cases}$$

The idea is to apply the Castelnuovo-Severi inequality to the diagram with $\pi_1 = w$, $\pi_2 = f$ the d-gonal map $X \longrightarrow \mathbb{P}^1$. We shall analyze below our particular case of cyclic coverings from the linear system point of view, and as the reader can remark, linear system-theoretic argument is also an efficient way in treating Castelnuovo-Severi. Assume X possesses a (base-point free) g^1_d which is invariant under the action of w. It can be seen easily that there is a divisor $D_0 \in g^1_d$ such that $wD_0 = D_0$. Putting

$$\begin{aligned} &\mathfrak{L} := \mathfrak{L}(D_0) := \{f \in \mathcal{M}(X) : \ (f) + D_0 \geq 0\}, \\ &\mathfrak{L}_0 := \{f \in \mathfrak{L} : \ w^* f = f\}, \ \mathfrak{L}_1 := \{f \in \mathfrak{L} : \ w^* f = -f\}, \end{aligned}$$

and letting $d_i := \dim \mathfrak{L}_i$, $i = 0, 1$, one has $d_0 + d_1 = 2$. So there are two cases we have to consider: 1) $d_0 = d_1 = 1$ and 2) $d_0 = 2$, $d_1 = 0$.

From the covering $w \colon X \longrightarrow X^+$ it follows that there exists a function $y \in \mathcal{M}(X) \backslash w^* \mathcal{M}(X^+)$ such that 1) y generates $\mathcal{M}(X)$ over $w^*\mathcal{M}(X^+)$; 2) $w^* y = -y$ and

$$(y) = \sum_{j=1}^{s} P_j + w^{-1}(E_0^+)$$

where P_j are fixed points of w on X, and E_0^+ is a divisor on X^+. Furthermore one can write

$$D_0 = \sum_{j=1}^{s} \alpha_j P_j + w^{-1}(D_0^+)$$

with $\alpha_j = 0$, or 1, $\sum \alpha_j \equiv d(\bmod 2)$ and a divisor D_0^+ on X^+.

For $f \in \mathcal{L}_1$ we have $w^*(f/y) = f/y$, so $f/y \in w^*\mathcal{M}(X^+)$. A computation shows

$$D_f := (f) + D_0 = \sum_{j=1}^{s} \alpha_{1j} P_j + w^{-1}(D_f^+),$$

where D_f^+ is an effective divisor on X^+, and α_{1j} is the least residue of $\alpha_j + 1 (\bmod 2)$. One concludes that the linear system corresponding to $\mathcal{L}_i$ consists of all divisors of the form $\sum_{j=1}^{s} \alpha_{ij} P_j + w^{-1}(D^+)$, linearly equivalent to D_0 with D^+ effective on X^+ (here we make a convention $\alpha_{0j} := \alpha_j$).

On the other hand, for $i = 0, 1$ we have obviously

$$D_0 = \sum_{j=1}^{s} \alpha_{ij} P_j + w^{-1}(D_i^+) - (y^i),$$

where $D_1^+ = E_0^+ + D_0^+ - D'$ and D' is a divisor on X^+ supported at $\{P_j^+ := w(P_j)\}$. Thus

$$\deg D_i^+ = (d - \sum_{j=1}^{s} \alpha_{ij})/2.$$

From the above we see that $\dim |D_i^+| = 0$ if and only if $d_i = 0$. Moreover, if $\dim |D_i^+| \geq 1$, then one can fix an effective divisor $\overline{D}_i^+$ from this linear system, and an $\overline{f} := \sum_{j=1}^{s} \alpha_{ij} P_j + w^{-1}(\overline{D}_i^+) - D_0 \in \mathcal{L}_i$. The map $f^+ \longmapsto \overline{f} w^* f^+$ gives a precise isomorphism $\mathcal{L}(\overline{D}_i^+) \simeq \mathcal{L}_i$. Thus $d_i = \dim |D_i^+|$.

The case $d_0 = d_1 = 1$ implies both divisors D_i^+, $i = 0, 1$ are effective, and hence one gets

$$s \leq 2d \tag{4.1}$$

This also can be seen by using the fibre product argument as follows. If $f \colon X \longrightarrow \mathbb{P}^1$ is a d-gonal map $w^* f = -f$, then w induces a 2:1-covering $\pi \colon \mathbb{P}^1 \longrightarrow \mathbb{P}^1$ so that we obtain a commutative diagram

$$\begin{array}{ccc} \mathbb{P}^1 \times_{\mathbb{P}^1} X^+ & \xrightarrow{w} & X^+ \\ \Big\downarrow f & & \Big\downarrow f' \\ \mathbb{P}^1 & \xrightarrow{\pi} & \mathbb{P}^1 \end{array}$$

and X is a normalization of $\mathbb{P}^1 \times_{\mathbb{P}^1} X^+$, where f' is the induced d-gonal function on X^+ from Descending Lemma. Hence by adjunction formula we have

$$g(X) \leq p_a(\mathbb{P}^1 \times_{\mathbb{P}^1} X^+) = g(X^+) + d - 1.$$

Immediately it is nothing but (4.1) by Riemann-Hurwitz and Fricke's formulas. This is a particular case of Castelnuovo-Severi.

In fact we conjecture that this case never happens for modular curves $X_0(N)$ and d odd. If true it would give a fair solution of the trigonal classification problem. It should be noted that this problem was settled recently by M. Baker, Y. Hasegawa and M. Shimura with the aid of computer and algebraic method exposed in §3.

4.2. Example. The most "difficult" case in the Ogg's list [17] is $N = 37$. In this case $g(X_0(37)) = 2,\ g(X_0^+(37)) = 1, s(37) = 2\ (d = 2)$. We have actually $\deg D_0^+ = 1,\ \deg D_1^+ = 0$, and so $d_0 = d_1 = 1$.

Therefore the major case in this approach is the second case ($d_0 = 2,\ d_1 = 0$). In characteristic zero it can be done as follows (*cf.* [11]). Take a prime divisor $q \mid N$. Assume that on $X_0(N/q)$ we also have the second case. Take a non-constant $f \in \mathfrak{L}_0(X_0(N/q))$. Its pull-back via $\varphi\colon X_0(N) \longrightarrow X_0(N/q)$, for simplicity denoted by the same letter, belongs to $\mathfrak{L}_0(X_0(N))$. Thus f is induced by a function on the extended upper-half plane which is invariant under $\Gamma_0(N/q)$, $\begin{pmatrix} 0 & -1 \\ N/q & 0 \end{pmatrix}$, $\begin{pmatrix} 0 & -1 \\ N & 0 \end{pmatrix}$, and so under $\begin{pmatrix} q & 0 \\ 0 & 1 \end{pmatrix}$ too. Such a function should be constant: a contradiction. Hence by the first case on $X_0(N/q)$ one obtains: $s(N/q) \leq 2d$.

In positive characteristic class number approach does not give a full answer. At some part one has to combine with the algebraic approach of §3. We restrict ourselves to the following situation. In the notations above assume moreover $q \nmid N' := N/q$ (certainly characteristic of the base field $\mathrm{char}\,(k) \nmid N$ as assumed before). In any characteristic an elliptic curve E is said to be CM curve, if its endomorphism ring is strictly larger than $\mathbb{Z}$. In terms of pairs let E be an elliptic curve defined over k, and let $V_{N'},\ V_q$ be cyclic subgroups of order N', q respectively. Assuming the invariance of the action of $w_N,\ w_{N'}$ on both $\mathfrak{L}$ spaces of $X_0(N),\ X_0(N')$ we have the following commutative diagram

$$\begin{array}{ccc}
(E, V_q + V_{N'}) & \stackrel{w_N}{\longmapsto} & (E/V_q + V_{N'}, E[q]/V_q + E[N']/V_{N'}) \\
\varphi_* \downarrow & & \downarrow \varphi_* \\
(E, V_{N'}) & & (E/V_q + V_{N'}, E[N']/V_{N'}) \\
\downarrow \wr & & \downarrow w_{N'} \\
(E/V_q, V_q + V_{N'}/V_q) & \simeq & (E/V_q + E[N'], E'[N'] + V_{N'}/E[N'])
\end{array}$$

where $E' := E/V_q + V_{N'}$. Actually one can construct a tower $(E, V_{N'}) \longmapsto (E/V_q, V_q + V_{N'}/V_q) \longmapsto (E/V_{q^2}, V_{q^2} + V_{N'}/V_{q^2}) \longmapsto \cdots$. Thus E is a CM curve: a contradiction. So again by the first case one gets $s(N') \leq 2d$.

At this point one can also argue altenatively by using reduction modulo q of $X_0(N)$ *à la* Igusa-Deligne-Rapoport ([3]): the reduction of $X_0(N)$ at q consists of two components, isomorphic to $X_0(N')$, intersecting transversally at supersingular points, and the involution w_N interchanges these components.

4.3. Remark. The effective bound in the class number approach is not so good, since by a well-known theorem of Goldfeld-Gross-Zagier one gets just an exponential bound. Howerver a (non-effective) version of the Brauer-Siegel theorem gives (asymptotically) "almost" quadratic bound, so in practice this approach is also efficient.

5. Applications. As noted above in characteristic zero S.-T. Yau (unpublished, *cf.* [24]), and P. Zograf ([25]) independently found an analytical approach based on the Selberg eigenvalue theorem which treats all congruence subgroups and gives a linear bound for their indexes in $PSL_2(\mathbb{Z})$. More generally, let Γ be a discrete subgroup of $PSL_2(\mathbb{R})$ having a fundamental domain of finite area $A(\Gamma\backslash\mathbb{H})$ *w.r.t.* the Poincaré metric on the upper half-plane $\mathbb{H}$. The Laplacian $\Delta := -4(\mathrm{Im} z)^2 \partial^2/\partial z \partial \overline{z}$ acting on $\mathbb{H}$ admits a unique self-adjoint non-negative extension in $L_2(\Gamma \setminus \mathbb{H})$: the automorphic Laplacian which we denote by Δ_Γ. In the case $\Gamma \setminus \mathbb{H}$ is not compact, the spectrum of Δ_Γ consists of a continuous part - the semiaxis $[1/4, \infty)$, and a discrete part which we order as $0 = \lambda_0 < \lambda_1 \leq \cdots$.

5.1. Yau-Zograf theorem. *In the notation above assume that the Riemann surface $\overline{\Gamma \setminus \mathbb{H}}$ admits a meromorphic function of degree $d > 0$. Then we have either $A(\Gamma \setminus \mathbb{H}) < 32\pi d$, or*

$$\lambda_1 < \frac{8\pi d}{A(\Gamma \setminus \mathbb{H})}.$$

In the case Γ_0 is a discrete subgroup of $PSL_2(\mathbb{R})$ commensurable with the full modular group $PSL_2(\mathbb{Z})$, one gets

Corollary 5.2. *For any congruence subgroup Γ of Γ_0 such that the Riemann surface $\overline{\Gamma \setminus \mathbb{H}}$ admits a meromorphic function of degree d we have*

$$3\chi(\Gamma_0) \cdot [\Gamma_0 : \Gamma] < 64d.$$

This is because of Selberg's eigenvalue theorem for congruence subgroup stating that $\lambda_1 \geq 3/16$. Selberg's proof at some part relies also to Weil's bound on the number of points on curves over finite fields. In fact A. Selberg conjectured that $\lambda_1 \geq 1/4$ for congruence subgroups. At present an almost "half way" was made recently by W. Luo, Z. Rudnick and P. Sarnak: $\lambda_1 \geq 171/784$.

Since $\chi(PSL_2(\mathbb{Z})) = 1/6$, we come to the bound mentioned in the Introduction: the index of any congruence subgroup of $PSL_2(\mathbb{Z})$ is less than $128d$, where d is the degree of a non-constant meromorphic function on the Riemann surface $\overline{\Gamma \setminus \mathbb{H}}$. Although this bound turned out to be surprisingly sharp for large values d, in practice the algebraic methods carry much more arithmetic information for the classification problem, say with low values d. It remains to note a mysterious "coincidence": the apperance of Weil's bound in both approaches.

5.3. Modular parametrization. The famous modularity conjecture (now a theorem of Breuil-Conrad-Diamond-Taylor) asserts that every elliptic curve E defined over $\mathbb{Q}$ is modular, *i.e.* there is a modular parametrization $\psi\colon X_0(N) \longrightarrow E$. The degree of ψ called the modular degree plays an important role in the whole stuff here. From what we said above it follows that there exists a function $B(d)$ on d such that

$$\phi(N) < B(\deg \psi),$$

where $B(d)$ can be taken from either of the approaches above. This is somewhat opposite to the one conjectured by L. Szpiro: $\deg \psi << N^{2+\varepsilon}$.

5.4. On congruence subgroups. In comparision with the Yau-Zograf theorem a natural question arising here is whether one can proceed algebraically with congruence subgroups of $PSL_2(\mathbb{Z})$? The answer is yes. It is clear that to this end it suffices to deal with modular curve $X(N)$ corresponding to the principal congruence subgroup

$$\Gamma(N) := \left\{ \gamma = \begin{pmatrix} a & b \\ c & d \end{pmatrix} \in PSL_2(\mathbb{Z}) : \ \gamma \equiv Id \ (\mathrm{mod}\ N) \right\}$$

Next we have to involve the following coding-theoretic trick ([21]). Recall that the curve $X(N)$ has a model $X_P(N)$ defined over the field K_N which is the unique quadratic subfield of the cyclotomic field $\mathbb{Q}(\zeta_N)$. Hence one can consider reduction of $X(N)$ at a prime $\mathfrak{p}$ in K_N lying over $p \nmid N$. Thus we get a smooth projective geometrically irreducible curve $X_N := X_P(N)/\mathfrak{p}$. Since $k(\mathfrak{p}) \subseteq \mathbb{F}_{p^2}$, we see that the trick of counting points is similar to that in Ogg's method: all points of X_N lying over supersingular values of j-invariant are rational over $\mathbb{F}_{p^2}$ and the number of such points $= [PSL_2(\mathbb{Z}) : \Gamma(N)](p-1)/12$. Now the proof is going through along the ideas of §3. This trick allows us also to have easily a complete classification of torsions over trigonal bases. Note that such a classification in the case of hyperelliptic bases was given in [12].

5.5. On d-gonal version of Radermacher's conjecture. The original conjecture of Radermacher asserts that (over complex numbers) there are only finitely many congruence subgroups of genus 0 in the modular group $PSL_2(\mathbb{Z})$. J. Thomson first, and P. Zograf later with effective estimate (*see* [26] and references therein) have proved that the number of congruence subgroups of $PSL_2(\mathbb{Z})$ of a given genus is finite. The following statement is a d-gonal version of Radermacher's conjecture in any characteristic which follows immediately from the methods exposed above.

Proposition 5.5.1. *There are only finitely many congruence subgroups $\Gamma \supset \Gamma(N)$ of $PSL_2(\mathbb{Z})$ with N not divisible by a prime p and such that the reduction of $\Gamma \setminus \mathbb{H}$ at a prime divisor $\mathfrak{p}$ of p in K_N is d-gonal.*

Note that in treating trigonal, tertragonal, ..., or hyperelliptic-elliptic (*i.e.* low-gonal) modular curves the arithmetic-algebraic methods of §§3-4 have more advantages.

5.5.2. A semi-continuity type question. It asks whether the following statement is true: if $X_0(N)/p$ is d-gonal for all p from a set of primes of big density (say, near to 1), then $X_0(N)/\mathbb{Q}$ is d-gonal?

5.6. CKR conjecture. Let N be a prime ≥ 23, and let $X := X_0(N)$. We know that X has only two cusps $\{c_0, c_\infty\}$. Consider the standard Albanese morphism *w.r.t.* c_∞ embedding X into its Jacobian J, and put

$$\mathcal{S}_\infty := X(\overline{\mathbb{Q}}) \cap J_{tors}(\overline{\mathbb{Q}}).$$

In [2] the authors conjectured that if $P \in \mathcal{S}_\infty$, then either $P \in \{c_0, c_\infty\}$, or X is hyperelliptic and P is a branch point. More recently the conjecture was settled independently by M. Baker, K. Ribet and A. Tamagawa. Curiously enough, at some stage of the proof it requires involving a known list of low-gonal modular curves (*see* Baker's thesis 1999, and [20] for details).

5.7. On Seshadri's constant and lattices. This is an important constant in algebraic geometry, and quite often very difficult to be computed, or estimated. I will give just a vague insight which seems to me that it might be closely related to the approach based on Selberg's eigenvalue theorem. For simplicity I assume $k = \mathbb{C}$. Let X be a smooth complex projective variety, L an ample line bundle on X, and $x \in X$ a fixed point. Consider the blowing-up of X at x

$$f: Y := Bl_x(X) \longrightarrow X$$

with exceptional divisor $E := f^{-1}(x)$. Set

$$\varepsilon(L, x) := \sup\{\varepsilon \geq 0 : \ f^* c_1(L) - \varepsilon \cdot [E] \ \text{ is nef}\}.$$

Now let $A := \mathbb{C}^g/\Lambda$, and let (A, D) be a principally polarized abelian variety. The divisor D corresponds to the Hermitian form H on Λ. We define the Seshadri constant $\varepsilon(A, \Theta)$ as $\varepsilon(\mathcal{O}_A(\Theta), x)$ at a point $x \in A$ (this value is common for all points of A by homogeneity). Put

$$m(A, D) := \min_{x \in \Lambda - \{0\}} H(x, x).$$

R. Lazarsfeld [10] has proved for a d-gonal curve C of genus g we have the following inequalities

$$\frac{\pi}{4} m(J(C), \Theta) \leq \varepsilon(J(C), \Theta) \leq \frac{gd}{g + d - 1}.$$

As in §5.5 let Γ_0 be a discrete subgroup of $PSL_2(\mathbb{R})$ commensurable with $PSL_2(\mathbb{Z})$. Assume that C corresponds to a Fuchsian subgroup of the first kind of Γ_0. Then one may conjecture that $m(J(C), \Theta) \geq c \cdot \lambda_1 \cdot \chi(\Gamma_0) \cdot [\Gamma_0 : \Gamma]$ for an absolute constant c. If true, besides obvious application to the d-gonality problem one would have an interesting application to the lattice theory.

6. Drinfel'd modular curves. In this section let $\mathbb{F}_q$ be a finite field with q elements of characteristic p; and let $A := \mathbb{F}_q[T]$ be the polynomial ring of one variable with coefficients in $\mathbb{F}_q$ which can be considered as the ring of regular functions on the affine line $\mathbb{P}^1 - \{\infty\}$. We denote by $K := \mathbb{F}_q(T)$ the corresponding rational function field, $K_\infty := \mathbb{F}_q((T^{-1}))$ its completion at ∞; $\mathbb{C}_\infty$ the completed algebraic closure of K_∞. The fraktur letters $\mathfrak{a}, \mathfrak{b}, \ldots$ will denote ideals of A; $\mathfrak{m}, \mathfrak{n}, \ldots$ will denote monic elements of A, and also (principal) ideals they generate); in particular $\mathfrak{p}, \mathfrak{q}, \ldots$ will be prime elements (i.e., monic non-constant irreducible elements) of A.

The group $GL_2(K_\infty)$ acts on the Drinfel'd upper half-plane $\Omega := \mathbb{P}^1(\mathbb{C}_\infty) - \mathbb{P}^1(K_\infty) = \mathbb{C}_\infty - K_\infty$ by fractional linear transformations. For $\mathfrak{n} \in A$ we denote by

$\Gamma_0(\mathfrak{n})$ the Hecke congruence subgroup $\left\{ \begin{pmatrix} a & b \\ c & d \end{pmatrix} \in GL_2(A) : \ c \equiv 0 \bmod \mathfrak{n} \right\}$, and let

$$Y_0(\mathfrak{n}) := \Gamma_0(\mathfrak{n}) \setminus \Omega, \quad X_0(\mathfrak{n}) := \Gamma_0(\mathfrak{n}) \setminus \overline{\Omega}$$

be the affine Drinfel'd modular curve and its natural compactification obtained by adding to $Y_0(\mathfrak{n})$ a finite number of points (the so-called cusps), where $\overline{\Omega} := \Omega \cup \mathbb{P}^1(K)$. It is well known that the modular curve $Y_0(\mathfrak{n})$ parametrizes isomorphism classes of pairs (ϕ, λ), where ϕ is a Drinfel'd module (of rank 2) over $\mathbb{C}_\infty$ and λ is a cyclic $\mathfrak{n}$-isogeny of ϕ, or equivalently isomorphism classes of pairs (ϕ, γ), where ϕ is a Drinfel'd module over $\mathbb{C}_\infty$ and γ is a cyclic $\mathfrak{n}$-kernel of ϕ (for definitions *cf.* below).

There is a natural and well-known analogy between elliptic curves and Drinfel'd modules of rank 2. Let's discuss briefly here scheme-theoretic aspects of Drinfel'd modular curves (for details *see* [5], [22], [8], [6]). Let L be a field equipped with an A-algebra structure $\imath \colon A \to L$, and let $L\{\tau\} = \mathrm{End}_L(\mathbb{G}_a)$ be the endomorphism ring of the additive group scheme over L, where τ is the endomorphism of $\mathbb{G}_{a,L}$ given by $\tau \colon x \mapsto x^q$ (the Frobenius endomorphism over $\mathbb{F}_q$). This is a skew polynomial ring with the commutation rule $\tau.c = c^q.\tau$ for $c \in L$. A Drinfel'd module (of rank 2) over L is a ring homomorphism

$$\phi \colon A \to L\{\tau\}, \quad a \mapsto \phi_a := \phi(a) = \imath(a) + \sum_{i=1}^{2n} a_i \tau^i$$

with $a_{2n} \neq 0$, where $n = \deg(a)$, for all $a \in A \setminus \{0\}$. Clearly it is defined by $\phi_T = \imath(T) + c_1\tau + c_2\tau^2, \quad c_1, c_2 \in L, \quad c_2 \neq 0$. The element $j(\phi) = c_1^{q+1}/c_2$ is called the j-invariant of ϕ.

A homomorphism from the Drinfel'd module ϕ to the Drinfel'd module ψ is a $\lambda \in L\{\tau\}$ such that $\lambda.\phi_a = \psi_a.\lambda$, $\forall a \in A$. If $\lambda \neq 0$, then λ is called an isogeny. If in addition λ is invertible, i.e. $\in L\{\tau\}^* = L^*$, then we have the notion of an isomorphism between ϕ and ψ. The isomorphism class of a Drinfel'd module (over the algebraic closure $\overline{L}$) is well determined by its j-invariant.

For a non-zero ideal $\mathfrak{a}$ of A the scheme of $\mathfrak{a}$-torsion (or $\mathfrak{a}$-division) points of ϕ is the subscheme $G_\mathfrak{a} := \mathrm{Ker}(\phi_\mathfrak{a}) := \bigcap_{a \in \mathfrak{a}} \{\mathrm{Ker}(\phi_a)\}$ of $\mathbb{G}_{a,L}$ which is a flat finite group scheme of degree ${}^{\#}(A/\mathfrak{a})^2$. It is étale iff "char"$L := \mathrm{Ker}\, \imath \notin V(\mathfrak{a})$, where $V(\mathfrak{a}) := \{\mathfrak{p} \in \mathrm{Spec}\, A : \ \mathfrak{p} | \mathfrak{a}\}$.

Let now S be an A-scheme equipped with $\imath^* \colon S \to \mathrm{Spec}\, A$, and $\imath \colon A \to \Gamma(S, \mathcal{O}_S)$ is the canonical ring homomorphism. A (rank 2) Drinfel'd module $(\mathcal{L}, \phi)$ over S is a pair consisting of a line bundle $\mathcal{L}$ over S and a ring homomorphism

$$\phi\colon A \to \mathrm{End}_S(\mathcal{L}), \quad a \mapsto \phi_a = \imath(a) + \sum_{i=1}^{2n} \alpha_i(a)\tau^i$$

into the endomorphism ring of the additive group scheme $(\mathcal{L}, +)$, where $\tau^i\colon \mathcal{L} \to \mathcal{L}^{\otimes q^i}$ is the i-th power of the Frobenius endomorphism, $n = \deg(a)$, $\alpha_i(a) \in \Gamma(S, \mathcal{L}^{\otimes(1-q^i)})$, and $\alpha_{2n}(a)$ is nowhere vanishing on S, $\forall a \in A \setminus \{0\}$.

Morphisms are morphisms of the corresponding S-group schemes, compatible with the A-action. The morphism $\imath^*$ is called the charateristic of S (or sometimes also the characteristic of $(\mathcal{L}, \phi)$). The ideal $\mathfrak{a}$ is said to be prime to the characteristic of S, if $\imath^*(S) \cap V(\mathfrak{a}) = \emptyset$.

The subscheme $G_{\mathfrak{a}}$ of $\mathfrak{a}$-torsion points on $(\mathcal{L}, \phi)$ is defined as the subgroup scheme $\mathrm{Ker}(\phi_{\mathfrak{a}}) := \bigcap_{a\in\mathfrak{a}} \{\mathrm{Ker}(\phi_a)\}$ of $\mathbb{G}_{a,S}$, where the intersection here means the intersection (fiber product)of subschemes. The subgroup scheme $G_{\mathfrak{a}}$ is also finite and flat over S, of degree $^{\#}(A/\mathfrak{a})^2$. It is étale over S iff $\mathfrak{a}$ is prime to the characteristic of S. For simplicity we shall assume this condition in what follows.

A level-$\mathfrak{a}$-structure on $(\mathcal{L}, \phi)$ is an isomorphism of group schemes

$$l\colon (\mathfrak{a}^{-1}/A)^2 \times S \to G_{\mathfrak{a}}$$

over S, compatible with the A-action.

The functor $\mathcal{M}(\mathfrak{a})$ assigning to each A-scheme S the set of isomorphism classes of (rank 2) Drinfel'd modules over S with a level-$\mathfrak{a}$-structure is representable by an affine scheme $M(\mathfrak{a})$ of finite type over A. Fibres of the morphism $M(\mathfrak{a}) \to \mathrm{Spec}\ A$ over $\mathrm{Spec}\ A[\mathfrak{a}^{-1}] := \mathrm{Spec}\ A \setminus V(\mathfrak{a})$ are smooth curves, and empty over $V(\mathfrak{a})$. The scheme $M(\mathfrak{a})$ has a canonical compactification – a unique smooth, proper scheme $\overline{M(\mathfrak{a})}$ over $\mathrm{Spec}\ A[\mathfrak{a}^{-1}]$, containing $M(\mathfrak{a})$ as an open dense subscheme, such that fibres of the morphism $\overline{M(\mathfrak{a})} \to \mathrm{Spec}\ A[\mathfrak{a}^{-1}]$ are smooth complete curves.

A level-$\mathfrak{a}$-Hecke-structure is a cyclic subgroup scheme Z of $G_{\mathfrak{a}}$ with isomorphism of subgroup schemes

$$l_0\colon (\mathfrak{a}^{-1}/A) \times S \to Z$$

over S, compatible with the A-action.

The functor $\mathcal{M}_0(\mathfrak{a})$ assigning to each A-scheme S the set of isomorphism classes of (rank 2) Drinfel'd modules over S with a level-$\mathfrak{a}$-Hecke-structure is, in general, not representable, but it has a coarse moduli scheme $M_0(\mathfrak{a})$ and let $\overline{M_0(\mathfrak{a})}$ denote its natural compactification. The generic fibres of morphisms $M_0(\mathfrak{a}) \to \operatorname{Spec} A$ and $\overline{M_0(\mathfrak{a})} \to \operatorname{Spec} A[\mathfrak{n}^{-1}]$ are models defined over K of the curves $Y_0(\mathfrak{a})$ and $X_0(\mathfrak{a})$ respectively. In the case $\mathfrak{a} = \mathfrak{n}$ is a prime, one has the following description. The schemes $M(\mathfrak{n})$ and $\overline{M(\mathfrak{n})}$ can be equipped with a natural action of $GL_2(A/\mathfrak{n})$. If B is the subgroup of $GL_2(A/\mathfrak{n})$ consisting of upper-triangular matrices, then $M_0(\mathfrak{n}) := B \setminus M(\mathfrak{n})$, $\overline{M_0(\mathfrak{n})} := B \setminus \overline{M(\mathfrak{n})}$.

Writting

$$\mathfrak{n} = \prod_{i=1}^{s} \mathfrak{p}_i^{e_i}$$

with different primes $\mathfrak{p}_i$, we put

$$\varepsilon(\mathfrak{n}) := q^n \prod_{i=1}^{s} (1 + q^{-\deg(\mathfrak{p}_i)}),$$

where $n := \deg(\mathfrak{n})$. We shall be interested in Drinfel'd modules over extensions L of K and $A/(\mathfrak{p})$ for a prime $\mathfrak{p} \nmid \mathfrak{n}$, so the assumption on primality to the characteristic of L is satisfied, and the above can be interpreted as follows. The $\mathfrak{n}$-torsion points $\operatorname{Ker}(\phi_\mathfrak{n})(\overline{L})$ of ϕ is a free rank 2 $A/\mathfrak{n}$-module. Its $\varepsilon(\mathfrak{n})$ different rank 1 $A/\mathfrak{n}$-submodules are the cyclic $\mathfrak{n}$-kernels of ϕ. Such a kernel is L-rational, if as a set it is stable under the action of the absolute Galois group of L. A cyclic $\mathfrak{n}$-isogeny of ϕ is an $\lambda \in \overline{L}\{\tau\}$ with degree equal $\deg(\mathfrak{n})$ and $\operatorname{Ker}(\lambda)(\overline{L})$ is a cyclic $\mathfrak{n}$-kernel of ϕ. Clearly every cyclic $\mathfrak{n}$-kernel γ gives rise to a unique cyclic $\mathfrak{n}$-isogeny λ up to isomorphism, and one can choose λ in $L\{\tau\}$ (i.e., λ is L-rational) iff γ is L-rational. Thus we come to the modular interpretation of $Y_0(\mathfrak{n})$ noted above.

7. d-gonality problem. As in the case of $X_0(N)$ the curve $X_0(\mathfrak{n})$ has a canonical model defined over K which has good reduction mod $\mathfrak{p}$ for $\mathfrak{p} \nmid \mathfrak{n}$. So a similar consideration of $X_0(\mathfrak{n})$ mod two different primes $\mathfrak{p}, \mathfrak{q}$ not dividing $\mathfrak{n}$ would give us a bound for the degree $\deg(\mathfrak{n})$ by means of the d-gonality of $X_0(\mathfrak{n})$. As will be shown below one can obtain a bound $\deg(\mathfrak{n}) \asymp 4\log_q d$.

Like $X_0(N)$ the curve $X_0(\mathfrak{n})$ over finite fields, say $\mathbb{F}_{q^2}$, by taking reduction mod (T) provided $T \nmid \mathfrak{n}$, has at least 2^s rational cusps, and many supersingular points. Indeed take the supersingular Drinfel'd module ϕ_0 over $A/(T)$ defined by $\phi_0(T) := \tau^2$. The number of cyclic $\mathfrak{n}$-kernels of ϕ_0 is $\varepsilon(\mathfrak{n})$, and all of them are $\mathbb{F}_{q^2}$-rational, since $\phi_0(T)$ acts on these kernels as the Frobenius endomorphism over $\mathbb{F}_{q^2}$ (the generator of the absolute Galois group $\operatorname{Gal}(\overline{\mathbb{F}}_q/\mathbb{F}_{q^2})$). It is known also that the automorphism group of ϕ_0 is $\mathbb{F}^*{}_{q^2}$, and the action of $\mathbb{F}^*{}_q$ is trivial. So an orbit under the action of $\mathrm{Aut}(\phi_0)$ could contain at most $(q+1)$ kernels, and we find at least $\frac{\varepsilon(\mathfrak{n})}{q+1}$ points on $X_0(\mathfrak{n})$ mod (T) which are rational over $\mathbb{F}_{q^2}$. Hence we get

$$2^s + \frac{\varepsilon(\mathfrak{n})}{q+1} \leq d'(q^2 + g'q + 1) \tag{7.1}$$

where $d'|d$ and $g' \leq (d/d' - 1)^2$ as in Theorem 3.2.

It should be noted here that there might be another way for counting supersingular points like in the classical (coding-theoretic) case by using standard covering trick (but one should care about ramifications).

Furthermore as in §3 the descending arguments applied to the covering of curves $X_0(\mathfrak{n})_K \to X_0(\mathfrak{m})_K$ for $\mathfrak{m}|\mathfrak{n}$ allow us to decompose $\mathfrak{n} = \mathfrak{n}_1\mathfrak{n}_2$ with $T \nmid \mathfrak{n}_1$, $(T+1) \nmid \mathfrak{n}_2$. Applying (7.1) to $\mathfrak{n}_1$, $\mathfrak{n}_2$ respectively, we come to an upper bound for the $\deg(\mathfrak{n}) \asymp 4\log_q d$.

For an improved method following the idea with asymptotical bound (3.1.3) with applications to the case of Drinfel'd modules of rank 2 we refer the reader to [16]. In fact this gives us an asymptotically sharper bound $\deg(\mathfrak{n}) \asymp 2\log_q d$, which is enough good in practice. It should be noted that the case of hyperelliptic $X_0(\mathfrak{n})$ (an analogue of Ogg's result [17]) was treated in [19].

8. Miscellaneous remarks. Comparing parallelly with §§4-5 one can have a list of analogous problems for the Drinfel'd modular case. Here we mention only the most notable among them.

8.1. Analogue of the automorphic Laplacian. Its motivation is obviously because of the Yau-Zograf approach in the case of classical modular curves. This problem seems quite important (and very difficult) for a further development of the topics exposed in this paper.

8.2. Analogue of CKR conjecture. It is an interesting problem to study torsion points on $X_0(\mathfrak{n})$. We feel that it would be desirable to have a model-theoretic approach which could unify the whole picture here.

8.3. Cryptographic aspects. Elliptic curves now play a very important role in public key cryptographic applications. What can we say about Drinfel'd modules (especially of rank 2)? A Drinfel'd module analogue of RSA is based on the inversion problem for Drinfel'd modules. By this we mean the following: given a finite field

L of characteristic p, a Drinfel'd module $\phi\colon A \longrightarrow L\{\tau\}$, and an element $a \in A$ such that $\phi(a)\colon L \longrightarrow L$ is an isomorphism; find $b \in A$ such that $\phi(b)\colon L \longrightarrow L$ is the inverse for $\phi(a)$. Similarly a Drinfel'd module version of the Diffie-Hellman public key distribution system is as follows. Given an element $\zeta \in L$, a Drinfel'd module $\phi\colon A \longrightarrow L\{\tau\}$ as above (the public file). Alice generates randomly an $a_{\mathrm{Al}} \in A$. She sends to Bob the element $\phi(a_{\mathrm{Al}})(\zeta)$. Bob generates a random $a_{\mathrm{Bo}} \in A$, and he sends to Alice the element $\phi(a_{\mathrm{Bo}})(\zeta)$. Alice can compute $\phi(a_{\mathrm{Al}})(\phi(a_{\mathrm{Bo}}(\zeta))$, while Bob can compute $\phi(a_{\mathrm{Bo}})(\phi(a_{\mathrm{Al}}(\zeta))$. The common key then is $\phi(a_{\mathrm{Al}}a_{\mathrm{Bo}})(\zeta)$. Certainly Eve knows about $\phi(a_{\mathrm{Al}})(\zeta)$, $\phi(a_{\mathrm{Bo}})(\zeta)$ (and the public file). So she faces with the discrete logarithm problem for Drinfel'd modules: given $x, y \in L$, find $a \in A$ such that $\phi(a)(x) = y$. Most recently T. Scanlon ([18]) has shown that public key cryptosystems based on Drinfel'd modules are not secure (he treated essentially the case $A = \mathbb{F}_p[T]$, $L = \mathbb{F}_q$). However one would like to consider these stuffs from the geometric point of view and by analogy with the ECC. First one has to derive a good class of Drinfel'd modules (of rank 2), say with finite "Mordell-Weil rank". Next the point is how to choose a "good" prime $\mathfrak{p}$, and then consider their reduction modulo $\mathfrak{p}$. We hope to say a bit more details in the near future.

Acknowledgements. I would like to thank the organizers for the kind invitation and excellent organization of the Symposium. My thanks are due to M.-H. Saito for a long time and fruitful collaboration. I am indebted to F. Momose for inspiring discussions on class number approach. My thanks are also due to J. Chao for pointing out a recent paper ([18]) and helpful discussions on the topics.

REFERNCES

[1] Castelnuovo, G., *Sulle serie algebriche di gruppi di punti appartenenti ad una curva algabrica*, Rend. d. R. Acad. Lincei (5) **15** (1906), 337-344.

[2] Coleman, R., Kaskel, B., Ribet, K., *Torsions points on $X_0(N)$*, Proc. Symp. Pure Math., **66**-1 (1999), 27–49.

[3] Deligne, P., Rapoport, M. *Les schémas de modules de courbes elliptiques*, in *Modular Functions of One Variable* II, Lect. Notes in Math. 349, 1973, 143–316.

[4] Denis, L., *Hauters canonique et modules de Drinfeld*, Math. Ann., **294** (1992), 213–223.

[5] Drinfel'd, V. G., *Elliptic modules* I, Math. USSR Sb., **23** (1974), 561–592.

[6] *Drinfeld Modules, Modular Schemes and Applications*, (Ed.: Gekeler, E.-U., van der Put, M., Reversat, M., Van Geel, J.), World Scientific, 1997.

[7] Edixhoven, B., *Rational torsion points over number fields (after Kamienny and Mazur)*, Sém. Bourbaki 1993-1994, n^0 782, Astérisque, **227** (1995), 209–227.

[8] Gekeler, E.-U., *Über Drinfeld'sche modulkurven vom Hecke-typ*, Comp. Math., **57** (1986), 219–236.

[9] Hartshorne, R., *Algebraic Geometry*, Graduate Texts in Math. 52, Springer-Verlag, 1977.

[10] Lazarsfeld, R., *Lengths of periods and Seshadri constants of abelian varieties,* Math. Res. Letters, **3** (1996), 439–447.

[11] Momose, F., *Personal communication*, March 1996.

[12] NguyenKhac, V., *Class numbers, d-gonality of modular curves and bounding torsions*, Proceedings of Algebraic Geometry Symposium in Sendai, January 1996, 111–118.

[13] NguyenKhac, V., *Uniformly bounding torsions for Drinfel'd modules of rank 2,* Talk at the International Algebraic Conference Dedicated to the Memory of A. G. Kurosh, Moscow State University, May 27–31, 1998.

[14] NguyenKhac, V., Saito, M.-H., *d-gonality of modular curves and bounding torsions,* Kyoto-Math 96-07, Kyoto University, 1996, 16 p.

[15] NguyenKhac, V., Shioda, T., *On the Castelnuovo-Weil lattices* I, Preprint, 2000.

[16] NguyenKhac, V., Yamada, S.-i., *On d-gonality of Drinfel'd modular curves and strong uniform boundedness conjecture,* Proc. Japan Acad., **77** (2001), Ser. A, N^0 7, 126–129.

[17] Ogg, A. P., *Hyperelliptic modular curves,* Bull. Soc. Math. France, **102** (1974), 449–462.

[18] Scanlon, T., *Publis key cryptosystems based on Drinfeld modules are insecure,* J. Cryptology, 9 April 2001, online.

[19] Schweizer, A., *Hyperelliptic Drinfeld modular curves,* in [6], 330–343.

[20] Tamagawa, A., *Ramification of torsion points on curves with ordinary semistable Jacobian varieties,* Duke J. Math., **106** (2001), 281–319.

[21] Tsfasman, M. A., Vladuţ, S. G., *Algebraic-geometric codes,* Kluwer Acad. Publ., 1990.

[22] Vladuţ, S. G., Manin, Yu. I., *Linear codes and modular curves,* J. Soviet. Math., **30** (1985), 2611–2643.

[23] Yang, P. C., Yau, S.-T., *Eigenvalues of the Laplacian of compact Riemann surfaces and minimal submanifolds*, Ann. Scu. Norm. Sup. Pisa **8** (1980), 55–63.

[24] Yau, S.-T., *An application of eigenvalue estimate to algebraic curves defined by congruence subgroups*, Math. Res. Letters, **3** (1996), 167–172.

[25] Zograf, P. G., *Small eigenvalues of automorphic Laplacians in spaces of cusp forms*, Zapiski Nauchnyh Seminarov LOMI im. V. A. Steklova AN SSSR, **134** (1984), 157–168 (in Russian; English transl.: J. Soviet Math., v. **36** (1987), 106–114).

[26] Zograf, P. G., *A spectral proof of Radermacher's conjecture for congruence subgroups of the modular group,* J. Reine Angew. Math., **441** (1991), 113–116.

ON THE ASYMPTOTIC BEHAVIOR OF ADMISSIBLE VARIATIONS OF MIXED HODGE STRUCTURE

GREGORY PEARLSTEIN

Introduction

Let $f : X \to S$ be a complex analytic family of compact Kähler manifolds. Then, by virtue of the Hodge decomposition theorem, each fiber $\mathcal{H}_s = H^k(X_s, \mathbb{C})$ of the vector bundle $\mathcal{H} = R^k_{f*}(\mathbb{Q}) \otimes \mathcal{O}_S$ carries a pure, polarized Hodge structure of weight k. Accordingly, one obtains a C^∞-decomposition

$$\mathcal{H} = \bigoplus_{p+q=k} \mathcal{H}^{p,q}$$

of $\mathcal{H}$ into a sum of C^∞ subbundles $\mathcal{H}^{p,q}$ with fiber $H^{p,q}(X_s)$ over the point $s \in S$.

Alternatively, such variations of Hodge structure can be studied in terms of the associated period map

$$\varphi : S \to \mathcal{D}/\Gamma$$

obtained by parallel translation of the data of $\mathcal{H}$ to a fixed reference fiber H via the Gauss-Manin connection of the underlying local system $\mathcal{H}_{\mathbb{Q}} = R^k_{f*}(\mathbb{Q})$. In particular, given a family of projective varieties $f : \bar{X} \to \bar{S}$ for which the generic fiber is smooth, the construction described above defines a variation of Hodge structure over the complement of the critical locus of f.

Replacing the family of smooth projective varieties considered above by a surjective, quasi-projective morphism $f : \bar{X} \to \bar{S}$, one obtains a variation of graded-polarized mixed Hodge structure. In this paper, we discuss some of our recent work regarding the asymptotic behavior of such degenerations of mixed Hodge structure.

Our presentation is organized as follows: In §1, we recall the definition of a variation of Hodge structure and discuss Schmid's Nilpotent Orbit Theorem [11]. In §2, we review the notion of a variation of graded-polarized mixed Hodge structure

I wish to thank S. Usui and A. Ohbuchi for organizing this very fine conference, and for their hospitality and generous support during my stay in Japan. I also thank the referee for his many helpful comments and suggestions.

and discuss the geometry of the associated classifying spaces $\mathcal{M}$. In §3, we recall the notion of an admissible variation of graded-polarized mixed Hodge structure following [14], and discuss a theorem of Deligne [4] [9] [7] [12] which shows how to attach a distinguished representation of $sl_2(\mathbb{C})$ to the limiting data of an admissible variation of graded-polarized mixed Hodge structure. In §4, we prove an analog [9] of Schmid's Nilpotent Orbit Theorem for admissible variations of graded-polarized mixed Hodge structure $\mathcal{H} \to \Delta^*$. In §5, we announce [10] an analog of Schmid's SL_2 Orbit Theorem for a special class of nilpotent orbits of graded-polarized mixed Hodge structure, and outline the proof.

§1. Degenerations of Hodge Structure

Let $f : X \to S$ be a family of compact Kähler manifolds and

$$\mathcal{H} = \bigoplus_{p+q=k} \mathcal{H}^{p,q} \tag{1.1}$$

denote the associated decomposition of $\mathcal{H} = R^k_{f*}(\mathbb{Q}) \otimes \mathcal{O}_S$ discussed in §1. In order to replace the decomposition (1.1) by a system of holomorphic objects, recall that the category of pure Hodge structures of weight k defined over a given noetherian subring A of $\mathbb{R}$ such that $A \otimes \mathbb{Q}$ is a field is equivalent to the category of finitely generated A-modules H_A equipped with a decreasing filtration F of $H_{\mathbb{C}} = H_A \otimes \mathbb{C}$ by complex subspaces such that

$$F^p \oplus \overline{F^{k-p+1}} = H_{\mathbb{C}} \tag{1.2}$$

for each index p. Indeed, given a pure Hodge structure of weight k, one defines

$$F^p = \bigoplus_{r \geq p} H^{r,k-r} \tag{1.3}$$

Conversely, given a filtration F which satisfies equation (1.2) for each index p, one defines $H^{p,q} = F^p \cap \bar{F}^q$.

Remark. A polarization of a pure Hodge structure H of weight k consists of a $(-1)^k$-symmetric bilinear form $Q : H_A \otimes H_A \to A$ such that (i) $Q(F^p, F^{k-p+1}) = 0$ for each index p; and (ii) $i^{p-q}Q(v, \bar{v}) > 0$ for any non-zero element $v \in H^{p,q}$.

Applying equation (1.3) pointwise to the fibers of $\mathcal{H}$, one obtains a decreasing filtration $\mathcal{F}$ of $\mathcal{H} = R^k_{f*}(\mathbb{Q}) \otimes \mathcal{O}_S$ by holomorphic subbundles, subject to the following horizontality condition

$$\nabla(\mathcal{F}^p) \subseteq \mathcal{F}^{p-1} \otimes \Omega^1_S \tag{1.4}$$

which reflects the fact that the Gauss–Manin connection ∇ acts on $\mathcal{H}$ as cup product with the associated Kodaira–Spencer map.

Definition. Let S be a complex manifold, and A be a noetherian subring of $\mathbb{R}$ such that $A \otimes \mathbb{Q}$ is a field. Then, a variation of pure polarized A-Hodge structure over S consists of a local system of finitely generated A-modules $\mathcal{H}_A$ over S equipped with a decreasing filtration $\mathcal{F}$ of $\mathcal{H} = \mathcal{H}_A \otimes \mathcal{O}_S$ by holomorphic subbundles and a flat, non-degenerate bilinear form $Q : \mathcal{H}_A \otimes \mathcal{H}_A \to A$ which satisfy the following two conditions:

(a) $\nabla(\mathcal{F}^p) \subseteq \mathcal{F}^{p-1} \otimes \Omega^1_S$ for each index p.
(b) For each point $s \in S$, the triple $((\mathcal{H}_A)_s, \mathcal{F}_s, Q_s)$ determines a pure, polarized Hodge structure of weight k defined over A.

Now, as noted in the introduction, such a variation of Hodge structure is equivalent to a holomorphic, horizontal, locally liftable map

$$\varphi : S \to \mathcal{D}/\Gamma \tag{1.5}$$

from S into the quotient of a suitable classifying space $\mathcal{D}$ by the action of the monodromy group Γ of $\mathcal{H}$ via parallel translation of the data of $\mathcal{H}$ to a fixed reference fiber $H = \mathcal{H}_{s_0}$.

Lemma 1.6 [5]. *The classifying space $\mathcal{D}$ appearing in equation* (1.5) *is a complex manifold upon which the real group $G_{\mathbb{R}} = Aut_{\mathbb{R}}(Q)$ acts transitively by biholomorphisms.*

Sketch of Proof. The first step is to properly define $\mathcal{D}$. Namely, $\mathcal{D}$ is to consist of all decreasing filtrations F of the ambient vector space $H_{\mathbb{C}}$ such that $(H_{\mathbb{R}}, F, Q)$ is a pure, polarized Hodge structure of weight k, such that $\dim F^p = \dim \mathcal{F}^p$ for each index p.

To endow the set $\mathcal{D}$ with a complex structure, one then exhibits $\mathcal{D}$ as an open subset of a complex homogeneous space $\check{\mathcal{D}}$ upon which the Lie group $G_{\mathbb{C}} = Aut_{\mathbb{C}}(Q)$ acts transitively by biholomorphisms. Namely, a decreasing filtration F of $H_{\mathbb{C}}$ belongs to $\check{\mathcal{D}}$ if and only if (a) $\dim F^p = \dim \mathcal{F}^p$; and (b) $Q(F^p, F^{k-p+1}) = 0$.

Example 1.7. Let $k = 1$, $H_{\mathbb{Z}} = \mathbb{Z}e_1 \oplus \mathbb{Z}e_2$, $Q(e_1, e_2) = 1$ and $f^p := \dim(F^p) = 2-p$ for $p = 0, 1, 2$. Then, the associated classifying space $\mathcal{D}$ is the upper half plane U upon which the Lie group $G_{\mathbb{R}} = SL_2(\mathbb{R})$ acts transitively via fractional linear transformations.

In 1973, W. Schmid proved the following result, which shows the period map of a variation of Hodge structure $\mathcal{H} \to \Delta^*$ defined over a punctured disk Δ^* to be asymptotic to a period map φ_{nilp} of a special type, called a nilpotent orbit:

Definition (Nilpotent Orbit). A nilpotent orbit of pure, polarized Hodge structure (modeled on $\mathcal{D}$) consists of a filtration $F \in \check{\mathcal{D}}$ and an element $N \in Lie(G_{\mathbb{R}})$ such that:

(i) $N(F^p) \subseteq F^{p-1}$ for each index p

(ii) There exists a positive constant α such that $\mathrm{Im}(z) > \alpha \implies e^{zN}.F \in \mathcal{D}$.

Nilpotent Orbit Theorem [11]. *Let $\mathcal{H} \to \Delta^*$ be a variation of pure, polarized Hodge structure with unipotent monodromy, and $F(z) : U \to \mathcal{D}$ be a lifting of the period map $\varphi(s) : \Delta \to \mathcal{D}/\Gamma$ of $\mathcal{H}$ to the upper half-plane U which makes the following diagram commute*

$$\begin{array}{ccc} U & \xrightarrow{F} & \mathcal{D} \\ {\scriptstyle s=\exp(2\pi i z)}\big\downarrow & & \big\downarrow \\ \Delta^* & \xrightarrow{\varphi} & \mathcal{D}/\Gamma \end{array}$$

Then, the map $\psi(z) = e^{-zN}.F(z)$ satisfies the periodicity condition $\psi(z+1) = \psi(z)$, and hence descends to a map $\psi : \Delta^* \to \check{\mathcal{D}}$. Moreover,

1. *The limiting Hodge filtration $F_\infty = \lim_{s\to 0} \psi(s)$ exists as an element of $\check{\mathcal{D}}$.*
2. *The pair (F_∞, N) is a nilpotent orbit.*
3. *There exists constants α, β and K such that $\mathrm{Im}(z) > \alpha \implies e^{zN}.F_\infty \in \mathcal{D}$, and*
$$d(F(z), e^{zN}.F_\infty) < K\mathrm{Im}(z)^\beta e^{-2\pi \mathrm{Im}(z)}$$
relative to any $G_{\mathbb{R}}$-invariant hermitian distance on $\mathcal{D}$.

Remark. To be coordinate free, the limiting Hodge filtration F_∞ constructed above should be viewed as an object attached to $T_0^*(\Delta)$.

In particular, as a consequence of Schmid's nilpotent orbit theorem, the problem of classifying the local monodromy of a variation of Hodge structure reduces to one of classifying all possible nilpotent orbits (F, N) modeled on a given classifying space $\mathcal{D}$. This in turn, is accomplished by Schmid's SL_2 Theorem, which asserts that every nilpotent orbit $e^{zN}.F_\infty$ is asymptotic to a nilpotent orbit $e^{zN}.\hat{F}_\infty$ which arises via a representation of $SL_2(\mathbb{R})$.

More precisely, recall that given a nilpotent endomorphism N of a finite dimensional vector space V, there exists a unique increasing monodromy weight filtration $W(N)$ of V such that for each index j:

(i) $N(W_j(N)) \subseteq W_{j-2}(N)$
(ii) N^j induces an isomorphism $W_j(N)/W_{j-1}(N) \to W_{-j}(N)/W_{-j-1}(N)$.

Then, as a consequence of Schmid's SL_2 Orbit Theorem, one has the following result:

Theorem 1.8 [11]. *Let (F, N) be a nilpotent orbit modeled on a classifying space $\mathcal{D}$ of pure Hodge structures of weight k. Then, for each index ℓ, F induces a* *pure Hodge structure of weight ℓ on each quotient $Gr^W_\ell = W_\ell/W_{\ell-1}$ of the shifted monodromy weight filtration $W_j = W(N)[-k]_j = W(N)_{j-k}$ via the rule*

$$F^p Gr^W_\ell = \frac{F^p \cap W_\ell + W_{\ell-1}}{W_{\ell-1}}$$

In particular, as consequence of Theorem (1.8), the limiting Hodge filtration F_∞ of a variation of pure, polarized Hodge structure $\mathcal{H} \to \Delta^*$ pairs with the corresponding monodromy weight filtration $W(N)[-k]$ to define a mixed Hodge structure, with respect to which N is a morphism of type $(-1,-1)$:

Definition (Mixed Hodge Structure). Let A be a noetherian subring of $\mathbb{R}$ such that $A \otimes \mathbb{Q}$ is a field. Then, a mixed Hodge structure (F, W) defined over A consists of a finitely generated A-module V_A equipped with an increasing filtration $W(V_A \otimes \mathbb{Q})$ of $V_A \otimes \mathbb{Q}$ together with a decreasing filtration F of $V_{\mathbb{C}} = V_A \otimes \mathbb{C}$ by complex subspaces such that F induces pure Hodge structure of weight k on each quotient space $Gr^W_k = W_k/W_{k-1}$, $W_k = W_k(V_A \otimes \mathbb{Q}) \otimes \mathbb{C}$, via the rule

$$F^p Gr^W_k = \frac{F^p \cap W_k + W_{k-1}}{W_{k-1}}$$

Likewise, a (-r,-r)-morphism of a mixed Hodge structure (F, W) is an A-linear map $L : V_A \to V_A$ such that $L(F^p) \subseteq F^{p-r}$ and $L(W_k) \subseteq W_{k-2r}$ for each pair of indices p and k.

§2. Variations of Mixed Hodge Structure

Let X be a complex algebraic variety. Then, by a theorem of Deligne [2], the cohomology groups $H^k(X, \mathbb{C})$ of X carry functorial mixed Hodge structures.

Example 2.1. Let M be a compact Riemann surface and S be a finite set of distinct points in M. Then, the associated mixed Hodge structure on $H^1(M \setminus S, \mathbb{C})$ is given by the following pair of filtrations:

$$W_0 = 0, \qquad W_1 = H^1(M, \mathbb{C}), \qquad W_2 = H^1(M \setminus S, \mathbb{C})$$
$$F^2 = 0, \qquad F^1 = \Omega^1(S), \qquad F^0 = H^1(M \setminus S, \mathbb{C})$$

where $\Omega^1(S)$ is the set of meromorphic 1-forms on M which are holomorphic on $M \setminus S$ and have at worst simple poles along S.

The graded quotients Gr^W_k of the mixed Hodge structure considered in Example 2.1 are polarized by the bilinear forms:

$$Q_2(\alpha, \beta) = 4\pi^2 \sum_{p \in S} \mathrm{Res}_p(\alpha)\mathrm{Res}_p(\beta), \qquad Q_1(\alpha, \beta) = \int_M \alpha \wedge \beta$$

More generally, a collection of bilinear forms $Q = \{Q_k\}$ is said to be a graded-polarization of the mixed Hodge structure (F, W) provided that for each index k, the pair (FGr_k^W, Q_k) is a pure, polarized Hodge structure of weight k.

In analogy with the pure case, the definition of a variation of graded-polarized mixed Hodge structure is as follows:

Definition. Let S be a complex manifold. Then, a variation of graded-polarized mixed Hodge structure $\mathcal{V} \to S$ consists of a $\mathbb{Q}$-local system $\mathcal{V}_{\mathbb{Q}}$ over S equipped with

(1) a rational, increasing weight filtration $\mathcal{W} = \mathcal{W}(\mathcal{V}_{\mathbb{Q}}) \otimes \mathbb{C}$ of $\mathcal{V}_{\mathbb{C}} = \mathcal{V}_{\mathbb{Q}} \otimes \mathbb{C}$;
(2) a decreasing filtration $\mathcal{F}$ of $\mathcal{V} = \mathcal{V}_{\mathbb{C}} \otimes \mathcal{O}_S$ by holomorphic subbundles;
(3) a collection of bilinear forms $Q_k : Gr_k^{\mathcal{W}}(\mathcal{V}_{\mathbb{Q}}) \otimes Gr_k^{\mathcal{W}}(\mathcal{V}_{\mathbb{Q}}) \to \mathbb{Q}$ of alternating parity $(-1)^k$;

such that

(a) $\mathcal{F}$ is horizontal with respect to the Gauss-Manin connection ∇ of $\mathcal{V}$, i.e. $\nabla(\mathcal{F}^p) \subseteq \mathcal{F}^{p-1} \otimes \Omega^1_S$;
(b) for each index k, the triple $(Gr^{\mathcal{W}}(\mathcal{V}_{\mathbb{Q}}), \mathcal{F}Gr_k^{\mathcal{W}}, Q_k)$ is a variation of pure, polarized Hodge structure of weight k.

Example 2.2. Let $X \to S$ be a family of smooth, complex algebraic curves punctured at $k \geq 2$ points. Then, the fibers $H^1(X_s, \mathbb{C})$ patch together to form a variation of graded-polarized mixed Hodge structure $\mathcal{V} \to S$.

As in the pure case, a variation of graded-polarized mixed Hodge structure $\mathcal{V} \to S$ is equivalent to a holomorphic, horizontal, locally liftable map

$$\varphi : S \to \mathcal{M}/\Gamma \tag{2.3}$$

from S into the quotient of a suitable classifying space $\mathcal{M}$ of graded-polarized mixed Hodge structure by action of the monodromy group Γ of $\mathcal{V}$ via parallel translation to a fixed reference fiber $V = \mathcal{V}_{s_o}$. Moreover, the classifying space $\mathcal{M}$ is a complex manifold upon which a real Lie group G acts transitively by biholomorphisms:

Theorem 2.4 [8], [15]. *The classifying space $\mathcal{M}$ appearing in equation* (2.3) *is a complex manifold upon which the group*

$$G = \{\, g \in GL(V)^W \mid Gr(g) \in Aut_{\mathbb{R}}(Q) \,\}$$

acts transitively by automorphisms, where $W = \mathcal{W}_{s_o}$ and $Gr(g) : Gr^W \to Gr^W$ denotes the map induced by an element of $GL(V)^W = \{\, g \in GL(V) \mid g(W_k) \subseteq W_k \,\}$ on Gr^W.

Sketch of Proof. As in the proof of Theorem (1.6), the first step is to properly define the classifying space $\mathcal{M}$. Namely, $\mathcal{M}$ is to consist of all decreasing filtrations F of V such that:

(i) (F, W) is a mixed Hodge structure which is graded-polarized by Q;
(ii) $\dim F^p Gr^W_k = \dim \mathcal{F}^p Gr^{\mathcal{W}}_k$ for each pair of indices p and k.

Likewise, as in the proof of Theorem (1.6), in order to endow the set $\mathcal{M}$ with the structure of a complex manifold, one exhibits $\mathcal{M}$ as an open subset of a dual space $\check{\mathcal{M}}$ upon which the complex Lie group

$$G_{\mathbb{C}} = \{\, g \in GL(V)^W \mid Gr(g) \in Aut_{\mathbb{C}}(Q) \,\}$$

acts transitively. Namely, a decreasing filtration F of V belongs to $\check{\mathcal{M}}$ if and only if (a) $\dim F^p = \dim \mathcal{F}^p$; (b) $Q(F^p Gr^W_k, F^{k-p+1} Gr^W_k) = 0$; and (c) $\dim F^p Gr^W_k = \dim \mathcal{F}^p Gr^{\mathcal{W}}_k$.

Example 2.5. Let $\mathcal{M}$ be the classifying space of graded-polarized mixed Hodge structure defined by the following data: (a) $\mathbb{Q}$-structure $V_{\mathbb{Q}} = \mathbb{Q}e_0 \oplus \mathbb{Q}e_1$; (b) Hodge numbers $h^{1,1} = h^{0,0} = 1$; (c) weight filtration $W_0 = W_1 = \mathbb{Q}e_0$, $W_2 = V$; (d) graded-polarizations $Q_{2j}([e_j], [e_j]) = 1$. Then, $\mathcal{M}$ is isomorphic to $\mathbb{C}$ via the map $F^1(z) = \mathrm{span}(e_1 + ze_0)$, and $F^2(z) = 0$, $F^0(z) = V$.

Remark. The classifying spaces $\mathcal{M}$ introduced above do not parametrize extension classes of graded-polarized mixed Hodge structure. Indeed, under the isomorphism given in Example 2.5, $(F(z), W) \equiv (F(z+n), W)$ for every $n \in \mathbb{Z}$.

In outline form, the geometry of the classifying spaces $\mathcal{M}$ defined above is as follows: As in the pure case, the fundamental vector bundles attached to $\mathcal{M}$ are the Hodge bundles

$$\mathcal{I}^{p,q} \to \mathcal{M} \tag{2.6}$$

defined by pointwise application the following theorem:

Theorem 2.7 [2]. *Let (F, W) be a mixed Hodge structure. Then, there exists a unique, functorial decomposition*

$$V = \bigoplus_{p,q} I^{p,q} \tag{2.8}$$

of the underlying vector space V such that:

(i) $F^p = \oplus_{r \geq p} I^{r,s}$ *for each index* p*;*
(ii) $W_k = \oplus_{r+s \leq k} I^{r,s}$ *for each index* k*;*
(iii) $\overline{I^{p,q}} = I^{q,p} \mod \oplus_{r<q,\, s<p} I^{r,s}$ *for each bi-index* (p, q).

In particular, a mixed Hodge structure (F, W) is said to be split over $\mathbb{R}$ if the associated bigrading (2.8) satisfies the symmetry condition

$$\overline{I^{p,q}_{(F,W)}} = I^{q,p}_{(F,W)} \tag{2.9}$$

for each bi-index (p, q).

Theorem 2.10 [6], [8]. *Let $\mathcal{M}$ be a classifying space upon which the Lie group G acts transitively. Then, the subset $\mathcal{M}_{\mathbb{R}}$ consisting of all points $F \in \mathcal{M}$ such that (F, W) is split over $\mathbb{R}$ is a C^∞ submanifold of $\mathcal{M}$ upon which the Lie group*

$$G_{\mathbb{R}} = G \cap GL(V_{\mathbb{R}})$$

acts transitively.

Moreover, as a consequence of the following result, there exists a natural C^∞ fibration $\mathcal{M} \to \mathcal{M}_{\mathbb{R}}$:

Theorem 2.11 [1]. *Let (F, W) be a mixed Hodge structure. Then, there exists a unique real element δ in*

$$\Lambda^{-1,-1}_{(F,W)} = \{\, \alpha \in gl(V) \mid \alpha(I^{p,q}_{(F,W)}) \subseteq \bigoplus_{r<p,\, s<q} I^{r,s}_{(F,W)} \ \forall\, (p,q) \,\} \tag{2.12}$$

such that $(e^{-i\delta}.F, W)$ is split over $\mathbb{R}$. Moreover, δ commutes with every morphism α of (F, W) [including morphisms of type $(-k, -k)$].

In analogy with the pure case, the classifying space $\mathcal{M}$ carries a natural mixed Hodge metric h defined by pointwise application of the following construction:

Lemma 2.13 [6], [9]. *Let (F, W) be a mixed Hodge structure. Then, given any graded-polarization Q of (F, W), there exists a unique hermitian inner product h on the underlying vector space V such that*

(i) *The decomposition $V = \oplus_{p,q} I^{p,q}_{(F,W)}$ is orthogonal;*
(ii) *If u and v belong to $I^{p,q}_{(F,W)}$, then $h(u, v) = i^{p-q} Q_{p+q}([u], [\bar{v}])$.*

More precisely, let $\mathcal{M}$ be a classifying space of graded-polarized mixed Hodge structure and $\mathbf{g}_{\mathbb{C}} = Lie(G_{\mathbb{C}})$. Then, each point $F \in \mathcal{M}$ induces a mixed Hodge structure $(F^{\cdot}\mathbf{g}_{\mathbb{C}}, W.\mathbf{g}_{\mathbb{C}})$ on $\mathbf{g}_{\mathbb{C}}$ and hence a direct sum decomposition

$$\mathbf{g}_{\mathbb{C}} = \bigoplus_{r,s} \mathbf{g}^{r,s}_{(F,W)} \tag{2.14}$$

via Theorem (2.7). Indeed:

$$\mathbf{g}^{r,s}_{(F,W)} = \{\, \alpha \in \mathbf{g}_{\mathbb{C}} \mid \alpha(I^{p,q}_{(F,W)}) \subseteq I^{p+r,q+s}_{(F,W)} \ \ \forall\,(p,q)\,\} \tag{2.15}$$

In particular

$$\mathbf{g}_{\mathbb{C}} = \mathbf{g}^F_{\mathbb{C}} \oplus q_F, \qquad q_F = \bigoplus_{r<0} \mathbf{g}^{r,s}_{(F,W)} \tag{2.16}$$

(as vector spaces) where $\mathbf{g}^F_{\mathbb{C}} = Lie(G^F_{\mathbb{C}})$ denotes the isotopy subalgebra of $\mathbf{g}_{\mathbb{C}}$ at F. Consequently, since $\mathcal{M}$ is an open subset of $\check{\mathcal{M}}$, the map

$$u \in q_F \mapsto e^u.F \in \check{\mathcal{M}} \tag{2.17}$$

restricts to a biholomorphism from a neighborhood U of zero in q_F onto a neighborhood of F in $\mathcal{M}$. Accordingly, $q_F \cong T_F(\mathcal{M})$ via the action

$$< u, g >= \frac{d}{dz} g(e^{zu}.F)\bigg|_{z=0} \tag{2.18}$$

of $u \in q_F$ on $g \in C^\infty_F(\mathcal{M})$. Utilizing the isomorphism $q_F \cong T_F(\mathcal{M})$ defined by equation (2.18), one then obtains a natural hermitian metric on $\mathcal{M}$ via the rule

$$h_F(\alpha,\beta) = Tr(\alpha\beta^*) \tag{2.19}$$

where β^* denote the adjoint of β with respect to the inner product defined by Lemma (2.13).

Finally, both the metric (2.19) and the Hodge bundles (2.6) share the following symmetry:

Theorem 2.20 [6], [8]. *Let $F \in \mathcal{M}$ and $g \in G_{\mathbb{R}} \cup \exp(\Lambda^{-1,-1}_{(F,W)})$. Then,*

— $I^{p,q}_{(g.F,W)} = g.I^{p,q}_{(F,W)}$.
— *The induced map $L_{g*} : T_F(\mathcal{M}) \to T_{g.F}(\mathcal{M})$ is an isometry.*

Moreover, given any element $g \in G$, let G^F denote the isotopy subgroup of F in G and

$$Lie_{-s} = \{\, \alpha \in gl(V) \mid \alpha(W_k) \subseteq W_{k-s} \ \ \forall k\,\} \tag{2.21}$$

Then, g admits a distinguished C^∞ decomposition of the form

$$g = g_{\mathbb{R}} e^{\lambda} f \tag{2.22}$$

where $g_{\mathbb{R}} \in G_{\mathbb{R}}$, $\lambda \in \Lambda^{-1,-1}_{(F,W)}$ and $f \in G^F \cap \exp(Lie_{-1})$.

§3. Admissibility Criteria

The proof of the nilpotent orbit theorem presented in §1 depends crucially upon the fact that the associated classifying spaces $\mathcal{D}$ have negative holomorphic sectional curvature along the horizontal subbundle $T^{horiz}(\mathcal{D})$ of $T(\mathcal{D})$. In contrast, the classifying spaces of graded-polarized mixed Hodge structure $\mathcal{M}$ introduced in §2 usually admit holomorphic, horizontal maps $f : \mathbb{C} \to \mathcal{M}$, and hence do not admit any such metric [cf. Example 2.5]. Accordingly, one has no reason to expect the limiting Hodge filtration of an abstract variation of graded-polarized mixed Hodge structure $\mathcal{V} \to \Delta^*$ to exist, as the following example shows:

Example 3.1. Via the isomorphism $\mathcal{M} \cong \mathbb{C}$ defined in Example 2.5, any holomorphic map $f : \Delta^* \to \mathbb{C}$ defines a variation of graded-polarized mixed Hodge structure $\mathcal{V} \to \Delta^*$ with trivial monodromy. In particular, the limiting Hodge filtration F_∞ of $\mathcal{V}$ exists if and only if $\lim_{z\to 0} f(z)$ exists as an element of $\mathbb{C}$.

For variations of geometric origin however, one can prove the existence of the limiting Hodge filtration via a generalization [14] of J. Steenkbrink's work on limits of Hodge structures [13].

Now, by Theorem (1.8), the limiting Hodge filtration of F_∞ of a variation of pure, polarized Hodge structure pairs with the associated monodromy weight filtration $W(N)[-k]$ to define a mixed Hodge structure. The analogous object in the mixed case is the relative weight filtration

$$ {}^rW = {}^rW(N, W) \tag{3.2} $$

defined by the following lemma:

Lemma 3.3 [14]. *Let W be an increasing filtration of a finite dimensional vector space V, and N be a nilpotent endomorphism of V which preserves W. Then, there exists at most one increasing filtration ${}^rW = {}^rW(N, W)$ of V such that*

(1) $N({}^rW_k) \subseteq {}^rW_{k-2}$.
(2) ${}^rW_j Gr^W_k = W(N : Gr^W_k \to Gr^W_k)_{j-k}$.

In particular, as shown by Deligne in [3], the relative weight filtration of a geometric variation of graded-polarized mixed Hodge structure exists via comparison with the ℓ-adic case. Thus a minimal set of conditions for an abstract variation of graded-polarized mixed Hodge structure $\mathcal{V} \to \Delta^*$ with unipotent monodromy to be akin to a variation of geometric origin are:

(I) The limiting Hodge filtration of $\mathcal{V}$ exists.
(II) The relative weight filtration ${}^rW$ of $\mathcal{V}$ exists.

Following [14], we shall therefore call an abstract variation of graded-polarized mixed Hodge structure $\mathcal{V} \to \Delta^*$ with unipotent monodromy *admissible* provided $\mathcal{V}$ satisfies conditions (I) and (II) above.

Theorem 3.4, [14]. *The limiting Hodge filtration of an admissible variation of graded-polarized mixed Hodge structure $\mathcal{V} \to \Delta^*$ with unipotent monodromy pairs with the associated relative weight filtration ${}^{r}W$ to define a mixed Hodge structure, relative to which the monodromy logarithm N of $\mathcal{V}$ is a morphism of type $(-1,-1)$.*

As in the pure case, a nilpotent orbit of graded-polarized mixed Hodge structure modeled on the classifying space $\mathcal{M}$ consists of a filtration $F \in \check{\mathcal{M}}$ and an element $N \in \mathbf{g}_{\mathbb{R}}$ such that

(a) $N(F^p) \subseteq F^{p-1}$ for each index p.
(b) There exists a positive constant α such that $\mathrm{Im}(z) > \alpha \implies e^{zN}.F \in \mathcal{M}$.

Such a nilpotent orbit (F, N) is then said to be admissible if and only if the corresponding relative weight filtration ${}^{r}W = {}^{r}W(N, W)$ exists. In particular, one has the following result:

Lemma 3.5. *Let $\mathcal{V} \to \Delta^*$ be an admissible variation of graded-polarized mixed Hodge structure with unipotent monodromy. Then,*

(1) *The limiting Hodge filtration F_∞ of $\mathcal{V}$ exists;*
(2) *(F_∞, N) is an admissible nilpotent orbit of graded-polarized mixed Hodge structure.*

Proof. Assertion (1) is included in the definition of the admissibility of $\mathcal{V}$. Assertion (2) is a direct consequence of Schmid's Nilpotent Orbit Theorem, applied to the induced variations of pure, polarized Hodge structure $\mathcal{F}Gr^{\mathcal{W}}$, and the existence of the relative weight filtration ${}^{r}W$ postulated in the admissibility of $\mathcal{V}$.

A full analog of Schmid's Nilpotent Orbit Theorem for the mixed case would assert that further:

(3) There exists constants α, β and K such that $\mathrm{Im}(z) > \alpha \implies e^{zN}.F_\infty \in \mathcal{M}$ and

$$d_{\mathcal{M}}(F(z), e^{zN}.F_\infty) < K\mathrm{Im}(z)^{\beta} e^{-2\pi \mathrm{Im}(z)}$$

where $F(z) : U \to \mathcal{M}$ is any monodromy equivariant lifting of the period map of $\mathcal{V}$ to the upper half-plane U, and $d_{\mathcal{M}}$ denotes the Riemannian distance function of $\mathcal{M}$ induced by the mixed Hodge metric h.

As a first step towards the proof of (3), we now recount Deligne's extension [4] of the following construction to admissible nilpotent orbits of mixed Hodge structure [cf. [9] for complete proofs]:

Theorem 3.6 [11]. *Let (F, N) be a nilpotent orbit of pure, polarized Hodge structure of weight k, and*

$$V = \bigoplus_{p,q} I^{p,q} \tag{3.7}$$

denote the bigrading of the underlying vector space V defined by the limiting mixed Hodge structure $(F, W(N)[-k])$. Then, there exists a unique representation $\rho : sl_2(\mathbb{C}) \to \mathfrak{g}_{\mathbb{C}}$ such that $\rho(n_-) = N$ and $H = \rho(h)$ acts on (3.7) as multiplication by $p+q-k$ on $I^{p,q}$, where

$$n_- = \begin{pmatrix} 0 & 1 \\ 0 & 0 \end{pmatrix}, \qquad h = \begin{pmatrix} -1 & 0 \\ 0 & 1 \end{pmatrix}, \qquad n_+ = \begin{pmatrix} 0 & 0 \\ 1 & 0 \end{pmatrix} \tag{3.8}$$

Proof. Apply Lemma (3.10) below to the pair (N, H).

To this end, we define a grading Y of an increasing filtration W of a finite dimensional vector space V to be a semisimple endomorphism Y of V with integral eigenvalues such that W_k is equal to the direct sum of eigenspace $E_k(Y)$ of Y with eigenvalue k, and W_{k-1} for each index k. In particular, given a mixed Hodge structure (F, W) with bigrading

$$V = \bigoplus_{p,q} I^{p,q}$$

the semisimple endomorphism of V which acts as multiplication by $p+q$ on $I^{p,q}$ defines a functorial grading

$$Y = Y_{(F,W)} \tag{3.9}$$

of the underlying weight filtration W. Moreover, as shown in [1], given any increasing filtration W of a complex vector space V, the set $\mathcal{Y}(W)$ consisting of all grading Y of W is an affine space upon which the Lie group $\exp(Lie_{-1})$ [cf. equation (2.21)] acts simply transitively via the adjoint action

$$g.Y = Ad(g)Y$$

Lemma 3.10. *Let N be a nilpotent endomorphism of finite dimensional complex vector space V. Then, given a grading H of $W(N)$ such that $[H, N] = -2N$, there exists a unique representation $\rho : sl_2(\mathbb{C}) \to gl(V)$ such that $\rho(n_-) = N$ and $\rho(h) = H$.*

Sketch of Proof. Via the semisimplicity of $sl_2(\mathbb{C})$, it will suffice to define the highest weight vectors of ρ. Namely, $v \in V$ is of highest weight k if v belongs to $E_k(H) \cap \ker(N^{k+1})$.

Definition. A grading ${}^rY$ of ${}^rW = {}^rW(N, W)$ is said to be compatible with N and W if and only if (a) ${}^rY$ preserves W; and (b) $[{}^rY, N] = -2N$.

Corollary 3.11. *Let* ${}^rY$ *be a grading of* ${}^rW = {}^rW(N,W)$ *which is compatible with* N *and* W*, and* Y *be a grading of* W *which commutes with* ${}^rY$*. Then, there exists a unique representation* ρ_Y *of* $sl_2(\mathbb{C})$ *on the ambient vector space* V *such that* ρ *commutes with* Y*, and*

$$\rho_Y(n_-) = N_0, \qquad \rho_Y(h) = {}^rY - Y$$

where $N = N_0 + N_{-1} + \cdots$ *relative to the eigenvalues of* $\operatorname{ad} Y$*.*

Proof. One simply lifts the representation of $sl_2(\mathbb{C})$ on Gr^W defined by application of Lemma (3.10) to the pair $Gr(N)$, $Gr({}^rY - Y)$ via the isomorphism $V \cong Gr^W$ defined by the grading Y.

Theorem 3.12, Deligne [4]. *Let* ${}^rY$ *be a grading of* ${}^rW = {}^rW(N,W)$ *which is compatible with* N *and* W*. Then, there exists a unique, functorial grading* $Y = Y(N, {}^rY)$ *of* W *which commutes with* ${}^rY$ *such that*

$$[N - N_0, N_0^+] = 0$$

where $N_0 = \rho_Y(n_-)$ *and* $N_0^+ = \rho_Y(n_+)$*.*

Sketch of Proof [9]. Let Y' be a grading of W which commutes with ${}^rY$ such that

$$[N - \rho_{Y'}(n_-), \rho_{Y'}(n_+)] \equiv 0 \mod Lie_{-s}$$

[cf. (2.21)]. Then, making use of the decomposition

$$V = \operatorname{Image}(\rho_{Y'}(n_-)) \oplus \ker(\rho_{Y'}(n_+))$$

one can show the existence of an element g in $\exp(Lie_{-s})$ which commutes with ${}^rY$ such that

$$[N - \rho_{Y''}(n_-), \rho_{Y''}(n_+)] \equiv 0 \mod Lie_{-(s+1)}$$

where $Y'' = g.Y'$.

In particular, since $Lie_{-s} = 0$ for some finite index s, given any grading Y_1 of W which commutes with ${}^rY$, the iterated application of the above procedure to Y_1 generates a sequence of gradings $Y_1, \ldots, Y_n$ terminating at the desired grading Y.

Remark. There is a typesetting error in the full proof of Theorem (3.12) given in [9]: On pages 238–239, the phrase "commutes with ${}^rW$" should be "commutes with ${}^rY$".

To apply Theorem (3.12) to the study of admissible variations of graded-polarized mixed Hodge structure $\mathcal{V} \to \Delta^*$, observe that by Theorem (3.12) the monodromy logarithm N of $\mathcal{V}$ is a $(-1,-1)$ morphism of the limiting mixed Hodge structure $(F_\infty, {}^rW)$, and hence

$$[{}^rY, N] = -2N$$

where ${}^rY = Y_{(F,{}^rW)}$ denotes grading of ${}^rW = {}^rW(N,W)$ defined by the $I^{p,q}$'s of $(F, {}^rW)$.

Theorem 3.13. *Let $e^{zN}.F$ be an admissible nilpotent orbit of graded-polarized mixed Hodge structure, and ${}^rY = Y_{(F,{}^rW)}$ be the grading of ${}^rW$ defined by the $I^{p,q}$'s of $(F,{}^rW)$. Then,*

(a) *${}^rY$ is compatible with N and W;*
(b) *The associated grading $Y = Y(F,W,N)$ of W defined by application of Theorem (3.12) to N and ${}^rY$ preserves F.*

Proof. Let $(F,{}^rW) \mapsto (e^{-i\delta}.F,{}^rW)$ denote the splitting operation defined by Theorem (2.11). Then,

$$Y(F,W,N) = e^{i\delta}.Y(e^{-i\delta}.F,W,N)$$

via the functoriality of Theorem (3.12). Accordingly, it suffices to prove Theorem (3.13) for nilpotent orbits of the form $e^{zN}.F$ with $(F,{}^rW)$ split over $\mathbb{R}$.

In particular, since the construction of Theorem (3.12) is compatible with the operations of tensor product and direct sum, it suffices to consider the following two subcases:

(i) $V = \oplus_p I^{p,p}$ and $N(I^{p,p}) \subseteq I^{p-1,p-1}$;
(ii) $V = I^{p,0} \oplus I^{0,p}$ and $N = 0$.

Indeed, every mixed Hodge structure $(F,{}^rW)$ which is split over $\mathbb{R}$ and has N as morphism of type $(-1,-1)$ arises from (i) and (ii) via the operations of direct sum and tensor product.

Case (i): Since $I^{p,p}$ is the $2p$-eigenspace of ${}^rY$, and Y commutes with ${}^rY$, Y must preserve $I^{p,p}$. Consequently, Y also preserves $F^p = \oplus_{r\geq p} I^{r,r}$.
Case (ii): Since $N = 0$ and ${}^rW = {}^rW(N,W)$ exists, W must be the trivial filtration $0 = W_{p-1} \subseteq W_p = V$ which has only the trivial grading $Y(v) = pv$.

Corollary 3.14. *Let $e^{zN}.F$ be an admissible nilpotent orbit of graded-polarized mixed Hodge structure with limiting mixed Hodge structure $(F,{}^rW)$ which is split over $\mathbb{R}$. Then, the associated grading $Y = Y(F,W,N)$ defined by Theorem (3.13) is defined over $\mathbb{R}$.*

Proof. Y is obtained from the gradings attached to cases (i) and (ii) above via the operations of tensor product and direct sum [cf. Theorem (3.13)]. As these gradings are defined over $\mathbb{R}$, so is Y.

Corollary 3.15. *Let $e^{zN}.F$ be an admissible nilpotent orbit of graded-polarized mixed Hodge structure with limiting mixed Hodge structure $(F,{}^rW)$ which is split over $\mathbb{R}$. Then, the representation $\rho = \rho_Y$ obtained by application of Corollary (3.11) to $Y = Y(F,W,N)$ and ${}^rY = Y_{(F,{}^rW)}$ restricts to a representation $\rho : sl_2(\mathbb{R}) \to \mathfrak{g}_{\mathbb{R}}$.*

Proof. Since N, Y and ${}^rY$ are defined over $\mathbb{R}$, so is ρ as $\rho(n_-) = N_0$ and $\rho(h) = {}^rY - Y$. Moreover, since N acts by infinitesimal isometries on Gr^W, so does ρ as both N and N_0 induce the same map on Gr^W.

§4. Nilpotent Orbit Theorem

Building on the results of §2 and §3, we now prove an analog of Schmid's nilpotent orbit theorem for an admissible variation of graded-polarized mixed Hodge structure $\mathcal{V} \to \Delta^*$ with unipotent monodromy.

As in the pure case, we begin by lifting the period map $\varphi(s) : \Delta^* \to \mathcal{M}/\Gamma$ of $\mathcal{V}$ to a map $F(z) : U \to \mathcal{M}$ from the upper half-plane U into $\mathcal{M}$ which makes the following diagram commute

$$\begin{array}{ccc} U & \xrightarrow{F} & \mathcal{D} \\ {\scriptstyle s=\exp(2\pi i z)}\downarrow & & \downarrow \\ \Delta^* & \xrightarrow{\varphi} & \mathcal{D}/\Gamma \end{array} \tag{4.1}$$

Accordingly, $F(z+1) = e^N.F(z)$ where N is the monodromy logarithm of $\mathcal{V}$, and hence the $\psi(z) = e^{-zN}.F(z)$ descends to a map

$$\psi(s) : \Delta^* \to \check{\mathcal{M}}$$

Moreover, as per our discussion of §3, $\mathcal{V}$ is admissible if and only if

(I) The limiting Hodge filtration $F_\infty = \lim_{s\to 0} \psi(s)$ exists as an element of $\check{\mathcal{M}}$;
(II) The relative weight filtration ${}^rW = {}^rW(N, W)$ exists.

In particular, if $\mathcal{V}$ is admissible and q is a vector space complement to the isotopy algebra of F_∞ in $\mathbf{g}_{\mathbb{C}}$, there exists a unique q-valued function $\Gamma(s)$ such that $\Gamma(0) = 0$ and

$$F(z) = e^{zN} e^{\Gamma(s)}.F_\infty \tag{4.2}$$

over some neighborhood O of zero in Δ. Indeed, since $G_{\mathbb{C}}$ acts transitively on $\check{\mathcal{M}}$, the map

$$u \mapsto e^u.F_\infty$$

restricts to a biholomorphism from some neighborhood of zero in q onto a neighborhood of F_∞ in $\check{\mathcal{M}}$, and hence $\psi(s) = e^{\Gamma(s)}.F_\infty$. Accordingly, upon setting

$$q_\infty = \bigoplus_{a<0} \mathbf{g}^{a,b}_{(F_\infty, {}^rW)} \tag{4.3}$$

we obtain the following result:

Theorem 4.4. *Let $\varphi(s)$ be the period map of an admissible variation of graded-polarized mixed Hodge structure $\mathcal{V} \to \Delta^*$ and $F(z) : U \to \mathcal{M}$ be a lifting of $\varphi(s)$ to the upper half-plane which makes (4.1) commute. Then, there exists a unique q_∞-valued holomorphic function $\Gamma(s)$ defined on a neighborhood O of zero in Δ such that $F(z) = e^{zN}e^{\Gamma(s)}.F_\infty$ and $\Gamma(0) = 0$.*

Proof. Since $(F_\infty, {}^rW)$ induces a mixed Hodge structure on $\mathbf{g}_\mathbb{C}$, the subalgebra (4.3) is a vector space complement to the isotopy algebra

$$\mathbf{g}_\mathbb{C}^{F_\infty} = \bigoplus_{a \geq 0} \mathbf{g}^{a,b}_{(F_\infty, {}^rW)}$$

To see that $\Gamma(0) = 0$, one simply notes that $e^{\Gamma(s)}.F_\infty = \psi(s)$ and $\psi(0) = F_\infty$.

Remark. For the multivariable version of Theorem (4.4) and applications to Higgs bundles and quantum cohomology, see [8].

Theorem (4.4) can be viewed as a group theoretic version of Schmid's nilpotent orbit theorem in that it expresses the period map of an admissible variation $\mathcal{V} \to \Delta^*$ as a deformation a nilpotent orbit $e^{zN}.F_\infty$ by a function of order $|s| = e^{-2\pi \mathrm{Im}(z)}$ as $\mathrm{Im}(z) \to \infty$.

To prove a full analog, with distance estimate, we now record the following lemmata, the proofs of which are elementary:

Lemma 4.5 [9]. *Let Y be a grading of W which is split over $\mathbb{R}$, and y be a positive real number. Then,*

$$\alpha \in \mathbb{R} \implies y^{\alpha Y} : \mathcal{M} \to \mathcal{M}$$

where $y^{\alpha Y} = \exp(\alpha \log(y) Y)$. Moreover, if $\alpha < 0$ then

$$F_1, F_2 \in \mathcal{M} \implies d_\mathcal{M}(y^{\alpha Y}.F_1, y^{\alpha Y}.F_2) \leq y^{-\alpha(L-1)} d_\mathcal{M}(F_1, F_2)$$

where $d_\mathcal{M}$ is the hermitian distance defined by the metric (2.13) and L is the length of W [i.e. the smallest positive integer L such that $Lie_{-L} = 0$, cf. (2.21)].

Lemma 4.6 [9]. *Let F_o be an element of $\mathcal{M}$, and $| * |$ be a norm on $\mathbf{g}_\mathbb{C}$. Then, there exists a neighborhood S of F_o in $\mathcal{M}$, a neighborhood U of zero in $\mathbf{g}_\mathbb{C}$, and a constant $K > 0$ such that*

$$F \in S,\ u \in U \implies e^u.F \in \mathcal{M}, \qquad d_\mathcal{M}(e^u.F, F) < K|u|$$

Nilpotent Orbit Theorem [9]. *Let $\varphi(s)$ be the period map of an admissible variation of graded-polarized mixed Hodge structure $\mathcal{V} \to \Delta^*$ with unipotent monodromy, and $F(z) : U \to \mathcal{M}$ be a lifting of $\phi(s)$ to the upper half-plane which makes (4.1) commute. Then,*

(1) *The limiting Hodge filtration F_∞ of $\mathcal{V}$ exists;*
(2) *(F_∞, N) is a nilpotent orbit of graded-polarized mixed Hodge structure;*
(3) *There exists constants α, β and K such that $\mathrm{Im}(z) > \alpha \implies e^{zN}.F_\infty \in \mathcal{M}$ and*

$$d_{\mathcal{M}}(F(z), e^{zN}.F_\infty) < K y^\beta e^{-2\pi y}, \qquad z = x + iy$$

Proof. The validity of assertions (1) and (2) was established in Lemma (3.5). To verify assertion (3), assume for the moment that the limiting mixed Hodge structure $(F_\infty, {}^rW)$ of $\mathcal{V}$ is split over $\mathbb{R}$. Then,

— ${}^rY = Y_{(F_\infty, {}^rW)}$ and $Y = Y(F_\infty, W, N)$ are defined over $\mathbb{R}$ (Corollary 3.14);
— $\rho(h) = {}^rY - Y$ is an element of $\mathbf{g}_{\mathbb{R}}$ (Corollary 3.15);
— $F_o = e^{iN}.F_\infty \in \mathcal{M}$ (Schmid's SL_2 Orbit Theorem, applied to Gr^W).

Moreover, $[{}^rY, N] = -2N$ since N is a $(-1,-1)$ morphism of $(F_\infty, {}^rW)$. Thus,

$$e^{iyN} = y^{-\frac{1}{2}{}^rY} e^{iN} y^{\frac{1}{2}{}^rY} \tag{4.7}$$

and hence

$$e^{iyN}.F_\infty = (y^{-\frac{1}{2}{}^rY} e^{iN} y^{\frac{1}{2}{}^rY}).F_\infty = y^{-\frac{1}{2}{}^rY} e^{iN}.F_\infty = y^{-\frac{1}{2}{}^rY}.F_o \tag{4.8}$$

as ${}^rY$ preserves F_∞.

Taking note of the fact that $F(z) = e^{zN} e^{\Gamma(s)}.F_\infty$ and $e^{xN} \in G_{\mathbb{R}}$ acts on $\mathcal{M}$ by isometries, it then follows from equations (4.7) and (4.8) that

$$\begin{aligned} d_{\mathcal{M}}(e^{zN} e^{\Gamma(s)}.F_\infty, e^{zN}.F_\infty) &= d_{\mathcal{M}}(e^{iyN} e^{\Gamma(s)}.F_\infty, e^{iyN}.F_\infty) \\ &= d_{\mathcal{M}}(y^{-\frac{1}{2}{}^rY} e^{iN} y^{\frac{1}{2}{}^rY} e^{\Gamma(s)}.F_\infty, y^{-\frac{1}{2}{}^rY}.F_o) \\ &= d_{\mathcal{M}}(y^{-\frac{1}{2}{}^rY} e^{\tilde{\Gamma}(z)}.F_o, y^{-\frac{1}{2}{}^rY}.F_o) \end{aligned} \tag{4.9}$$

upon setting

$$e^{\tilde{\Gamma}(z)} = Ad(e^{iN}) Ad(y^{\frac{1}{2}{}^rY}) e^{\Gamma(s)} \tag{4.10}$$

and recalling that ${}^rY$ preserves F_∞.

Now,

$$[{}^rY, Y] = 0 \implies y^{-\frac{1}{2}{}^rY} = y^{-\frac{1}{2}({}^rY - Y)} y^{-\frac{1}{2}Y}$$

Consequently,

$$d_{\mathcal{M}}(y^{-\frac{1}{2}{}^rY}e^{\tilde{\Gamma}(z)}.F_o, y^{-\frac{1}{2}{}^rY}.F_o) = d_{\mathcal{M}}(y^{-\frac{1}{2}Y}e^{\tilde{\Gamma}(z)}.F_o, y^{-\frac{1}{2}Y}.F_o)$$

because ${}^rY - Y \in \mathbf{g}_{\mathbb{R}}$. Thus, by Lemma (4.5):

$$d_{\mathcal{M}}(e^{zN}e^{\Gamma(s)}.F_\infty, e^{zN}.F_\infty) \leq y^{\frac{1}{2}(L-1)}d_{\mathcal{M}}(e^{\tilde{\Gamma}(z)}.F_o, F_o) \tag{4.11}$$

Invoking Lemma (4.6), one therefore has

$$d_{\mathcal{M}}(e^{zN}e^{\Gamma(s)}.F_\infty, e^{zN}.F_\infty) \leq Ky^{\frac{1}{2}(L-1)}|\tilde{\Gamma}(z)| \tag{4.12}$$

To complete the proof in the split over $\mathbb{R}$ case, one simply decomposes $\Gamma(s)$ according to the eigenvalues of $ad\,{}^rY$ in order to obtain a bound on the asymptotic behavior of $|\tilde{\Gamma}(z)|$. Namely,

$$|\tilde{\Gamma}(z)| = O(y^{\frac{1}{2}c}e^{-2\pi y}) \tag{4.13}$$

as $y \to \infty$, where c denotes the maximal eigenvalue of $ad\,{}^rY$ on q_∞. Inserting (4.13) into (4.12), one obtains the desired estimate

$$d_{\mathcal{M}}(e^{zN}e^{\Gamma(s)}.F_\infty, e^{zN}.F_\infty) \leq Ky^{\beta}e^{-2\pi y}$$

Suppose now that $(F_\infty, {}^rW)$ is not split over $\mathbb{R}$. Then, by Theorem (2.11), there exists an element

$$\delta \in \Lambda^{-1,-1}_{(F_\infty, {}^rW)}$$

such that

$$(\hat{F}_\infty, {}^rW) = (e^{-i\delta}.\hat{F}_\infty, {}^rW)$$

is split over $\mathbb{R}$. In particular, upon setting ${}^rY = Y_{(\hat{F}_\infty, {}^rW)}$, $Y = Y(\hat{F}_\infty, W, N)$ and proceeding as above, one obtains the general proof mutatis mutandis.

§5. The SL$_2$ Orbit Theorem

In simplest terms, Schmid's SL_2 orbit theorem asserts that every nilpotent orbit $e^{zN}.F$ of pure, polarized Hodge structure is asymptotic to a 1-parameter map $U \to \mathcal{D}$ which arises form a representation of SL_2 on the associated Lie group $G_{\mathbb{R}}$.

More precisely, one calls a nilpotent orbit $\theta(z) = e^{zN}.\hat{F}$ an SL_2 orbit if and only if there exists a point $F_o \in \mathcal{D}$ and a representation $\psi : SL_2(\mathbb{R}) \to G_{\mathbb{R}}$ such that

$$\theta(g.i) = \psi(g).F_o \tag{5.1}$$

for every element $g \in SL_2(\mathbb{R})$.

SL_2 Orbit Theorem [11]. *Let $e^{zN}.F$ be a nilpotent orbit of pure, polarized Hodge structure. Then, there exists (i) an SL_2 orbit $e^{zN}.\hat{F}$; and (ii) a real-analytic function $g(y) : (a, \infty) \to G_{\mathbb{R}}$, such that $y > a \implies e^{iyN}.F \in \mathcal{D}$, and*

(a) $e^{iyN}.F = g(y)e^{iyN}.\hat{F}$;
(b) *$g(y)$ and $g^{-1}(y)$ have convergent power series expansions at ∞ of the form $(1+\sum_{k>0} A_k y^{-k})$ where A_k is an element of $\mathbf{g}_{\mathbb{R}} \cap \ker(ad\, N)^{k+1}$ which maps $W(N)_\ell$ into $W(N)_{\ell+k-1}$ for each index ℓ.*

In this section, we outline the proof of the following extension of Schmid's SL_2 Orbit Theorem:

Theorem 5.2 [10]. *Let $e^{zN}.F$ be an admissible nilpotent orbit of graded-polarized mixed Hodge structure for which the underlying weight filtration W is of length 2 (i.e. $0 = W_{k-2} \subset W_{k-1} \subset W_k = V$). Then, there exists (i) an admissible SL_2 orbit $e^{zN}.\hat{F}$; (ii) a real-analytic function $g(y) : (a, \infty) \to G_{\mathbb{R}}$; and (iii) an element $\zeta \in \mathbf{g}_{\mathbb{R}} \cap \ker(ad\, N) \cap \Lambda^{-1,-1}_{(\hat{F}, {}^rW)}$, such that $y > a \implies e^{iyN}.F \in \mathcal{M}$, and*

(a) $e^{iyN}.F = g(y)e^{iyN}.\hat{F}$;
(b) *$e^{-\zeta}g(y)$ and $g^{-1}(y)e^{\zeta}$ have convergent power series expansions at ∞ of the form $(1 + \sum_{k>0} A_k y^{-k})$ with $A_k \in \ker(ad\, N)^{k+1}$.*

Remark 1. In analogy with Theorem (2.20), in order to extend the SL_2 orbit theorem to nilpotent orbits of weight length $L > 2$, one must replace the $G_{\mathbb{R}}$-valued function $g(y)$ by the product of a $G_{\mathbb{R}}$-valued function $g_{\mathbb{R}}(y)$ and an $\exp(\Lambda^{-1,-1})$-valued function $e^{\lambda(y)}$.

Remark 2. Although every mixed Hodge structure (F, W) of weight length 2 is split over $\mathbb{R}$, and hence equivalent to a direct sum of pure Hodge structures, the corresponding statement for nilpotent orbits is false (i.e. not every such nilpotent orbit is a direct sum of pure orbits).

Remark 3. The presentation below is condensed from [10].

In particular, as consequence of Theorem (5.2) and some auxiliary calculations, one obtains:

Corollary 5.3. *The holomorphic sectional curvature of an admissible nilpotent orbit $e^{zN}.F$ of weight length 2 is negative, and bounded away from zero whenever $y = Im(z)$ is sufficiently large.*

Corollary 5.4. *Let $e^{zN}.F$ be an admissible nilpotent orbit of weight length 2. Then,*

$$\lim_{y \to \infty} e^{-iyN}.Y_{(e^{iyN}.F, W)} = Y(F, W, N)$$

where $Y(F,W,N)$ *is the grading of* W *constructed by Theorem* (3.13).

As in the pure case, the proof of Theorem (5.2) begins with the construction of a connection on a suitable principal bundle $P \to X$.

More precisely, let W be a weight filtration of arbitrary length L, and $\mathcal{M} = \mathcal{M}(W, Q, h^{p,q})$ be a classifying space of graded-polarized mixed Hodge structure upon which the Lie group G acts transitively. Define

— $\tilde{\mathcal{Y}}(W)$ to be the set of all gradings Y of W such that $(Y - \bar{Y})(W_k) \subseteq W_{k-2}$ for each index k.
— $\mathcal{D}$ to be the direct product of classifying spaces of pure, polarized Hodge structure to which $\mathcal{M}$ projects via the map $F \mapsto FGr^W$.
— $\tilde{\mathcal{X}} = \mathcal{D} \times \tilde{\mathcal{Y}}(W)$.
— $\tilde{G}$ to be the Lie group consisting of all elements $g \in G$ which induce real automorphisms of W_k/W_{k-2} for each index k.

Then, there exists a unique projection map $\pi : \tilde{\mathcal{X}} \to \mathcal{M}$ with C^∞-section

$$\sigma(F) = (FGr^W, Y_{(F,W)}) \tag{5.5}$$

and fiber $\pi^{-1}(F) \cong Lie(\tilde{G}^F)_{-2}$. (Namely, $\pi(\{H^{p,q}\}, Y)$ is the filtration $F \in \mathcal{M}$ obtained by lifting the graded filtration $FGr^W = \oplus_{a \geq p} H^{a,b}$ to the ambient vector space V via the grading Y.)

Lemma 5.6. *The space* $\pi : \tilde{\mathcal{X}} \to \mathcal{M}$ *is a holomorphic affine bundle upon which the Lie group* $\tilde{G}$ *acts both transitively and equivariantly.*

In particular, given a point $x \in \tilde{\mathcal{X}}$, the associated sequence

$$\tilde{G}^x \to \tilde{G} \to \tilde{G}/\tilde{G}^x \cong \tilde{\mathcal{X}} \tag{5.7}$$

defines a principal bundle $P(x)$ over $\tilde{\mathcal{X}}$.

Lemma 5.8. *Let* F_o *be an element of* $\mathcal{M}_{\mathbb{R}}$ *[cf. 2.10] and* $x_0 = \sigma(F_0)$. *Then, the vector space decomposition*

$$\tilde{\mathbf{g}} = \tilde{\mathbf{g}}^{x_o} \oplus (\tilde{\mathbf{g}} \cap \{ \bigoplus_{(r,s)\neq(0,0)} \mathbf{g}^{r,s}_{(F_o,W)} \}) \tag{5.9}$$

of $\tilde{\mathbf{g}} = Lie(\tilde{G})$ *is invariant under the action of* $Ad(\tilde{G}^{x_o})$, *and hence defines a connection* ∇ *on* $P(x_o)$.

Corollary 5.10. *Let F_o be an element of $\mathcal{M}_{\mathbb{R}}$, $x_o = \sigma(F_o)$ and ∇ denote the associated connection on $P(x_o)$ defined by Lemma (5.8). Then, given*

- *a nilpotent orbit $e^{zN}.F$ such that $e^{iyN}.F \in \mathcal{M}$ whenever $y > a$;*
- *an element $h_o \in \tilde{G}$ which maps x_o to $\sigma(e^{iaN}.F)$;*

there exists a unique lifting $h(y) : (a,\infty) \to \tilde{G}$ of $\sigma(e^{iyN}.F)$ to $P(x_0)$ which is tangent to ∇ and satisfies the initial condition $h(a) = h_o$.

Continuing the analogy with the pure case, the next step in the proof of Theorem (5.2) is to derive a system of differential equations which govern the function $h(y)$ defined by Corollary (5.10). To this end, given $F \in \mathcal{M}$ let

$$\mathbf{g}_{\mathbb{C}} = \eta_+ \oplus \eta_0 \oplus \eta_- \oplus \Lambda^{-1,-1}$$

denote the vector space decomposition of $\mathbf{g}_{\mathbb{C}}$ defined by the subalgebras

$$\eta_+ = \bigoplus_{r\geq 0, s<0} \mathbf{g}^{r,s}_{(F,W)} \qquad \eta_- = \bigoplus_{r<0, s\geq 0} \mathbf{g}^{r,s}_{(F,W)}$$

$$\eta_0 = \mathbf{g}^{0,0}_{(F,W)} \qquad \Lambda^{-1,-1} = \bigoplus_{r<0, s<0} \mathbf{g}^{r,s}_{(F,W)}$$

and observe that $F \in \mathcal{M}_{\mathbb{R}} \implies \bar{\eta}_+ = \eta_-$, $\bar{\eta}_0 = \eta_0$ and $\overline{\Lambda^{-1,-1}} = \Lambda^{-1,-1}$. Consequently, the linear operator $L : \mathbf{g}_{\mathbb{C}} \to \mathbf{g}_{\mathbb{C}}$ defined by the rule

$$\mathrm{L}|_{\eta_+} = +i, \qquad \mathrm{L}|_{\eta_0} = 0, \qquad \mathrm{L}|_{\eta_- \oplus \Lambda^{-1,-1}} = -i$$

restricts to a linear map $\mathrm{L} : \tilde{\mathbf{g}} \to \tilde{\mathbf{g}}$ whenever $F \in \mathcal{M}_{\mathbb{R}}$.

Theorem 5.11. *Let $h(y)$ be the $\tilde{G}$-valued function attached to the base point $F_o \in \mathcal{M}_{\mathbb{R}}$ and the nilpotent orbit $e^{zN}.F$ via Corollary (5.10). Then,*

$$h^{-1}\frac{dh}{dy} = -\mathrm{L}Ad(h^{-1}(y))N \tag{5.12}$$

where L *is the linear operator defined by $F = F_o$.*

Now, in order to reduce equation (5.12) to a more tractable form, define

$$\alpha(y) = -2h^{-1}\frac{dh}{dy} \qquad \beta(y) = Ad(h^{-1}(y))N \tag{5.13}$$

Then,

$$-2\frac{d\beta}{dy} = [\beta(y), \alpha(y)] \tag{5.14}$$

Moreover, on account of the horizontality of $N = \bar{N}$ along $e^{iyN}.F$ and the fact that $h^{-1}(y)$ maps $I^{p,q}_{(e^{iyN}.F,W)}$ to $I^{p,q}_{(F_o,W)}$:

$$\beta(y) = \beta^{1,-1}(y) + \beta^{0,0}(y) + \beta^{-1,1}(y) + \beta_+(y) + \beta_-(y) \tag{5.15}$$

where

$$\beta_+(y) = \sum_{k>0} \beta^{0,-k}(y), \qquad \beta_-(y) = \sum_{k>0} \beta^{-1,-k}(y) \tag{5.16}$$

and $\sum \beta(y)^{r,s}$ denote the Hodge components of $\beta(y)$ relative to (F_o, W).

Inserting equations (5.13) and (5.14) into equation (5.12), it then follows that

$$\alpha(y) = \alpha^{1,-1}(y) + \alpha^{-1,1}(y) + \alpha_+(y) + \alpha_-(y) \tag{5.17}$$

where

$$\begin{array}{ll} \alpha^{1,-1}(y) = 2i\beta^{1,-1}(y) & \alpha^{-1,1} = -2i\beta^{-1,1}(y) \\ \alpha_+(y) = 2i\beta_+(y) & \alpha_-(y) = -2i\beta_-(y) \end{array} \tag{5.18}$$

In particular, upon coupling (5.14) with equations (5.15)–(5.16) and (5.17)–(5.18), one finds that

$$\frac{d}{dy}\begin{pmatrix}\alpha_- \\ \alpha_+\end{pmatrix} = \begin{pmatrix} \frac{1}{2}ad\,Z(y) & -ad\,X^-(y) \\ -ad\,X^+(y) & -\frac{1}{2}ad\,Z(y)\end{pmatrix}\begin{pmatrix}\alpha_- \\ \alpha_+\end{pmatrix} + [\alpha_+, \alpha_-] \tag{5.19}$$

where

$$X^+(y) = \alpha^{1,-1}(y), \qquad Z(y) = 2i\beta^{0,0}(y), \qquad X^-(y) = \alpha^{-1,1}(y) \tag{5.20}$$

form a system of solutions to the monopole equations of su_2, i.e.

$$\begin{gathered} -2\frac{d}{dy}X^+(y) = [Z(y), X^+(y)], \qquad 2\frac{d}{dy}X^-(y) = [Z(y), X^-(y)] \\ -\frac{d}{dy}Z(y) = [X^+(y), X^-(y)] \end{gathered} \tag{5.21}$$

Returning now to the proof of Theorem (5.2) proper, let $e^{zN}.F$ be an admissible nilpotent orbit of graded-polarized mixed Hodge structure for which the underlying weight filtration W is of length 2, and

$$(\hat{F}, {}^rW) = (e^{-i\delta}.F, {}^rW) \tag{5.22}$$

denote the associated splitting of $(F, {}^rW)$ defined by Theorem (2.11). Then, the map $\theta(z) = e^{zN}.\hat{F}$ is an admissible SL_2 orbit of graded-polarized mixed Hodge structure, with base point

$$F_o = e^{iN}.\hat{F} \in \mathcal{M}_{\mathbb{R}} \tag{5.23}$$

and representation $\rho = \psi_* : sl_2(\mathbb{R}) \to \mathbf{g}_{\mathbb{R}}$ obtained by application of Corollary (3.15) to $\theta(z)$. Moreover, because W is of length 2, $\tilde{G} = G_{\mathbb{R}}$ and the associated function $\beta(y)$ is of the form

$$\beta(y) = \beta^{1,-1}(y) + \beta^{0,0}(y) + \beta^{-1,1}(y) + \beta^{0,-1}(y) + \beta^{-1,0}(y)$$

Now, in order to collect the associated system of equations (5.18)–(5.21) into a more manageable form, let $\{e, f\}$ be a real basis of $\mathbb{C}^2$ and

$$X^- = \begin{pmatrix} 0 & 1 \\ 0 & 0 \end{pmatrix}, \qquad Z = \begin{pmatrix} -1 & 0 \\ 0 & 1 \end{pmatrix}, \qquad X^+ = \begin{pmatrix} 0 & 0 \\ 1 & 0 \end{pmatrix}$$

denote the standard matrix representation of su_2 relative to the basis

$$\nu_0 = e - if, \qquad \nu_1 = e + if$$

Then, given a finite dimensional complex vector space V and a finite dimensional representation U of $sl_2(\mathbb{C})$, contraction against the Casimir element

$$\Omega = 2X^+X^- + 2X^-X^+ + Z^2$$

of $su_2 \subset sl_2(\mathbb{C})$ defines a natural paring

$$Q : Hom(sl_2(\mathbb{C}), gl(V)) \otimes Hom(U, gl(V)) \to Hom(U, gl(V))$$

via the rule

$$Q(A,B)(u) = 2[A(X^+), B(X^-.u)] + 2[A(X^-), B(X^+.u)] + [A(Z), B(Z.u)]$$

Moreover, upon defining

$$\Phi(y)X^+ = X^+(y), \qquad \Phi(y)Z = Z(y), \qquad \Phi(y)X^- = X^-(y)$$

and

$$\Psi(y)\nu_1 = \alpha^{0,-1}(y), \qquad \Psi(y)\nu_0 = \alpha^{-1,0}$$

equations (5.18)–(5.21) can be written in the compact form

$$-8\Phi'(y) = Q(\Phi, \Phi), \qquad -2\Psi'(y) = Q(\Phi, \Psi) \tag{5.24}$$

To complete the proof of Theorem (5.2), one must now take up the arduous task of finding explicit series solutions to the non-linear system of equations (5.24) [cf. §6 of [1]]. Once this task is complete (approximately 20–30 pages of intricate calculation), one then obtains the desired function $g(y)$, and the constant ζ via the equations

$$h(y) = g(y)y^{-\frac{1}{2}\rho(h)}$$

and

$$e^{\zeta} = \lim_{y \to \infty} g(y)$$

REFERENCES

[1]. Cattani E., Kaplan A., and Schmid W., *Degeneration of Hodge structures*, Ann. of Math. **123** (1986), 457–535.

[2]. Deligne P., *Théorie de Hodge I*, Actes, Congrés Intern. Math. Nice (1970), 425–430; II, Publ. Math I.H.E.S. **40** (1971), 5–58; III, Publ. Math I.H.E.S. **44** (1974), 5–77.

[3]. ———, *La conjecture de Weil, II.*, Inst. Hautes Études Sci. Publ. Math. **52** (1980), 137–252.

[4]. ———, *Private communication* (1995.).

[5]. Griffiths P., *On the periods of certain rational integrals*, Ann. Math. **90** (1969), 460-541.

[6]. Kaplan A., *Notes on the moduli spaces of Hodge structures* (1995).

[7]. Kaplan A., and Pearlstein G., *Singularities of Variations of Mixed Hodge Structure*, Preprint: math.AG/0007040.

[8]. Pearlstein G., *Variations of Mixed Hodge Structure, Higgs Fields and Quantum Cohomology*, Manuscripta Math. **102** (2000), 269–310.

[9]. ———, *Degenerations of Mixed Hodge Structure*, Duke Math. Jour. **110** (2001), 217–251.

[10]———, *SL_2 Orbits and Degenerations of Mixed Hodge Structure*, In preparation.

[11].Schmid W., *Variation of Hodge Structure: The Singularities of the Period Mapping.*, Invent. Math. **22** (1973), 211 – 319.

[12].Schwarz C., *Relative monodromy weight filtrations*, Math. Zeit. **236** (2001), 11–21.

[13].Steenbrink J., *Limits of Hodge Structure*, Invent. Math. **31** (1976), 229–257.

[14].Steenbrink J., and Zucker S., *Variation of Mixed Hodge Structure I.*, Invent. Math. **80** (1985), 489 – 542.

[15].Usui S., *Variation of mixed Hodge structure arising from family of logarithmic deformations. II. Classifying space.*, Duke Math. Jour. **51** (1984), 851–875.

Department of Mathematics, University of California, Irvine CA 92697–3875

DEGENERATION OF $SL(n)$-BUNDLES ON A REDUCIBLE CURVE

XIAOTAO SUN

INTRODUCTION

It is a classic idea in algebraic geometry to use degeneration method. In particular, it achieved successes recently in the studying of moduli spaces of vector bundles (See [Gi], [GL1], [GL2], [NR] and [S1]). In connection of string theory, it needs also to study the degeneration of moduli spaces of G-bundles for any reductive algebraic group G (See [F1],[F2]).

Let $\mathcal{X} \to B$ be a proper flat family of curves of genus g such that $\mathcal{X}_b$ $(b \neq 0)$ smooth and $\mathcal{X}_0$ a semistable curve. It was known that there exists a family

$$\mathcal{M}_{\mathcal{X}}(G)_0 \to B_0 = B \setminus \{0\}$$

of moduli spaces of semistable G-bundles. The question becomes that for what geometric objects on $\mathcal{X}_0$ such that the moduli space of them gives a compactification $\mathcal{M}_{\mathcal{X}}(G) \to B$ of $\mathcal{M}_{\mathcal{X}}(G)_0 \to B_0 = B \setminus \{0\}$. We can consider G as a subgroup of some $GL(n)$, and think a semistable G-bundle as a semistable vector bundle of rank n with some additional conditions. Thus we may think (not strictly) the moduli space of semistable G-bundles as a subscheme of the moduli space of semistable vector bundles of rank n. On the other hand, there is a natural choice of geometric objects, the torsion free sheaves, on $\mathcal{X}_0$. The moduli space of semistable torsion free sheaves gives a natural degeneration of moduli spaces of semistable vector bundles on $\mathcal{X}_b$ when b goes to 0. Then a possible approach to the problem is finding the correct torsion free G-sheaves on $\mathcal{X}_0$. However, the problem remains almost complete open except for special groups like $G = GL(n)$, $Sp(n)$ and $O(n)$ (See the introduction of [F1]). G. Faltings studied the cases that $G = Sp(n)$ and $O(n)$ but left the case $G = SL(n)$ open (See [F2]). In this paper, we will treat the case $G = SL(n)$ when $\mathcal{X}_0$ has two smooth irreducible components intersecting at one node x_0.

The work supported by a grant of NFSC for outstanding young researcher at contract number 10025103.

Let $\mathcal{U}_{\mathcal{X}} \to B$ be the family of moduli spaces of semistable torsion free sheaves and $\mathcal{SU}^0_{\mathcal{X}} \to B_0$ be the family of moduli spaces of semistable vector bundles with fixed determinant $\mathcal{L}$. Let $f : \mathcal{SU}_{\mathcal{X}} \to B$ be the Zariski closure of $\mathcal{SU}^0_{\mathcal{X}} \subset \mathcal{U}_{\mathcal{X}}$ in $\mathcal{U}_{\mathcal{X}}$. Then the problem becomes to give a moduli interpretation of $f^{-1}(0)$. Namely, to define a suitable moduli functor $\mathcal{SU}^{\natural}_{\mathcal{X}_0}$ such that $f^{-1}(0)$ universally corepresents $\mathcal{SU}^{\natural}_{\mathcal{X}_0}$. It is obvious that the above question can also be asked for moduli of bundles on higher dimensional variety. In the study of moduli spaces of bundles on surfaces (See [GL1] and [GL2]), Gieseker and Li have noted that the closed points $[\mathcal{F}] \in f^{-1}(0)$ satisfy the condition

$$(*) \qquad det(\mathcal{F}|_{\mathcal{X}_0\setminus\{x_0\}}) = \mathcal{L}|_{\mathcal{X}_0\setminus\{x_0\}}.$$

In general, it may not be true that a semistable sheaf $\mathcal{F}$ satisfying condition $(*)$ has to be a point of $f^{-1}(0)$. For example, when $\mathcal{X}_0$ is irreducible, the set of semistable sheaves satisfying the condition $(*)$ will have bigger dimension than $f^{-1}(0)$. However, in our case when $\mathcal{X}_0$ has two smooth irreducible components intersecting at one node, the points of $f^{-1}(0)$ are precisely the semistable sheaves satisfying condition $(*)$ (See Lemma 2.2). In fact, we defined a moduli functor $\mathcal{SU}^{\natural}_{\mathcal{X}_0}$, which is represented by a closed subscheme $\mathcal{SU}_{\mathcal{X}_0} \subset \mathcal{U}_{\mathcal{X}_0}$ of the moduli space of semistable torsion free sheaves on $\mathcal{X}_0$ (See Theorem 1.6). Moreover, we proved that $\mathcal{SU}_{\mathcal{X}_0}$ is a reduced, seminormal variety whose closed points are precisely the s-equivalent classes of semistable sheaves on $\mathcal{X}_0$ satisfying $(*)$. These was done in Section 1 (See Theorem 1.6). In Section 2, we showed that the above moduli problem has good specialzation (See [NS] for the notation), and the degeneration of moduli spaces of semistable $SL(n)$-bundles is $\mathcal{SU}_{\mathcal{X}_0}$ when b goes to 0.

There exist another degeneration of moduli spaces of $GL(n)$-bundles, the so called Gieseker type degeneration (see [Gi], [NSe], [Ka1] and [Ka2]), it was expected to study the degeneration of G-bundles by using Gieseker type degeneration. Thus it may be interesting to see our result from this view of point.

The paper comes from a conversation with Jun Li, who told me the condition $(*)$ in his joint works with Gieseker. The emails with D.S. Nagaraj and C.S. Seshadri concerning Proposition (4.1) of [NS] were helpful for the observation of Lemma 1.4. I thank them very much.

§1 Moduli space of semistable sheaves with fixed determinant

Let X be a projective curve of genus g with two smooth irreducible components X_1 and X_2 of genus g_1 and g_2, which intersect at a node x_0 of X. Let $X^0 = X \setminus \{x_0\}$ and $\mathcal{O}_X(1)$ be a fixed ample line bundle on X. Fix integers $r > 0$, d and $\chi = d + r(1-g)$.

For any sheaf E on X, $P(E,n) = \chi(E(n)) := dim\mathrm{H}^0(E(n)) - dim\mathrm{H}^1(E(n))$ is called Hilbert polynomial of E. If E is a torsion free sheaf on X, and let r_i denote the rank of restriction of E to X_i $(i = 1, 2)$. The rank of E is defined to be

$$rk(E) := \frac{1}{deg(\mathcal{O}_X(1))} \cdot \lim_{n\to\infty} \frac{P(E,n)}{n}.$$

Let $c_i = deg(\mathcal{O}_X(1)|_{X_i})$, $a_i = \frac{c_i}{c_1+c_2}$, then

$$P(E,n) = (c_1r_1 + c_2r_2)n + \chi(E), \quad rk(E) = a_1r_1 + a_2r_2.$$

Definition 1.1. *A sheaf E on X is called semistable (resp. stable) if, for any subsheaf $E_1 \subset E$, one has*

$$\chi(E_1) \leq \ (resp. \ <) \ \frac{\chi(E)}{rk(E)} \cdot rk(E).$$

The moduli functor $\mathcal{U}_X^\natural : (\mathbb{C} - schemes) \to (sets)$ was defined as

$$\mathcal{U}_X^\natural(S) = \left\{ \begin{array}{l} \mathcal{O}_S\text{-flat semistable sheaves } E \text{ on } X \times S \\ \text{of rank } r \text{ and } \chi(E|_{X\times\{s\}}) = \chi \text{ for any } s \in S \end{array} \right\}$$

and it is known that there exists a projective scheme $\mathcal{U}_X$, which universally corepresents the functor $\mathcal{U}_X^\natural$.

For any integer N and polynomial $P(n) = r(c_1+c_2)n+\chi$, let $\mathcal{W} = \mathcal{O}_X(-N)^{P(N)}$ and $\mathcal{W}_S = \mathcal{O}_{X\times S}(-N)^{P(N)}$. Recall that $Quot_{P(n)}(\mathcal{W})$ is Grothendieck quotient scheme, whose S-valued points may be described as the set of quotients

$$\mathcal{W}_S \twoheadrightarrow F$$

on $X \times S$, where F is flat over S and its Hilbert polynomial is $P(n)$. Let $\mathcal{R}^{ss} \subset Quot_{P(n)}(\mathcal{W})$ (resp. $\mathcal{R}^s$) be the open set where the sheaf F is semistable (resp. stable) and $\mathcal{W}_S \twoheadrightarrow F$ induces an isomorphism $H^0(\mathcal{W}_S(N)) \cong H^0(F(N))$. Since the set of semistable sheaves with fixed Hilbert polynomial is bounded, we can assume that N is chosen large enough so that: every semistable sheaf with Hilbert polynomial $P(n)$ appears as a point in $\mathcal{R}^{ss}$. The group $SL(P(N))$ acts on $Quot_{P(n)}(\mathcal{W})$ and thus on $\mathcal{R}^{ss}$. The moduli space $\mathcal{U}_X$ was constructed as a good quotient

$$\mathcal{U}_X = \mathcal{R}^{ss}//SL(P(N)).$$

More precisely, the following were known (See [Si] for the more general results)

Theorem 1.2. *([Se], [Si]) Let $\mathcal{U}_X = \mathcal{R}^{ss}//SL(P(N))$ be the good quotient. Then*

(1) *There exists a natural transformation $\mathcal{U}_X^\natural \to \mathcal{U}_X$ such that $\mathcal{U}_X$ universally corepresents $\mathcal{U}_X^\natural$.*

(2) *$\mathcal{U}_X$ is projective, and its geometric points are*

$$\mathcal{U}_X(\mathbb{C}) = \left\{ \begin{array}{l} s\text{-equivalent classes of semistable} \\ \text{sheaves of rank } r \text{ and degree } d \end{array} \right\}.$$

(3) *There is an open subset $\mathcal{U}_X^s \subset \mathcal{U}_X$, with inverse image equal to $\mathcal{R}^s$, whose points represent isomorphism classes of stable sheaves. Locally in the étale topology on $\mathcal{U}_X^s$, there exists a universal sheaf $\mathcal{E}^{univ}$ such that if $\mathcal{E} \in \mathcal{U}_X^\natural(S)$ whose fibres $\mathcal{E}_s$ are stable, then the pull-back of $\mathcal{E}^{univ}$ via $S \to \mathcal{U}_X^s$ is isomorphic to $\mathcal{E}$ after tensoring with the pull-back of a line bundle on S.*

Let $\mathcal{R}_0 \subset \mathcal{R}^{ss}$ be the dense open set of locally free sheaves. For each $F \in \mathcal{R}_0$, let $F_1 = F|_{X_1}$, $F_2 = F|_{X_2}$, we have $\chi(F_1) + \chi(F_2) = \chi + r$ and (by semistability)

$$a_1\chi \leq \chi(F_1) \leq a_1\chi + r, \quad a_2\chi \leq \chi(F_2) \leq a_2\chi + r. \tag{1.1}$$

Thus $\mathcal{R}_0$ is the disjoint union of $\mathcal{R}_0^{\chi_1,\chi_2} = \{F \in \mathcal{R}_0 |\ \chi(F|_{X_i}) = \chi_i\}$ where $\chi_1+\chi_2 = \chi + r$ satisfying (1.1). Let $\mathcal{R}^{\chi_1,\chi_2}$ be the Zariski closure of $\mathcal{R}_0^{\chi_1,\chi_2}$ in $\mathcal{R}^{ss}$, then

$$\mathcal{R}^{ss} = \bigcup_{\chi_1,\chi_2} \mathcal{R}^{\chi_1,\chi_2}$$

has at most $r+1$ irreducible components. Let $\mathcal{U}_{X_i}$ be the moduli space of semistable vector bundles F_i of rank r and $\chi(F_i) = \chi_i$. It was known that for all possible choices of χ_1, χ_2 satisfying (1.1), $\mathcal{R}^{\chi_1,\chi_2}$ is not empty if $\mathcal{U}_{X_1}$ and $\mathcal{U}_{X_2}$ are not empty.

$$\mathcal{U}_X = \bigcup_{\chi_1,\chi_2} \mathcal{U}^{\chi_1,\chi_2} = \bigcup_{\chi_1,\chi_2} \mathcal{R}^{\chi_1,\chi_2}//SL(P(N))$$

has $\sharp\{(\chi_1,\chi_2)\}$ irreducible components.

Let $\mathcal{L}_0$ be a line bundle of degree d on X, we define a subfunctor $S\mathcal{U}_X^\natural$ of $\mathcal{U}_X^\natural$ by

$$S\mathcal{U}_X^\natural(S) = \left\{ \begin{array}{l} \mathcal{O}_S\text{-flat semistable sheaves } E \text{ on } X \times S \text{ of rank } r \text{ satisfying} \\ det(E|_{X_S^0}) = p_X^*(\mathcal{L}_0)|_{X_S^0} \text{ and } \chi(E|_{X\times\{s\}}) = \chi \text{ for any } s \in S \end{array} \right\}$$

We will prove that there exists a closed subscheme $S\mathcal{U}_X \subset \mathcal{U}_X$ which is reduced, and universally corepresents the moduli functor $S\mathcal{U}_X^\natural$. To do that, we first prove that the local deformation functor of it is pro-representable. Let Λ be the category of Artinian local $\mathbb{C}$-algebras. Morphisms in Λ are local homomorphisms of $\mathbb{C}$-algebras. For any $A \in \Lambda$, we denote $X \times Spec(A)$ (resp. $X^0 \times Spec(A)$) by X_A (resp. X_A^0). At any point $e_0 = (\mathcal{W} \twoheadrightarrow E_0)$ of the $Quot$ scheme $Quot_{P(n)}(\mathcal{W})$, we have the local deformation functor

$$G(A) = \{A\text{-flat quotients } (\mathcal{W}_A \twoheadrightarrow E), \text{ with } \chi(E(n)) = P(n)\}$$

such that $G(\mathbb{C}) = \{e_0\}$. It is well known that $\hat{\mathcal{O}}_{Quot,e_0}$ pro-represents G. Let $\mathcal{L}_0$ be a fixed line bundle on X such that $det(E_0|_{X^0}) = \mathcal{L}_0|_{X^0}$. We can define a subfunctor of G by

$$F(A) = \{(\mathcal{W}_A \twoheadrightarrow E) \in G(A) \,|\, det(E|_{X_A^0}) = p_{X^0}^* \mathcal{L}_0\},$$

where p_{X^0} (resp. p_X, p_A) denotes the projection to X^0 (resp. X, $Spec(A)$).

Proposition 1.3. *The functor F is pro-representable.*

Proof. Let $A_1 \to A$ and $A_2 \to A$ be morphisms in Λ, and consider the map

$$F(A_1 \times_A A_2) \to F(A_1) \times_{F(A)} F(A_2). \tag{1.2}$$

By Theorem 2.11 of [Sc], it is enough to show that the map (1.2) is bijective when $A_2 \to A$ is a small extension and the tangent space t_F is a finite dimensional vector space. But the bijectivity of (1.2) implies that t_F has a vector space structure compatible with that of t_G (See Remark 2.13 of [Sc]). Thus $dim_{\mathbb{C}}(t_F) < \infty$ and we only need to check the bijectivity of (1.2).

Let $(\mathcal{W}_{A_1} \twoheadrightarrow E_1) \in F(A_1)$, $(\mathcal{W}_A \twoheadrightarrow E) \in F(A)$ $(\mathcal{W}_{A_2} \twoheadrightarrow E_2) \in F(A_2)$ such that the restriction morphisms $E_1 \xrightarrow{u_1} E$, $E_2 \xrightarrow{u_2} E$ induce isomorphisms

$$(\mathcal{W}_{A_1} \twoheadrightarrow E_1)|_{X_A} \cong (\mathcal{W}_A \twoheadrightarrow E), \quad (\mathcal{W}_{A_2} \twoheadrightarrow E_2)|_{X_A} \cong (\mathcal{W}_A \twoheadrightarrow E).$$

Let $B = A_1 \times_A A_2$, then, since G is pro-representable, there exists a unique $(\mathcal{W}_B \twoheadrightarrow \mathcal{E}) \in G(B)$ such that

$$(\mathcal{W}_B \twoheadrightarrow \mathcal{E})|_{X_{A_1}} \cong (\mathcal{W}_{A_1} \twoheadrightarrow E_1), \quad (\mathcal{W}_B \twoheadrightarrow \mathcal{E})|_{X_{A_2}} \cong (\mathcal{W}_{A_2} \twoheadrightarrow E_2),$$

where $\cong$ means equality as points of $Quot$ scheme. In particular, there exist morphism $\mathcal{E} \xrightarrow{q_1} E_1$, $\mathcal{E} \xrightarrow{q_2} E_2$ such that $u_1 \cdot q_1 = u_2 \cdot q_2$ and q_1, q_2 induce isomorphisms $\mathcal{E}|_{X_{A_1}} \cong E_1$, $\mathcal{E}|_{X_{A_2}} \cong E_2$. By restricting everything to X_B^0 and taking wedge product, we have (for $i = 1, 2$)

$$det(\mathcal{E}^0) \xrightarrow{det(q_i^0)} det(E_i^0), \quad det(E_i^0) \xrightarrow{det(u_i^0)} det(E^0)$$

satisfying $det(u_1^0) \cdot det(q_1^0) = det(u_2^0) \cdot det(q_2^0)$. Thus, by Corollary 3.6 of [Sc],

$$det(\mathcal{E}^0) \cong det(E_1^0) \times_{det(E^0)} det(E_2^0).$$

To prove that $det(\mathcal{E}^0) \cong p_{X^0}^*(\mathcal{L})$, we imitate the arguments of uniqueness in the proof of Proposition 3.2. Since $p_{X^0}^*(\mathcal{L})|_{X_{A_i}^0} \cong det(E_i^0)$, we have morphisms

$$p_{X^0}^*(\mathcal{L}) \xrightarrow{p_1} det(E_1^0), \quad p_{X^0}^*(\mathcal{L}) \xrightarrow{p_2} det(E_2^0)$$

which induce the isomorphisms and thus a commutative diagram

$$\begin{array}{ccc} det(E_2^0) & \xrightarrow{det(u_2^0)} & det(E^0) \\ {\scriptstyle p_2}\uparrow & & \\ p_{X^0}^*(\mathcal{L}) & & \uparrow{\scriptstyle \theta} \\ {\scriptstyle p_1}\downarrow & & \\ det(E_1^0) & \xrightarrow{det(u_1^0)} & det(E^0) \end{array}$$

where θ is an automorphism of $det(E^0)$. If there exists a morphism

$$det(E_2^0) \xrightarrow{\theta_2} det(E_2^0)$$

such that

$$\begin{array}{ccc} det(E_2^0) & \xrightarrow{\theta_2} & det(E_2^0) \\ {\scriptstyle det(u_2^0)}\downarrow & & \downarrow{\scriptstyle det(u_2^0)} \\ det(E^0) & \xrightarrow{\theta} & det(E^0) \end{array}$$

is commutative, then θ_2 has to be an isomorphism (since $A_2 \to A$ is a small extension) by Lemma 3.3 of [Sc]. Thus we can modify the morphism $p_{X^0}^*(\mathcal{L}) \xrightarrow{p_2} det(E_2^0)$ to

$$\widetilde{p}_2 : p_{X^0}^*(\mathcal{L}) \xrightarrow{p_2} det(E_2^0) \xrightarrow{\theta_2^{-1}} det(E_2^0),$$

so that $det(u_1^0) \cdot p_1 = det(u_2^0) \cdot \widetilde{p}_2$. Thus $det(\mathcal{E}^0) \cong p_{X^0}^*(\mathcal{L})$ and we are done if the lift of θ is always possible. But this is equivalent to the surjectivity of the canonical map

$$H^0(X_{A_2}^0, \mathcal{O}_{X_{A_2}^0}) \to H^0(X_A^0, \mathcal{O}_{X_A^0}),$$

which is true since for any finite dimensional $\mathbb{C}$-algebra A, we have

$$H^0(X^0_A, \mathcal{O}_{X^0_A}) = H^0(X^0, \mathcal{O}_{X^0}) \otimes_{\mathbb{C}} A.$$

For any point $e_0 = (\mathcal{W} \twoheadrightarrow E_0)$, we observe that F has the same singularity with G at e_0. To see it, we define a functor $T : \Lambda \to \mathbf{Set}$ by

$$T(A) = \{\text{Isomorphism classes of } A\text{-flat torsion free } (\mathcal{O}_{X,x_0} \otimes A)\text{-modules}\}$$

such that $T(\mathbb{C}) = \{E_0 \otimes \mathcal{O}_{X,x_0}\}$. There is a morphism of functors $\phi : G \to T$ defined by

$$\phi((\mathcal{W} \twoheadrightarrow E_A)) = E_A \otimes (\mathcal{O}_{X,x_0} \otimes A).$$

It is known that ϕ is formally smooth (See Theorem 4.1 of [F2], or Proposition (4.1) of [NS]). Here we remark that its restriction to the subfunctor F is also formally smooth.

Lemma 1.4. *The morphism $\phi : F \to T$ is formally smooth.*

Proof. Let $B \to A$ be a small extension, one need to check the surjectivity of

$$F(B) \to F(A) \times_{T(A)} T(B).$$

For any $(\mathcal{W}_A \twoheadrightarrow E_A) \in F(A)$ and $N \in T(B)$ satisfying $E_A \otimes (\mathcal{O}_{X,x_0} \otimes A) \cong N \otimes_B A$, we can find an open cover of X consists two affine sets U_1, U_2 such that $x_0 \in U_2 \setminus U_1$ and $U_1 \otimes Spec(A)$, $(U_2 \setminus \{x_0\}) \otimes Spec(A)$ trivializing the vector bundle $E^0_A := E_A|_{X^0_A}$ (See Proposition (4.1) of [NS]). Thus the gluing data of E^0_A is a matrix $M \in GL(\mathcal{O}_X(U_1 \cap U_2) \otimes A)$ and E_A is obtained by gluing E^0_A and $N \otimes_B A$. Then the lift $E_B \in F(B)$ was obtained by lifting the gluing data (See Proposition (4.1) of [NS]). Since $SL(\mathcal{O}_X(U_1 \cap U_2) \otimes B) \to SL(\mathcal{O}_X(U_1 \cap U_2) \otimes A)$ is surjective, it is clear that we can choose a lift $\widetilde{M} \in GL(\mathcal{O}_X(U_1 \cap U_2) \otimes B)$ of M such that the resulting sheaf $E_B \in F(B)$. This proves the lemma.

Corollary 1.5. *The functor F is pro-represented by a reduced semi-normal complete local $\mathbb{C}$-algebra.*

Proof. This follows Lemma 1.4 and the study of functor T in [F2] and [Se].

To construct the closed subscheme $\mathcal{SU}_X$, which will universally corepresent the moduli functor $\mathcal{SU}^\natural_X$, we recall some constructions in [S1] and [S2]. Let $\pi : \widetilde{X} \to X$ be the normalization of X and $\pi^{-1}(x_0) = \{x_1, x_2\}$, then $\widetilde{X}$ is a disjoint union of X_1 and X_2 (we will identify x_1, x_2 with x_0 when we work on X). A GPB (E, Q) of rank r on $\widetilde{X}$ is a vector bundle E of rank r on $\widetilde{X}$ (its restriction to X_i is denoted by

E_i), together with a quotient $E_{x_1} \oplus E_{x_2} \to Q$ of dimension r. We have constructed the moduli space

$$\mathcal{P} := \coprod_{\chi_1+\chi_2=\chi+r} \mathcal{P}_{\chi_1,\chi_2}$$

of s-equivalence classes of semistable GPB (E,Q) on $\widetilde{X}$ of rank r and $\chi(E) = \chi + r$ (See [S2]), where $\mathcal{P}_{\chi_1,\chi_2} = \{(E,Q) \in \mathcal{P} \,|\, \chi(E_i) = \chi_i\}$ and χ_1, χ_2 satisfy (1.1). There are also finite morphisms (See [S2])

$$\phi_{\chi_1,\chi_2} : \mathcal{P}_{\chi_1,\chi_2} \to \mathcal{U}_X^{\chi_1,\chi_2} \subset \mathcal{U}_X$$

such that $\phi = \coprod \phi_{\chi_1,\chi_2} : \mathcal{P} \to \mathcal{U}_X$ is the normalization of $\mathcal{U}_X$. The morphism ϕ is defined such that $\phi([(E,Q)]) := F$ satisfies the exact sequence

$$0 \to F \to \pi_* E \to {}_{x_0}Q \to 0.$$

A straightforward generalization of Lemma 5.7 in [S1] shows that there exists a morphism

$$Det : \mathcal{P}_{\chi_1,\chi_2} \to J_{X_1}^{\chi_1-r(1-g_1)} \times J_{X_2}^{\chi_2-r(1-g_2)}$$

such that $Det([(E,Q)]) = (det(E_1), det(E_2))$. Let $L_i = \mathcal{L}_0|_{X_i}$ $(i = 1,2)$ and

$$L_1(n_1) := L_1 \otimes \mathcal{O}_{X_1}(n_1 x_1), \quad L_2(n_2) := L_2 \otimes \mathcal{O}_{X_2}(n_2 x_2),$$

where $n_i = \chi_i - r(1-g_i) - deg(L_i)$. We have the closed subschemes

$$\mathcal{P}_{\chi_1,\chi_2}^{n_1,n_2} := Det^{-1}((L_1(n_1), L_2(n_2)),$$

where n_1, n_2 are determined uniquely by χ_1, χ_2 satisfying (1.1) since $\mathcal{L}_0$ was fixed. Thus we denote the closed subsets $\phi_{\chi_1,\chi_2}(\mathcal{P}_{\chi_1,\chi_2}^{n_1,n_2}) \subset \mathcal{U}_X^{\chi_1,\chi_2}$ by $\mathcal{SU}_X^{\chi_1,\chi_2}$, and define $\mathcal{SU}_X$ to be the closed subset

$$\bigcup_{\chi_1+\chi_2=\chi+r} \mathcal{SU}_X^{\chi_1,\chi_2} \subset \mathcal{U}_X$$

with the reduced scheme structure. Then we have

Theorem 1.6. *Let $\mathcal{SU}_X$ be the closed subscheme of $\mathcal{U}_X$ defined above. Then*

(1) *The natural transformation in Theorem 1.2 induces a transformation*

$$\mathcal{SU}_X^{\natural} \to \mathcal{SU}_X$$

such that $\mathcal{SU}_X$ *universally corepresents* $\mathcal{SU}_X^{\natural}$.

(2) $\mathcal{SU}_X$ *is a projective, seminormal variety of dimension* $(r^2-1)(g-1)$. *The number of irreducible components of* $\mathcal{SU}_X$ *is the same with that of* $\mathcal{U}_X$, *and its geometric points are*

$$\mathcal{SU}_X(\mathbb{C}) = \left\{ \begin{array}{l} s\text{-equivalent classes of semistable sheaves } \mathcal{E} \\ \text{of rank } r \text{ and degree } d \text{ with } det(\mathcal{E}|_{X^0}) = \mathcal{L}_0|_{X^0} \end{array} \right\}.$$

(3) *There is an open subset* $\mathcal{SU}_X^s \subset \mathcal{SU}_X$ *whose points represent isomorphism classes of stable sheaves with fixed determinant* $\mathcal{L}_0$ *on* $X^0 = X\backslash\{x_0\}$. *Locally in the étale topology on* $\mathcal{SU}_X^s$, *there exists a universal sheaf* $\mathcal{E}^{univ}$ *such that if* $\mathcal{E} \in \mathcal{SU}_X^{\natural}(S)$ *whose fibres* $\mathcal{E}_s$ *are stable, then the pull-back of* $\mathcal{E}^{univ}$ *via* $S \to \mathcal{SU}_X^s$ *is isomorphic to* $\mathcal{E}$ *after tensoring with the pull-back of a line bundle on* S.

Proof. Firstly, it is easy to check (2). In fact, the projectivity and the number of components follow from the construction of $\mathcal{SU}_X$. The semi-normality of $\mathcal{SU}_X$ follows from Corollary 1.5. To show that the set $\mathcal{SU}_X(\mathbb{C})$ consists of the sheaves F satisfying $det(F|_{X^0}) = \mathcal{L}_0|_{X^0}$, we only need a fact that any line bundles $\mathcal{L}_1$, $\mathcal{L}_2$ on a smooth projective curve Y satisfy $\mathcal{L}_1|_{Y\backslash\{y\}} = \mathcal{L}_2|_{Y\backslash\{y\}}$ if and only if $\mathcal{L}_1 = \mathcal{L}_2 \otimes \mathcal{O}_Y(ky)$ for some integer k.

To prove (1), let G be the functor represented by $\mathcal{R}^{ss}$ and F be the subfunctor defined by $F(S) = \{(\mathcal{W}_S \twoheadrightarrow \mathcal{E}) \in G(S) \,|\, det(\mathcal{E}|_{X_S^0}) = p_{X^0}^*\mathcal{L}_0\}$. Let $Z \subset \mathcal{R}^{ss}$ be the inverse image of $\mathcal{SU}_X$ and $\mathcal{I}_Z$ denote the ideal sheaf of Z. It is enough to show that for any $(\mathcal{W}_S \twoheadrightarrow \mathcal{E}) \in F(S)$, the morphism $\varphi_S : S \to \mathcal{R}^{ss}$ factors through $\varphi_S : S \to Z \subset \mathcal{R}^{ss}$. Namely, one has to prove $\varphi_S^*(\mathcal{I}_Z) = 0$, where $\varphi_S^* : \mathcal{O}_{\mathcal{R}^{ss}} \to \varphi_{S*}\mathcal{O}_S$. This is a local problem, it is enough to show that for any $s \in S$ the morphism

$$\varphi_{S,s} : \hat{S} = Spec(\hat{\mathcal{O}}_{S,s}) \to \mathcal{R}^{ss}$$

factors through $Spec(\hat{\mathcal{O}}_{Z,\varphi_S(s)})$. By Proposition 1.3 and Corollary 1.5, there exists a complete noetherian local $\mathbb{C}$-algebra R and $\xi = \varprojlim(\xi_n)$, where

$$\xi_n = (\mathcal{W}_{Spec(R/m^n)} \twoheadrightarrow \mathcal{E}_n) \in F(Spec(R/m^n)),$$

pro-represents the local deformation functor of F at s. Thus

$$\varphi_{S,s} : \hat{S} = Spec(\hat{\mathcal{O}}_{S,s}) \to \mathcal{R}^{ss}$$

factors through $f : Spec(R) \to \mathcal{R}^{ss}$, where the pullback $f^*(\mathcal{W}_{\mathcal{R}^{ss}} \twoheadrightarrow \mathcal{E}^{univ})$ of the universal quotient is $\xi = \varprojlim(\xi_n)$, which is in fact an element of $F(Spec(R))$ in our case (it is not true for general theory). Thus $f = f_\xi : Spec(R) \to \mathcal{R}^{ss}$ was given by an element $\xi \in F(Spec(R))$. Now we can use Lemma 1.7 below, which implies that $f_\xi^*(\mathcal{I}_Z) = \sqrt{0}$, to prove that f_ξ (and thus $\varphi_{S,s}$) factors through Z since R is reduced.

Having shown (1) and (2), the proof of claim (3) is the same with that of [Si].

Lemma 1.7. *Let $\xi \in F(S)$ and $f_\xi : S \to \mathcal{R}^{ss}$ be the induced morphism by ξ, then $f_\xi(\wp) \in Z$ for any point $\wp \in S$ (including non-closed points).*

Proof. Let $\xi = (\mathcal{W}_S \twoheadrightarrow \mathcal{F}) \in F(S)$. When S is reduced, we have a stratification $S = \coprod S_a$ of S, where $S_a := \{s \in S | \mathbf{a}(\mathcal{F}_s) = a\}$ are locally closed subschemes ($\mathbf{a}(\mathcal{F}_s)$ was defined by $\mathcal{F}_s \otimes \hat{\mathcal{O}}_{X,x_0} = \hat{\mathcal{O}}_{X,x_0}^{\oplus \mathbf{a}(\mathcal{F}_s)} \oplus \hat{m}_{x_0}^{\oplus(r-\mathbf{a}(F_s))}$), then using Lemma 2.7 of [S1] (for S_a) we get $f_\xi(S_a) \subset Z$. thus proves the lemma. For general S, we use the flattening stratifications $S = \coprod S_i$ of Mumford (See Lecture 8 of [Mu]) such that the sheaves used in the construction of Lemma 2.7 of [S1] are flat on S_i, then Lemma 2.7 goes through, thus the lemma is proved for general S.

§2 Degeneration of moduli space of semistable $SL(r)$-bundles

Let D be a complete discrete valuation ring with maximal idea $m_D = (t)D$ and $\mathcal{X} \to B = Spec(D)$ a flat family of proper connected curves. Assume that the generic fibre $\mathcal{X}_\eta$ is smooth and the closed fibre $\mathcal{X}_0$ is the curve X discussed in Section 1. Fix a relative ample line bundle $\mathcal{O}_{\mathcal{X}}(1)$ on $\mathcal{X}$ such that $\mathcal{O}_{\mathcal{X}}(1)|_{\mathcal{X}_0} = \mathcal{O}_X(1)$. Let $\mathcal{W}_{\mathcal{X}} = \mathcal{O}_{\mathcal{X}}(-N)^{\oplus P(N)}$ and $Quot_{P(n)}(\mathcal{W}_{\mathcal{X}}) \to B$ be the relative Grothendieck quotient scheme. For any B-scheme S, we will write $\mathcal{X}_S$ (resp. $\mathcal{W}_{\mathcal{X}_S}$) for $\mathcal{X} \times_B S$ (resp. $\mathcal{O}_{\mathcal{X}\times_B S}(-N)^{\oplus P(N)}$). Let $\mathcal{R}^{ss}_{\mathcal{X}}$ be the open set of semistable sheaves whose quotient map induces isomorphism $\mathcal{O}_S^{P(N)} \cong H^0(\mathcal{E}(N))$. It is well-known that the relative moduli space of semistable torsion free sheaves is the relative good quotient

$$\begin{array}{ccc} \mathcal{R}^{ss}_{\mathcal{X}} & \longrightarrow & \mathcal{U}_{\mathcal{X}} := \mathcal{R}^{ss}_{\mathcal{X}}//SL(P(N)) \\ {\scriptstyle f}\downarrow & & \downarrow{\scriptstyle \tilde{f}} \\ B & = & B \end{array}$$

Let $\mathcal{L}$ be a line bundle on $\mathcal{X}$ such that $\mathcal{L}|_{\mathcal{X}_0} = \mathcal{L}_0$ and $\mathcal{L}_\eta = \mathcal{L}|_{\mathcal{X}_\eta}$. Let

$$\mathcal{R}^{ss}_{\mathcal{X}_\eta}(\mathcal{L}_\eta) \subset \mathcal{R}^{ss}_{\mathcal{X}_\eta}$$

be the closed subscheme of sheaves with fixed determinant $\mathcal{L}_\eta$, and

$$S\mathcal{U}_{\mathcal{X}_\eta} := \mathcal{R}^{ss}_{\mathcal{X}_\eta}(\mathcal{L}_\eta)//SL(P(N))$$

be the moduli space of semistable bundles on $\mathcal{X}_\eta$ with fixed determinant $\mathcal{L}_\eta$. Let

$$\mathcal{R}^{ss}_{\mathcal{X}}(\mathcal{L}) := \overline{\mathcal{R}^{ss}_{\mathcal{X}_\eta}(\mathcal{L}_\eta)} \subset \mathcal{R}^{ss}_{\mathcal{X}}, \quad \mathcal{SU}_{\mathcal{X}} := \overline{\mathcal{SU}_{\mathcal{X}_\eta}} \subset \mathcal{U}_{\mathcal{X}} \tag{2.1}$$

be the Zariski closure of $\mathcal{R}^{ss}_{\mathcal{X}_\eta}(\mathcal{L}_\eta)$ and $\mathcal{SU}_{\mathcal{X}_\eta}$ in $\mathcal{R}^{ss}_{\mathcal{X}}$ and $\mathcal{U}_{\mathcal{X}}$ respectively. Then $\mathcal{SU}_{\mathcal{X}}$ is the relative good quotient of $\mathcal{R}^{ss}_{\mathcal{X}}(\mathcal{L})$. We will prove that

$$\widetilde{f} : \mathcal{SU}_{\mathcal{X}} \to B$$

universally corepresents a moduli functor $\mathcal{SU}^{\natural}_{\mathcal{X}}$, which is defined as

$$\mathcal{SU}^{\natural}_{\mathcal{X}}(S) = \left\{ \begin{array}{c} \mathcal{O}_S\text{-flat semistable sheaves } \mathcal{E} \text{ on } \mathcal{X}_S := \mathcal{X} \times_B S \text{ of} \\ \text{rank } r \text{ and degree } d \text{ satisfying } det(\mathcal{E}|_{\mathcal{X}^0_S}) = p^*_{\mathcal{X}^0}\mathcal{L} \end{array} \right\},$$

where $\mathcal{X}^0 := \mathcal{X} \setminus \{x_0\}$ and $p_{\mathcal{X}^0} : \mathcal{X}^0_S := \mathcal{X}^0 \times_B S \to \mathcal{X}^0$ denotes the projection. Actually, we will prove that the functor

$$\mathcal{R}^{\natural}_{\mathcal{X}}(\mathcal{L})(S) = \left\{ \begin{array}{l} \mathcal{O}_S\text{-flat quotients } (\mathcal{W}_{\mathcal{X}_S} \twoheadrightarrow \mathcal{E}_S) \\ \text{on } \mathcal{X}_S \text{ such that } \mathcal{E}_S \in \mathcal{SU}^{\natural}_{\mathcal{X}}(S). \end{array} \right\}$$

is represented by the closed subscheme $\mathcal{R}^{ss}_{\mathcal{X}}(\mathcal{L}) \subset \mathcal{R}^{ss}_{\mathcal{X}}$.

Lemma 2.1. *The functor $\mathcal{R}^{\natural}_{\mathcal{X}}(\mathcal{L})$ is represented by the closed subscheme $\mathcal{R}^{ss}_{\mathcal{X}}(\mathcal{L}) \subset \mathcal{R}^{ss}_{\mathcal{X}}$ and the restriction of universal quotient on $\mathcal{X} \times_B \mathcal{R}^{ss}_{\mathcal{X}}$.*

Proof. For any $(\mathcal{W}_{\mathcal{X}_S} \twoheadrightarrow \mathcal{E}_S) \in \mathcal{R}^{\natural}_{\mathcal{X}}(\mathcal{L})(S)$, there is a unique morphism

$$\varphi_S : S \to \mathcal{R}^{ss}_{\mathcal{X}}$$

such that pullback of the universal quotient $(\mathcal{W}_{\mathcal{X}_{\mathcal{R}^{ss}_{\mathcal{X}}}} \twoheadrightarrow \mathcal{E}^{univ}_{\mathcal{R}^{ss}_{\mathcal{X}}})$ is $(\mathcal{W}_{\mathcal{X}_S} \twoheadrightarrow \mathcal{E}_S)$. It is enough to show that φ_S is factorized through

$$\begin{array}{ccccc} S & \xrightarrow{\varphi_S} & \mathcal{R}^{ss}_{\mathcal{X}}(\mathcal{L}) & \longrightarrow & \mathcal{R}^{ss}_{\mathcal{X}} \\ \downarrow & & {\scriptstyle f_{\mathcal{L}}}\downarrow & & {\scriptstyle f}\downarrow \\ B & = & B & = & B \end{array}$$

This is true at the generic fibre, we only need to check the case when S is defined over $0 \in B$. In the proof of Theorem 1.6 (1), we have show that φ_S is factorized through (note that we are at the case of $\mathcal{X} \times_B S = \mathcal{X}_0 \times S$)

$$Z = \{(\mathcal{W} \twoheadrightarrow \mathcal{E}) \in \mathcal{R}^{ss}_{\mathcal{X}_0} \,|\, det(\mathcal{E}|_{X^0}) = p^*_{X^0}(\mathcal{L}_0)\}.$$

Thus the lemma follows the equality: $f^{-1}_{\mathcal{L}}(0) := \mathcal{R}^{ss}_{\mathcal{X}}(\mathcal{L})_0 = Z$, which we will prove in the next lemma.

Lemma 2.2. $f_{\mathcal{L}}^{-1}(0) := \mathcal{R}_{\mathcal{X}}^{ss}(\mathcal{L})_0 = Z$.

Proof. We first prove that $\mathcal{R}_{\mathcal{X}}^{ss}(\mathcal{L})_0 \subset Z$. Let $\xi_0 = (\mathcal{W} \twoheadrightarrow \mathcal{E}_0) \in \mathcal{R}_{\mathcal{X}}^{ss}(\mathcal{L})_0$, we need to show that $det(\mathcal{E}_0|_{X^0}) = p_{X^0}^*(\mathcal{L}_0)$.

By the definition of $\mathcal{R}_{\mathcal{X}}^{ss}(\mathcal{L})$, there exists a complete discrete valuation ring $\widetilde{D}$ dominant D, and a morphism $T := Spec(\widetilde{D}) \xrightarrow{\varphi} \mathcal{R}_{\mathcal{X}}^{ss}(\mathcal{L})$ over B such that

$$\varphi(\widetilde{\eta}) \in \mathcal{R}_{\mathcal{X}_\eta}^{ss}(\mathcal{L}_\eta), \quad \varphi(\widetilde{0}) = \xi_0.$$

Namely, there is a $\mathcal{O}_T$-flat torsion free sheaf $\mathcal{E}$ on $\mathcal{X} \times_B T$ such that $det(\mathcal{E}|_{\mathcal{X}_\eta}) = \mathcal{L}_\eta$ and $\mathcal{E}|_{\mathcal{X}_0} = \mathcal{E}_0$.

Let $\varpi : \widetilde{\mathcal{X}}_T \to \mathcal{X}_T$ be the desingularization of $\mathcal{X}_T$ at x_0. The exceptional divisor $\varpi^{-1}(x_0) = \sum E_i$ is a chain of (-2)-curves, and the special fibre $\widetilde{\mathcal{X}}_0$ of $\widetilde{\mathcal{X}}_T \to T$ at $\widetilde{0}$ is $X_1 + X_2 + \sum E_i$, and $X_1 \cap (X_2 + \sum E_i) = \{x_1\}$, $X_2 \cap (X_1 + \sum E_i) = \{x_2\}$. Identifying $\widetilde{\mathcal{X}}_T \setminus \varpi^{-1}(x_0) \cong \mathcal{X}_T \setminus \{x_0\} = \mathcal{X}_T^0$, we can extend $det(\mathcal{E}|_{\mathcal{X}_T^0})$ into a line bundle $\overline{det(\mathcal{E}|_{\mathcal{X}_T^0})}$ on $\widetilde{\mathcal{X}}_T$, which satisfies that $(\varpi^* p_{\mathcal{X}}^* \mathcal{L})|_{\widetilde{\mathcal{X}}_\eta} = \overline{det(\mathcal{E}|_{\mathcal{X}_T^0})}|_{\widetilde{\mathcal{X}}_\eta}$. Thus one has

$$\overline{det(\mathcal{E}|_{\mathcal{X}_T^0})} = (\varpi^* p_{\mathcal{X}}^* \mathcal{L}) \otimes \mathcal{O}_{\widetilde{\mathcal{X}}_T}(V),$$

where $V \subsetneq \widetilde{\mathcal{X}}_0$ is a vertical divisor. Since $\mathcal{O}_{\widetilde{\mathcal{X}}_T}(\widetilde{\mathcal{X}}_0)$ is trivial, there are integers n_1, n_2 such that $\mathcal{O}_{\widetilde{\mathcal{X}}_T}(V)|_{X_1} = \mathcal{O}_{X_1}(n_1 x_1)$, $\mathcal{O}_{\widetilde{\mathcal{X}}_T}(V)|_{X_2} = \mathcal{O}_{X_2}(n_2 x_2)$. Hence

$$det(\mathcal{E}_0|_{X^0}) = det(\mathcal{E}|_{\mathcal{X}_T^0})|_{X^0} = \overline{det(\mathcal{E}|_{\mathcal{X}_T^0})}|_{X^0} = p_{X^0}^*(\mathcal{L}_0).$$

We are left to prove $Z \subset \mathcal{R}_{\mathcal{X}}^{ss}(\mathcal{L})_0$. Since the locus $\mathcal{R}_0 \subset f^{-1}(0) = \mathcal{R}_{\mathcal{X}_0}^{ss}$ of locally free sheaves is a dense open subset of $\mathcal{R}_{\mathcal{X}_0}^{ss}$, we only need to check that

$$\mathcal{R}_0 \cap Z \subset \mathcal{R}_{\mathcal{X}}^{ss}(\mathcal{L})_0$$

Let $\xi_0 = (\mathcal{W} \twoheadrightarrow \mathcal{E}_0) \in \mathcal{R}_0 \cap Z$, then $\mathcal{E}_0$ is a vector bundle on $\mathcal{X}_0 = X$ and $det(\mathcal{E}_0|_{X^0}) = p_{X^0}^*(\mathcal{L}_0)$, which implies that (for $i = 1, 2$)

$$det(\mathcal{E}_0)|_{X_i} = det(\mathcal{E}_0|_{X_i}) = \mathcal{L}_0|_{X_i} \otimes \mathcal{O}_{X_i}(n_i x_0), \quad n_1 + n_2 = 0.$$

Let $\widetilde{\mathcal{L}} = \mathcal{L} \otimes \mathcal{O}_{\mathcal{X}}(X_1 + (n_1+1)X_2)$ and $\widetilde{\mathcal{L}}_0 = \widetilde{\mathcal{L}}|_{\mathcal{X}_0}$. Then, since $\mathcal{O}_{\mathcal{X}}(X_1 + X_2) \cong \mathcal{O}_{\mathcal{X}}$, we have $\mathcal{O}_{\mathcal{X}}(X_1 + (n_1+1)X_2)|_{X_i} = \mathcal{O}_{X_i}(n_i x_0)$ and $\widetilde{\mathcal{L}}_0|_{X_i} = det(\mathcal{E}_0)|_{X_i}$, which implies that

$$det(\mathcal{E}_0) = \widetilde{\mathcal{L}}_0 = \widetilde{\mathcal{L}}|_{\mathcal{X}_0}.$$

Let $d_i = deg(\mathcal{E}_0|_{X_i})$ and $J^{d_i}_{X_i}$ $(i = 1, 2)$ be the Jacobian variety of line bundles of degree d_i on X_i. Let $J(\mathcal{X}_\eta)$ be the Jacobian varirty of line bundles of degree d on $\mathcal{X}_\eta$. Then $J(\mathcal{X}_\eta)$ can be compactified into a relative Jacobian variety

$$J(\mathcal{X}) \to B$$

such that its special fibre $J(\mathcal{X})_0 = J^{d_1}_{X_1} \times J^{d_2}_{X_2}$ (Otherwise, we can modify $J(\mathcal{X})$ through an isomorphism by tensoring a suitable line bundle $\mathcal{O}_{\mathcal{X}}(k_1X_1 + k_2X_2)$ on $\mathcal{X}$). Let $\mathcal{R}^0_{\mathcal{X}} \subset \mathcal{R}^{ss}_{\mathcal{X}}$ be the locus of locally free sheaves. By taking determinant, we have a morphism

$$Det : \mathcal{R}^0_{\mathcal{X}} \setminus \bigcup_{\chi_i \neq d_i + r(1-g_i)} \mathcal{R}^{\chi_1,\chi_2}_0 \to J(\mathcal{X}).$$

The line bundle $\widetilde{\mathcal{L}}$ on $\mathcal{X}$ gives a section $\sigma : B \to J(\mathcal{X})$ such that $\sigma(0) = \sigma(B) \cap J(\mathcal{X})_0$. Let $\mathcal{R}^\sigma := Det^{-1}(\sigma(B)) \to B$ and $\mathcal{E}^\sigma = \mathcal{E}^{univ}|_{\mathcal{X}\times_B \mathcal{R}^\sigma}$. Then

$$det(\mathcal{E}^\sigma) = p^*_{\mathcal{X}}(\widetilde{\mathcal{L}}) \otimes p^*_{\mathcal{R}^\sigma}(\mathcal{K})$$

for some line bundle $\mathcal{K}$ on $\mathcal{R}^\sigma$, and $\xi_0 \in \mathcal{R}^\sigma$. Thus, for any open set $\xi_0 \in U$ such that $\mathcal{K}$ is trivial on $U \to B$, we have $U \cap \mathcal{R}^{ss}_{\mathcal{X}}(\mathcal{L}_\eta) \neq \emptyset$. Thus

$$\xi_0 \in \overline{\mathcal{R}^{ss}\mathcal{X}(\mathcal{L}_\eta)},$$

which proves the lemma.

Let $\Lambda_{\mathcal{X}}$ be the category of Artinian local $D-$algebras. For any point

$$e = (\mathcal{W} \twoheadrightarrow \mathcal{E}_0) \in \mathcal{R}^{ss}_{\mathcal{X}}(\mathcal{L})_0,$$

let $F_{\mathcal{X}} : \Lambda_{\mathcal{X}} \to \mathbf{Set}$ denote the local deformation functor of $\mathcal{R}^{ss}_{\mathcal{X}}(\mathcal{L})$ at e. Similarly, we define a functor $T_{\mathcal{X}} : \Lambda \to \mathbf{Set}$ by

$$T_{\mathcal{X}}(A) = \{\text{Isomorphism classes of } A\text{-flat torsion free } (\mathcal{O}_{X,x_0} \otimes_D A)\text{-modules}\}$$

such that $T(\mathbb{C}) = \{\mathcal{E}_0 \otimes \mathcal{O}_{X,x_0}\}$. There is a morphism of functors

$$\phi : F_{\mathcal{X}} \to T_{\mathcal{X}}$$

defined by $\phi((\mathcal{W}_{\mathcal{X}} \twoheadrightarrow \mathcal{E}_{\mathcal{X}_A})) = \mathcal{E}_{\mathcal{X}_A} \otimes (\mathcal{O}_{\mathcal{X},x_0} \otimes_D A)$. Similar with Lemma 1.4, we have

Lemma 2.3. *The morphism $\phi : F_{\mathcal{X}} \to T_{\mathcal{X}}$ is formally smooth.*

Proof. The same arguments with Lemma 1.4, we just remark an easy fact: Let B be a flat D-algebra and M be a B-module, flat over D. Assume that $M \otimes_D D/m$ is a free B/mB-module. Then M is a free B-module.

Let $\mathbf{X} = (x_{ij})_{(r-a)\times(r-a)}$ and $\mathbf{Y} = (y_{ij})_{(r-a)\times(r-a)}$ be the $(r-a)\times(r-a)$ matrics, and

$$\mathcal{Z} = \mathrm{Spec}\frac{D[\mathbf{X},\mathbf{Y}]}{(\mathbf{X}\cdot\mathbf{Y}-t, \mathbf{Y}\cdot\mathbf{X}-t)}.$$

Let $0 \in \mathcal{Z}$ be the point defined by the idea $(\mathbf{X},\mathbf{Y},t)\mathcal{O}_{\mathcal{Z}}$. Then we have

Lemma 2.4. *Let $e = (\mathcal{W} \twoheadrightarrow \mathcal{E}_0) \in \mathcal{R}^{ss}_{\mathcal{X}}(\mathcal{L})_0$ such that $\mathbf{a}(\mathcal{E}_0) = a$. Then there exist ℓ_1 and ℓ_2 such that*

$$\hat{\mathcal{O}}_{\mathcal{R}^{ss}_{\mathcal{X}}(\mathcal{L}),e}[[u_1,\cdots,u_{\ell_1}]] \cong \hat{\mathcal{O}}_{\mathcal{Z},0}[[v_1,\cdots,v_{\ell_2}]].$$

Proof. This is the consequence of Lemma 2.3 and the results of [F2] and [NS].

Theorem 2.5. *Let $\mathcal{X} \to B$ be a regular scheme with closed fibre $\mathcal{X}_0 = X$ and a fixed relative ample line bundle $\mathcal{O}_{\mathcal{X}}(1)$. Let $\mathcal{L}$ be a line bundle on $\mathcal{X}$ such that $\mathcal{L}|_{\mathcal{X}_0} = \mathcal{L}_0$ and let $\widetilde{f} : \mathcal{SU}_{\mathcal{X}} \to B$ be defined as (2.1). Then*

1. *There is a natural transformation $\mathcal{SU}^{\natural}_{\mathcal{X}} \to \mathcal{SU}_{\mathcal{X}}$ such that $\widetilde{f} : \mathcal{SU}_{\mathcal{X}} \to B$ universally corepresents $\mathcal{SU}^{\natural}_{\mathcal{X}}$.*
2. *$\widetilde{f} : \mathcal{SU}_{\mathcal{X}} \to B$ is a flat family of projective varieties of dimension $(r^2-1)(g-1)$, whose general fibres $\widetilde{f}^{-1}(b)$ are moduli spaces $\mathcal{SU}_{\mathcal{X}_b})$ of semistable vector bundles of rank r and degree d with fixed determinant $\mathcal{L}_b = \mathcal{L}|_{\mathcal{X}_b}$, and its closed fibre $\widetilde{f}^{-1}(0)$ is the variety $\mathcal{SU}_X$ in Theorem 1.6.*

Proof. (1) follows from Lemma 2.1 and the same arguments of [Si]. (2) is clear.

REFERENCES

[F1] G. Faltings, *A proof for the Verlinde formula*, J. Algebraic Geom. **3** (1994), 347–374.

[F2] G. Faltings, *Moduli-stacks for bundles on semistable curves*, Math. Ann. **304** (1996), 489–515.

[Gi] D. Gieseker, *A degeneration of the moduli space of stable bundles*, J. Differential Geom. **19** (1984), 173–206.

[GL1] D. Gieseker and J. Li, *Irreducibility of moduli of rank 2 vector bundles on algebraic surfaces*, J. Differential Geom. **40** (1994), 23–104.

[GL2] D. Gieseker and J. Li, *Moduli of high rank vector bundles over surfaces*, J. Amer. Math. Soc. **9** (1996), 107–151.

[Ka1] Ivan Kausz, *A modular compactification of the general linear group*, Documenta Math. **5** (200), 553–594.

[Ka2] Ivan Kausz, *A Gieseker type degeneration of moduli stacks of vector bundles on curves*, Preprint (2001), 1–59.

[Li] K. Z. Li, *Lectures on moduli theory (in Chinese, unpublished)* (2001).

[Mu] D. Mumford, *Lectures on curves on an algebraic surface*, Annals of Math. Studies **59** (1966).

[NR] M.S.Narasimhan and T.R. Ramadas, *Factorisation of generalised theta functions I*, Invent. Math. **114** (1993), 565–623.

[NS] D.S. Nagaraj and C.S. Seshadri, *Degenerations of the moduli spaces of vector bundles on curves I*, Proc. Indian Acad. Sci.(Math. Sci.) **107** (1997), 101–137.

[NSe] D.S. Nagaraj and C.S. Seshadri, *Degenerations of the moduli spaces of vector bundles on curves II*, Proc. Indian Acad. Sci.(Math. Sci.) **109** (1999), 165–201.

[Sc] M. Schlessinger, *Functors of Artin Rings*, Trans. of AMS. **130** (1968), 208–222.

[Se] C.S. Seshadri, *Fibrés vectoriels sur les courbes algébriques*, Astérisque **96** (1982).

[Si] C. Simpson, *Moduli of representations of the fundamental group of a smooth projective variety I*, I.H.E.S. Publications Mathématiques **79** (1994), 47–129.

[S1] X. Sun, *Degeneration of moduli spaces and generalized theta functions*, J. Algebraic Geom. **9** (2000), 459–527.

[S2] X. Sun, *Factorization of generalized theta functions at reducible case*, math. AG/0004111 (2000), 1–33.

INSTITUTE OF MATHEMATICS, ACADEMIA SINICA, BEIJING 100080, CHINA
E-mail address: xsun@math08.math.ac.cn

DEPARTMENT OF MATHEMATICS, THE UNIVERSITY OF HONG KONG, POKFULAM ROAD, HONG KONG
E-mail address: xsun@maths.hku.hk

Refined Brill-Noether Locus
and
Non-Abelian Zeta Functions for Elliptic Curves

Lin WENG
Graduate School of Mathematics, Kyushu University, Japan

In this paper, new local and global non-abelian zeta functions for elliptic curves are defined using moduli spaces of semi-stable bundles. To understand them, we also introduce and study certain refined Brill-Noether locus in the moduli spaces. Examples of these new zeta functions and a justification of using only semi-stable bundles are given too.

1. Refined Brill-Noether Locus

1.1. Moduli Space of Semi-Stable Bundles

1.1.1 Indecomposable Bundles

Let E be an elliptic curve defined over $\overline{\mathbf{F}_q}$, an algebraic closure of the finite field $\mathbf{F}_q$ with q-elements.

Recall that a vector bundle V on E is called indecomposable if V is not the direct sum of two proper subbundles, and that every vector bundle on E may be written as a direct sum of indecomposable bundles, where the summands and their multiplicities are uniquely determined up to isomorphism. Thus to understand vector bundles, it suffices to study the indecomposable ones. To this end, we have the following result of Atiyah [At]. In the sequel, for simplicity, we always assume that the characteristic of $\mathbf{F}_q$ is strictly bigger than the rank of V.

Theorem. (Atiyah) *(a) For any $r \geq 1$, there is a unique indecomposable vector bundle I_r of rank r over E, all of whose Jordan-Hölder constituents are isomorphic to $\mathcal{O}_E$. Moreover, the bundle I_r has a canonical filtration*

$$\{0\} \subset F^1 \subset \ldots \subset F^r = I_r$$

with $F^i = I_i$ and $F^{i+1}/F^i = \mathcal{O}_E$;

(b) For any $r \geq 1$ and any integer a, relative prime to r and each line bundle λ over E of degree a, there exists up to isomorphism a unique indecomposable bundle $W_r(a;\lambda)$ over E of rank r with λ the determinant;

(c) The bundle $I_r(W_{r'}(a;\lambda)) = I_r \otimes W_{r'}(a;\lambda)$is indecomposable and every indecomposable bundle is isomorphic to $I_r(W_{r'}(a;\lambda))$ for a suitable choice of r, r', λ. Every bundle V over E is a direct sum of vector bundles of the form $I_{r_i}(W_{r'_i}(a_i;\lambda_i))$, for suitable choices of r_i, r'_i and λ_i. Moreover, the triples (r_i, r'_i, λ_i) are uniquely specified up to permutation by the isomorphism type of V.

Here note in particular that $W_r(1,\lambda) \simeq \lambda$, and that indeed $I_r(W_{r'}(a;\lambda))$ is the unique indecomposable bundle of rank rr' such that all of whose successive quotients in the Jordan-Hölder filtration are isomorphic to $W_{r'}(a;\lambda)$.

1.1.2. Semi-Stable Bundles

As above, let V be a vector bundle over E. Define its slope $\mu(V)$ by $\mu(V) := \deg(V)/\mathrm{rank}(V)$. Then, following Mumford ([Mu]), V is called stable (resp. semi-stable), if for any proper subbundle W of V, $\mu(W) < \mu(V)$ (resp. $\mu(W) \leq \mu(V)$). For example, $I_r(W_{r'}(a;\lambda))$ is semi-stable with $\mu(I_r(W_{r'}(a;\lambda))) = a/r'$.

Theorem. (Atiyah) *(a) Every bundle V over E is isomorphic to a direct sum $\oplus_i V_i$ of semi-stable bundles, where $\mu(V_i) > \mu(V_{i+1})$;*
(b) Let V be a semi-stable bundle over E with slope $\mu(V) = a/r'$ where r' is a positive integer and a is an integer relatively prime to r'. Then V is a direct sum of bundles of the form $I_r(W_{r'}(a;\lambda))$, where λ is a line bundle of degree a. *1.1.3. Moduli Space of Semi-Stable Bundles*

Let V be a semi-stable vector bundle, then we may associate it a Jordan-Hölder filtration, which is far from being unique. However the associated graded bundle, denoted as $\mathrm{gr}(V)$, is unique. Following Seshadri, two semi-stable vector bundles V and V' are called S-equivalent, denoted by $V \sim_S V'$, if $\mathrm{gr}(V) \simeq \mathrm{gr}(V')$.

Now set

$$\mathcal{M}_{E,r}(\lambda) = \{V : \mathrm{semi-stable}, \mathrm{det}V = \lambda, \mathrm{rank}(V) = r\}/\sim_S .$$

Then from the above classification we have the following well-known

Theorem. (Atiyah, Mumford-Seshadri) *With respect to a fixed pair (r,λ), there exists a natural projective algebraic variety structure on $\mathcal{M}_{E,r}(\lambda)$. Moreover, if $\lambda \in \mathrm{Pic}^0(E)$, then $\mathcal{M}_{E,r}(\lambda)$ is simply the projective space $\mathbf{P}^{r-1}_{\overline{\mathbf{F}}_q}$.*

1.2. Refined Brill-Noether Locus

1.2.1. Rational Points

Now let E be an elliptic curve defined over a finite field $\mathbf{F}_q$. Then over $\overline{E} = E \times_{\mathbf{F}_q} \overline{\mathbf{F}_q}$, from 1.1.3, we have the moduli spaces $\mathcal{M}_{\bar{E},r}(\lambda)$ (resp. $\mathcal{M}_{\bar{E},r}(d)$) of semi-stable bundles of rank r with determinant λ (resp. degree d) over $\bar{E}$. As algebraic varieties, we may consider $\mathbf{F}_q$-rational points of these moduli spaces. Clearly, by definition, these rational points of moduli spaces correspond exactly to these classes of semi-stable bundles which themselves are defined over $\mathbf{F}_q$. (In the case for

$\mathcal{M}_{\bar{E},r}(\lambda)$, λ is assumed to be rational over $\mathbf{F}_q$.) Thus for simplicity, we simply write $\mathcal{M}_{E,r}(\lambda)$ or $\mathcal{M}_{E,r}(d)$ for the corresponding subsets of $\mathbf{F}_q$-rational points. For example, we then simply write $\mathrm{Pic}^0(E)$ for $\mathrm{Pic}^0(E)(\mathbf{F}_q)$. And by an abuse of notation, we often call these subsets the moduli spaces of semi-stable bundles too.

1.2.2. Standard Brill-Noether Locus

Note that if V is semi-stable with strictly positive degree d, then $h^0(E,V)=d$. Hence the standard Brill-Noether locus is either the whole space or empty. In this way, we are lead to study the case when $d=0$.

For this, recall that for $\lambda\in \mathrm{Pic}^0(E)$,

$$\mathcal{M}_{E,r}(\lambda)=\{V:\mathrm{semi-stable},\mathrm{rank}(V)=r,\det(V)=\lambda\}/\sim_S$$

is identified with

$$\{V=\oplus_{i=1}^r L_i:\otimes_i L_i=\lambda, L_i\in \mathrm{Pic}^0(E), i=1,\dots,r\}/\sim_{\mathrm{iso}}\simeq \mathbf{P}^{r-1}$$

where $/\sim_{\mathrm{iso}}$ means modulo isomorphisms.

Now introduce the standard Brill-Noether locus

$$W_{E,r}^a(\lambda):=\{[V]\in \mathcal{M}_{E,r}(\lambda):h^0(E,\mathrm{gr}(V))\geq a\}$$

and its 'stratification' by

$$W_{E,r}^a(\lambda)^0:=\{[V]\in W_{E,r}(\lambda):h^0(E,\mathrm{gr}(V))=a\}=W_{E,r}^a(\lambda)\backslash\cup_{b\geq a+1}W_{E,r}^b(\lambda).$$

One checks easily that $W_{E,r}^a(\lambda)\simeq \mathbf{P}^{(r-a)-1}$. Thus in particular, we have the following

Lemma. *With the same notation as above,*

$$W_{E,r+1}^{a+1}(\lambda)\simeq W_{E,r}^a(\lambda),\qquad \text{and}\qquad W_{E,r+1}^{a+1}(\lambda)^0\simeq W_{E,r}^a(\lambda)^0.$$

1.2.3. Refined Brill-Noether Locus

The Brill-Noether theory is based on the consideration of h^0. But in the case for elliptic curves, for arithmetic consideration, such a theory is not fine enough: not only h^0 plays an important role, the automorphism groups are important as well. Based on this, we introduce, for a fixed $(k+1)$-tuple non-negative integers $(a_0;a_1,\dots,a_k)$, the subvariety of $W_{E,r}^{a_0}$ by setting

$$\begin{aligned}W_{E,r}^{a_0;a_1,\dots,a_k}(\lambda):=\{[V]\in W_{E,r}^{a_0}(\lambda):\mathrm{gr}(V)=\mathcal{O}_E^{(a_0)}\oplus\oplus_{i=1}^k L_i^{(a_i)},&\\ \otimes_i L_i^{\otimes a_i}=\lambda,\ L_i\in \mathrm{Pic}^0(E),&\\ i=1,\dots,k\}.&\end{aligned}$$

Moreover, we define the associated 'stratification' by setting

$$W_{E,r}^{a_0;a_1,\ldots,a_k}(\lambda)^0 := \{[V] \in W_{E,r}^{a_0;a_1,\ldots,a_k}(\lambda),\ \#\{\mathcal{O}_E, L_1, \ldots, L_k\} = k+1\}.$$

From definition, we easily have the following

Lemma. *With the same notation as above,*

$$W_{E,r+1}^{a_0+1;a_1,\ldots,a_k}(\lambda) \simeq W_{E,r}^{a_0;a_1,\ldots,a_k}(\lambda)$$

and

$$W_{E,r+1}^{a_0+1;a_1,\ldots,a_k}(\lambda)^0 \simeq W_{E,r}^{a_0;a_1,\ldots,a_k}(\lambda)^0.$$

Moreover,

$$\mathcal{M}_{E,r}(\lambda) = \cup_{a_0;a_1,\ldots,a_k} W_{E,r}^{a_0;a_1,\ldots,a_k}(\lambda)^0,$$

where the union is a disjoint one.

In fact, the structures of $W_{E,r}^{a_0;a_1,\ldots,a_k}(\lambda)$ can be given explicitly: They are products of (copies of) projective bundles over E and (copies of) projective spaces.

Proposition. *With the same notation as above, regroup* $(a_0; a_1, \ldots, a_k)$ *as* $(a_0; b_1^{(s_1)}, \ldots, b_l^{(s_l)})$ *with the condition that* $b_1 > b_2 > \ldots > b_l$ *and* $s_1, s_2, \ldots, s_l \in \mathbf{Z}_{>0}$, *then*

(1) if $b_l = 1$,

$$W_{E,r}^{a_0;a_1,\ldots,a_k}(\lambda) \simeq \prod_{i=1}^{l-1} \mathbf{P}_E^{s_i-1} \times \mathbf{P}^{s_l};$$

(2) if $b_l > 1$,

$$W_{E,r}^{a_0;a_1,\ldots,a_k}(\lambda) \simeq \prod_{i=1}^{l} \mathbf{P}_E^{s_i}.$$

Proof. This is because we have the following two facts about the quotient of t products of elliptic curves:

(1) The quotient space $E^{(n)}/S_n$ is isomorphic to the $\mathbf{P}^{n-1}$-bundle over E; and

(2) The quotient of $E^{(n-1)}/S_n$ is isomorphic to $\mathbf{P}^{(n-1)}$. Here we embed $E^{(n-1)}$ as a subspace of $E^{(n)}$ under the map:

$$(x_1, \ldots, x_n) \mapsto (x_1, \ldots, x_{n-1}, x_n)$$

with $x_n = \lambda - (x_1 + x_2 + \ldots + x_{n-1})$.

We end this subsection with the following intersection theoretical discussion. For simplicity, let $\lambda = \mathcal{O}_E$. Then from above, we have the refined Brill-Noether loci $W_{E,r}^{a_0;a_1,\ldots,a_k}(\mathcal{O}_E)$ which are isomorphic to products of (copies of) projective bundles over E and (copies of) projective spaces. Thus it would be very interesting to see the intersections of these special subvarieties in $\mathcal{M}_{E,r}(\mathcal{O}_E) = \mathbf{P}^{r-1}$. For this

purpose, define the so-called Brill-Noether tautological ring $\mathbf{BN}_{E,r}(\mathcal{O}_E)$ to be the subring generated by all the associated refined Brill-Noether loci. For examples,
(1) If $r = 2$, then this ring contains two elements: 1-dimensional one $W_{E,2}^{2;0}(\mathcal{O}_E) = \{[\mathcal{O}_E \oplus \mathcal{O}_E]\}$ and the whole $\mathbf{P}^1$;
(2) If $r = 3$, then (generators of) this ring contains five elements: 2 of 0-dimensional objects: $W_{E,3}^{3;0}(\mathcal{O}_E) = \{[\mathcal{O}_E^{(3)}]\}$ and $W_{E,3}^{1;2}(\mathcal{O}_E) = \{[\mathcal{O}_E \oplus T_2^{(2)}] : T_2 \in E_2\}$ containing 4 elements; 2 of 1-dimensional objects: $W_{E,3}^{1;1,1} = \{[\mathcal{O}_E \oplus L \oplus L^{-1}] : L \in \mathrm{Pic}^0(E)\} \simeq \mathbf{P}^1$, a degree 2 projective line contained in $\mathbf{P}^2 = \mathcal{M}_{E,3}(\mathcal{O}_E)$; and $W_{E,3}^{0;2,1} = \{[L^{(2)} \oplus L^{-2}] : L \in \mathrm{Pic}^0(E)\}$ a degree 3 curve which is isomorphic to E; and finally the whole space. Moreover, the intersection of $W_{E,3}^{1;1,1} = \mathbf{P}^1$ and $W_{E,3}^{0;2,1} = E$ are supported on 0-dimensional locus $W_{E,3}^{1;1,1}$, with the multiplicity 3 on the single point locus $W_{E,3}^{3;0}(\mathcal{O}_E)$ and 1 on the complement of the points in $W_{E,3}^{1;1,1}$.

2. Measure Refined Brill-Noether Locus Arithmetically

2.1. Invariant α

In the rest of this section, we use the same notation as in 1.2.

To measure the Brill-Noether locus, we introduce the following arithmetic invariant $\alpha_{E,r}(\lambda)$ by setting

$$\alpha_{E,r}(\lambda) := \sum_{V \in [V] \in \mathcal{M}_{E,r}(\lambda)} \frac{q^{h^0(E,V)}}{\#\mathrm{Aut}(V)}.$$

Also set

$$\alpha_{E,r}^{a_0+1;a_1,\ldots,a_k}(\lambda) := \sum_{V \in [V] \in W_{E,r}^{a_0+1;a_1,\ldots,a_k}(\lambda)^0} \frac{q^{h^0(E,V)}}{\#\mathrm{Aut}(V)}.$$

Before going further, we remark that above, we write $V \in [V]$ in the summation. This is because in each S-equivalence class $[V]$, there are usually more than one vector bundles V. For example, $[\mathcal{O}_E^{(4)}]$ consists of $\mathcal{O}_E^{(4)}$, $\mathcal{O}_E^{(2)} \oplus I_2$, $I_2 \oplus I_2$, $\mathcal{O}_E \oplus I_3$, and I_4 by the result of Atiyah cited in 1.1.

Thus, by Lemma 1.2.3, we have the following

Lemma. *With the same notation as above,*

$$\alpha_{E,r}(\lambda) = \sum_{(a_0;a_1,\ldots,a_k);k} \alpha_{E,r}^{a_0;a_1,\ldots,a_k}(\lambda).$$

We end this section with the following

Conjecture. *For all* $\lambda \in \mathrm{Pic}^0(E)$,

$$\alpha_{E,r}(\lambda) = \alpha_{E,r}(\mathcal{O}_E).$$

2.2. Invariants β and γ

Due to the importance of automorphism groups, following Harder-Narasimhan, and Desale-Ramanan, we introduce the following β-series invariants $\beta_{E,r}(d)$, $\beta_{E,r}(\lambda)$ and $\beta_{E,r}^{a_0;a_1,\ldots,a_k}(\lambda)$ by setting

$$\beta_{E,r}(d) := \sum_{V \in [V] \in \mathcal{M}_{E,r}(d)} \frac{1}{\#\mathrm{Aut}(V)},$$

$$\beta_{E,r}(\lambda) := \sum_{V \in [V] \in \mathcal{M}_{E,r}(\lambda)} \frac{1}{\#\mathrm{Aut}(V)},$$

and

$$\beta_{E,r}^{a_0;a_1,\ldots,a_k}(\lambda) := \sum_{V \in [V] \in W_{E,r}^{a_0;a_1,\ldots,a_k}(\lambda)^0} \frac{1}{\#\mathrm{Aut}(V)}.$$

Corresponding to the Conjecture 2.1.1 for α, for β, we have the following deep

Theorem. ([HN] & [DR]) *For all* $\lambda, \lambda' \in \mathrm{Pic}^d(E)$,

$$\beta_{E,r}(\lambda) = \beta_{E,r}(\lambda').$$

Moreover,

$$N_1 \cdot \beta_{E,r}(\lambda) = \frac{N_1}{q-1} \cdot \prod_{i=2}^r \zeta_E(i) - \sum_{\Sigma_1^k r_i = r, \Sigma_i d_i = d, \frac{d_1}{r_1} > \ldots > \frac{d_k}{r_k}, k \geq 2} \prod_i \beta_{E,r_i}(d_i) \frac{1}{q^{\Sigma_{i<j}(r_j d_i - r_i d_j)}}.$$

Here N_1 denotes $\#E (:= \#E(\mathbf{F}_q))$ and $\zeta_E(s)$ denotes the Artin zeta function for elliptic curve $E/\mathbf{F}_q$.

Thus, we are lead to introducing the γ-series invariants $\gamma_{E,r}(\lambda)$ and $\gamma_{E,r}^{a_0+1;a_1,\ldots,a_k}(\lambda)$ by setting

$$\gamma := \alpha - \beta.$$

That is to say,

$$\gamma_{E,r}(\lambda) := \sum_{V \in [V] \in \mathcal{M}_{E,r}(\lambda)} \frac{q^{h^0(E,V)} - 1}{\#\mathrm{Aut}(V)},$$

and

$$\gamma_{E,r}^{a_0;a_1,\ldots,a_k}(\lambda) := \sum_{V\in[V]\in W_{E,r}^{a_0;a_1,\ldots,a_k}(\lambda)^0} \frac{q^{h^0(E,V)}-1}{\#\mathrm{Aut}(V)}.$$

Clearly, by the above Theorem, the Conjecture 2.1.2 for α is equivalent to the following

Conjecture. *For all* $\lambda \in \mathrm{Pic}^0(E)$,

$$\gamma_{E,r}(\lambda) = \gamma_{E,r}(\mathcal{O}_E).$$

The advantage of this Conjecture is that now the support of the summation is over $W_{E,r}^1(\lambda)$, a codimension 1 projective subspace.

Similarly, we have the following

Lemma. *With the same notation as above, for* $\lambda \in \mathrm{Pic}^0(E)$,

$$\beta_{E,r}(\lambda) = \sum_{a_0;a_1,\ldots,a_k} \beta_{E,r}^{a_0;a_1,\ldots,a_k}(\lambda)$$

and

$$\gamma_{E,r}(\lambda) = \sum_{a_0;a_1,\ldots,a_k} \gamma_{E,r}^{a_0;a_1,\ldots,a_k}(\lambda).$$

3. New Non-Abelian Zeta Functions for Elliptic Curves

3.1. Non-Abelian Local Zeta Functions

3.1.1. Definition

Let E be an elliptic curve defined over $\mathbf{F}_q$, the finite field with q elements. Then we have the associated ($\mathbf{F}_q$-rational points of) moduli spaces $\mathcal{M}_{E,r}(d)$. By definition, the *rank* r *non-abelian zeta function* $\zeta_{E,r,\mathbf{F}_q}(s)$ *of* E is defined by setting

$$\zeta_{E,r,\mathbf{F}_q}(s) := \sum_{V\in[V]\in\mathcal{M}_{E,r}(d),d\geq 0} \frac{q^{h^0(E,V)}-1}{\#\mathrm{Aut}(V)} q^{-s\cdot d(V)}, \qquad \mathrm{Re}(s) > 1.$$

Here $d(V)$ denotes the degree of V.

Remark. We call the above infinite sum the rank r non-abelian zeta function because when $r = 1$ the above summation $\zeta_{E,1}(s)$ coincides with the classical Artin zeta function for the elliptic curve E over $\mathbf{F}_q$. In other words, the classical Artin zeta function $\zeta_E(s)$ may be written as

$$\zeta_E(s) = \sum_{L\in \mathrm{Pic}^d(E),d\geq 0} \frac{q^{h^0(E,L)}-1}{\#\mathrm{Aut}(L)} q^{-s\cdot d(L)}, \qquad \mathrm{Re}(s)>1,$$

where $d(L)$ denotes the degree of L.

3.1.2. Basic Properties

With the above definition, by a direct yet long calculation, we have, by setting $t=q^{-s}$ and $Z_{E,r,\mathbf{F}_q}(t) := \zeta_{E,r,\mathbf{F}_q}(s)$, the following

Fundamental Identity. *Let E be an elliptic curve defined over $\mathbf{F}_q$. Then for any $r\in\mathbf{Z}_{>0}$,*

$$Z_{E,r,\mathbf{F}_q}(t) := \gamma_{E,r}(0)\cdot \frac{P_{E,r,\mathbf{F}_q}(t)}{(1-t^r)(1-q^rt^r)}.$$

Here

$$\begin{aligned} &P_{E,r,\mathbf{F}_q}(t) \\ &\quad =1+\sum_{i=1}^{r-1}(q^i-1)\frac{\beta_{E,r}(i)}{\gamma_{E,r}(0)}\cdot t^i + \Big((q^r-1)\frac{\beta_{E,r}(0)}{\gamma_{E,r}(0)} - (q^r+1)\Big)\cdot t^r \\ &\qquad +\sum_{i=1}^{r-1}(q^i-1)\frac{\beta_{E,r}(i)}{\gamma_{E,r}(0)}\cdot q^{r-i}t^{2r-i} + q^rt^{2r}. \end{aligned}$$

Remark. We in this paper choose not to write down the detailed elementary calculations, despite the fact that some of them are very long and sometimes a bit complicated.

As a direct consequence of this Fundamental Identity, we have the following

Theorem. *Let E be an elliptic curve defined over $\mathbf{F}_q$, the finite field with q elements. Then the associated rank r non-abelian zeta function $\zeta_{E,r,\mathbf{F}_q}(s)$ satisfies the following basic properties:*

*(1) (**Rationality**) $Z_{E,r,\mathbf{F}_q}(t)$ may be written as the quotient of two polynomials;*

*(2) (**Functional Equation**) $\zeta_{E,r,\mathbf{F}_q}(s) = \zeta_{E,r,\mathbf{F}_q}(1-s)$.*

From here we may conclude that after suitable arrangement, the product of two reciprocal roots of $P_{E,r,\mathbf{F}_q}(t)$, a degree $2r$ polynomial, are always equal to q, the cardinal number of the base field. Moreover, we know that up to the term $\gamma_{E,r}(0)$, from Theorem 2.1.2, the coefficients of these local non-abelian zeta functions can be computed.

3.2. Non-Abelian Global Zeta Functions

3.2.1. Definition.

Now let $\mathbf{E}$ be an elliptic curve defined over a number field F. For a fixed positive integer r, set S_{bad} to be the union of all infinite places (of F), all finite places where $\mathbf{E}$ have bad reductions, or where the characteristics of the residue fields are less than r. By definition, a (finite) places v of F is good, if v is not in S_{bad}.

Thus, in particular, for good places v, by taking reduction of $\mathbf{E}$ at v, we have the associated regular elliptic curve $\mathbf{E}_v$ defined over $F(v) \simeq \mathbf{F}_{q_v}$, the residue field of F at v, where q_v denotes cardinal number of $F(v)$. Then by applying the construction of 3.1, we get the rank r local non-abelian zeta functions $\zeta_{\mathbf{E}_v,r,\mathbf{F}_{q_v}}(s)$. In particular, we further obtain, by the rationality, the corresponding polynomials $P_{\mathbf{E}_v,r,\mathbf{F}_q}(t)$ of degree $2r$ (with 1 as constant terms).

Definition. *Let $\mathbf{E}$ be an elliptic curve defined over a number field F. Then for any positive integer r, define its associated rank r global non-abelian zeta function $\zeta_{\mathbf{E},r,F}(s)$ by setting*

$$\zeta_{\mathbf{E},r,F}(s) := \prod_{v:\mathrm{good}} P_{\mathbf{E}_v,r,\mathbf{F}_{q_v}}(q_v^{-s})^{-1}.$$

Here q_v denotes the cardinal number of the residue field of F at v.

Clearly, if $r = 1$, this then recovers the famous Hasse-Weil zeta function for elliptic curves, for which we have the celebrated BSD conjecture. Thus it seems to be quite natural for us to call the above Euler product a global non-abelian zeta function for elliptic curve.

Surely the biggest problem we are now facing in this algebraic part of our non-abelian zeta function is to give the precise region over which the Euler product in the definition converges. For this we have to use our refined Brill-Noether theory discussed in Section 1.

3.2.2. Estimations for β and γ

We now want to give estimations for invariants β and γ. So let E be an elliptic curve defined over a finite field $\mathbf{F}_q$ as before.

First we study $\beta_{E,r}$. For this, following Harder-Narasimhan, we interpret the Tamagawa number of $\mathrm{SL}(n)$ is 1 as follows:

Proposition. (=[DR Proposition 1.1]) *Let $\zeta_E(s)$ be the Artin zeta function of E. Then for any fixed $\lambda \in \mathrm{Pic}^d(E)$,*

$$\sum_{V:\mathrm{rank}(V)=r,\det(V)=\lambda} \frac{1}{\#\mathrm{Aut}(V)} = \frac{1}{q-1}\prod_{k=2}^{r} \zeta_E(k).$$

Therefore, for a fixed $\lambda \in \mathrm{Pic}^d(E)$,

$$\sum_{V:\mathrm{rank}(V)=r,\det(V)=\lambda} \frac{1}{\#\mathrm{Aut}(V)} = O(q^{-1}).$$

This implies in particular that

$$\beta_{E,r}(\lambda) = O(q^{-1}).$$

Thus, by Hasse's result ([Ha]) on $N_1 = \#\mathrm{Pic}^d(E)$, we know $N_1 = O(q)$. Therefore we complete the following

Proposition I. *With the same notation as above,* $\beta_{E,r}(d) = O(1)$, *as* $q \to \infty$.

Next we study $\gamma_{E,r}(0)$. As our final purpose is to give a good estimation for the coefficients of $P_{E,r,\mathbf{F}_q}(t)$, hence by the Fundamental Identity in 3.1.2, it suffices to give a lower bound for $\gamma_{E,r}(0)$. For this, we consider semi-stable vector bundles V with $\mathrm{gr}(V) = \mathcal{O}_E \oplus \oplus_{i=1}^{r-1} L_i$ with $L_i \in \mathrm{Pic}^0(E)$ and $\#\{\mathcal{O}_E, L_1, \dots, L_{r-1}\} = r$. Clearly then $V = \mathrm{gr}(V)$ and there are totally $O(N_1^{r-1})$ or better $O(q^{r-1})$ of them, as $q \to \infty$ by the above mentioned result of Hasse. On the other hand, easily, we have $h^0(V) = 1$ and $\#\mathrm{Aut}(V) = (q-1)^r$. Therefore

$$\sum_{V:\mathrm{gr}(V)=\mathcal{O}_E\oplus\oplus_{i=1}^{r-1},L_i\in\mathrm{Pic}^0(E),\#\{\mathcal{O}_E,L_1,\dots,L_{r-1}\}=r} \frac{q^{h^0(E,V)}-1}{\#\mathrm{Aut}(V)} = O(1).$$

This then implies the following

Proposition II. *With the same notation as above,* $\gamma_{E,r}(d) = O(1)$, *as* $q \to \infty$.

3.2.3. Convergence of Global Non-Abelian Zeta Functions

As above, let E be an elliptic curve defined over $\mathbf{F}_q$. Then by the Fundamental Identity and the Functional Equation for local non-abelian zeta functions of elliptic curves,

$$P_{E,r,\mathbf{F}_q}(t) = \prod_{i=1}^{r} (1 + A_i t + q t^2)$$

with $A_i \in \mathbf{R}$. So, as $q \to \infty$, by Proposition I and II in 3.2.1,

$$\sum_i A_i = O(q);$$

$$\sum_{i<j} A_i A_j = O(q^2);$$

$$\cdots\cdots\cdots\cdots\cdots$$

$$\prod_{i=1}^{r} A_i = O(q^r).$$

So, $A_i = O(q), i = 1, \ldots, r$. As a direct consequence, we then have the following

Theorem. *Let* $\mathbf{E}$ *be an elliptic curve defined over a number field* F. *Then all rank* r *global non-abelian zeta functions* $\zeta_{\mathbf{E},r,F}(s)$, *defined by (infinite) Euler products converge when* $\mathrm{Re}(s) > 2$.

Remark. When $r = 1$, by the above mentioned result of Hasse, the Hasse-Weil zeta functions, i.e., $\zeta_{\mathbf{E},1,F}(s)$ in our notation, converge when $\mathrm{Re}(s) > \frac{3}{2}$. While this result is better than ours in the case when $r = 1$, we are also quite satisfied with our one as our Theorem is the best possible for general r at this stage. (See e.g. 4.1.1 below.)

Now, a natural question is whether our rank r global non-abelian zeta functions have meromorphic extensions and satisfy the functional equation. We believe that it should be the case. In fact we have the following

Working Hypothesis. *By introducing also factors for bad places, the completed rank* r *non-abelian zeta functions* $\xi_{\mathbf{E},r,F}(s)$ *for elliptic curves have meromorphic continuation to the whole complex plane and satisfy the functional equation*

$$\xi_{\mathbf{E},r,F}(s) = \xi_{\mathbf{E},r,F}(1 + \frac{1}{r} - s).$$

Clearly, we then would also hope for certain type of such zeta functions, the inverse Mellin transform would lead to modular forms of fractional weight $1 + \frac{1}{r}$. Unfortunately, we have not yet obtained any examples to support this speculation. But if it holds, we then have a systematical way to construct fractional weight modular forms, for which, except in the case of half integers we know very little.

4. Examples and Justifications

4.1. Lower Rank Non-Abelian Zeta Functions

4.1.1. Rank Two

Let E be an elliptic curve defined over the finite field $\mathbf{F}_q$. If rank r is two, we need then only to calculate $\beta_{E,2}(0), \beta_{E,2}(1)$ and $\gamma_{E,r}(0)$.

We first consider $\beta_{E,2}(0)$. Then by our discussion on Brill-Noether locus, it suffices to calculate $\beta_{E,2}(\mathcal{O}_E)$. Now

$$\mathcal{M}_{E,2}(\mathcal{O}_E) = W_{E,2}^{2;0}(\mathcal{O}_E)^0 \cup W_{E,2}^{0;2}(\mathcal{O}_E)^0 \cup W_{E,2}^{0;1,1}(\mathcal{O}_E)^0.$$

Clearly,

$$W_{E,2}^{2;0}(\mathcal{O}_E)^0 = \{[V] : \mathrm{gr}(V) = \mathcal{O}_E^{(2)}\}$$

consisting of just 1 element;

$$W_{E,2}^{0;2}(\mathcal{O}_E)^0 = \{[V] : \mathrm{gr}(V) = T_2^{(2)}, T_2 \in E_2, T_2 \neq \mathcal{O}_E\}$$

consisting of 3 elements coming from non-trivial $T_2 \in E_2$, 2-torsion subgroup of E; while

$$W_{E,2}^{0;1,1}(\mathcal{O}_E)^0 = \{[V] : \mathrm{gr}(V) = L \oplus L^{-1}, L \in \mathrm{Pic}^0(E), L \neq L^{-1}\}$$

is simply the complement of the above 4 points in $\mathbf{P}^1$. With this, one checks that

$$\beta_{E,2}(0) = \Big(\frac{1}{(q^2-1)(q^2-q)} + \frac{1}{(q-1)q}\Big) + 3 \cdot \Big(\frac{1}{(q^2-1)(q^2-q)} + \frac{1}{(q-1)q}\Big)$$
$$+\big(q+1-(3+1)\big) \cdot \frac{1}{(q-1)^2}$$
$$= \frac{q+3}{q^2-1}.$$

And hence

$$\beta_{E,2}(0) = N_1 \cdot \frac{q+3}{q^2-1}.$$

As for $\beta_{E,2}(1)$, it is very simple: Any degree one rank two semi-stable bundle is stable. Moreover, by the result of Atiyah cited in 1.1, there is exactly one stable rank two bundle whose determinant is the fixed line bundle. Thus

$$\beta_{E,2}(1) = N_1 \cdot \frac{1}{q-1}.$$

Finally, we study $\gamma_{E,2}(0)$. We want to check Conjecture 2.1.2. Clearly if $\lambda \neq \mathcal{O}_E$, then $\gamma_{E,2}(\lambda)$ is supported on

$$W_{E,2}^{1;1}(\lambda) = \{[V] : \mathrm{gr}(V) = \mathcal{O}_E \oplus \lambda\}$$

consisting only one element with $V = \mathrm{gr}(V) = \mathcal{O}_E \oplus \lambda$. So

$$\gamma_{E,2}(\lambda) = \frac{q-1}{(q-1)^2} = \frac{1}{q-1}.$$

On the other hand, $\gamma_{E,2}(\mathcal{O}_E)$ is supported on

$$W_{E,2}^{2;0}(\mathcal{O}_E) = \{[V] : \mathrm{gr}(V) = \mathcal{O}_E^{(2)}\}$$

consisting only one element too. However, now in the single class $[V]$ with $\mathrm{gr}(V) = \mathcal{O}_E^{(2)}$, there are two elements, i.e., $\mathcal{O}_E^{(2)}$ and I_2. So

$$\gamma_{E,2}(\mathcal{O}_E) = \frac{q^2-1}{(q^2-1)(q^2-q)} + \frac{q-1}{(q-1)q} = \frac{1}{q-1} = \beta_{E,1}(\mathcal{O}_E).$$

Thus we have checked all conjectures in the case $r = 2$.

Proposition. *With the same notation as above, in the case $r = 2$, all conjectures in this paper are confirmed. In particular,*

$$Z_{E,2,\mathbf{F}_q}(t) = \frac{N_1}{q-1} \cdot \frac{1+(q-1)t+(2q-4)t^2+(q^2-q)t^3+q^2t^4}{(1-t^2)(1-q^2t^2)}.$$

We reminder the reader that in this case, $P_{E,2,\mathbf{F}_q}(t)$ is independent of E and with integer coefficients. Thus for global rank two non-abelian zeta function, we obtain an absolute Euler product, say in the case when the base field is $\mathbf{Q}$,

$$E_2(s) = \prod_{p\geq 3,\mathrm{prime}} \frac{1}{1+(p-1)p^{-s}+(2p-4)p^{-2s}+(p^2-p)p^{-3s}+p^2p^{-4s}},$$

$$\mathrm{Re}(s) > 2.$$

Clearly we expect more from such a beautiful Euler product. In fact, recently, in a private communication, Andrew Booker shows that the produt could be meromophically extended to the region $\mathrm{Re}(s) > 1$. *4.1.2. Rank Three*

First, we check that

$$\sum_{V,\mathrm{gr}(V)=\mathcal{O}_E^{(2)}} \frac{1}{\#\mathrm{Aut}(V)} = \sum_{W,\mathrm{gr}(W)=\mathcal{O}_E^{(3)}} \frac{q^{h^0(E,V)}-1}{\#\mathrm{Aut}(V)}.$$

By the fact that $\mathrm{Aut}(\mathcal{O}_E \oplus I_2) = (q-1)^2q^3$, the above identity is equivalent to

$$\frac{1}{(q^2-1)(q^2-q)} + \frac{1}{(q-1)q} = \frac{q^3-1}{(q^3-1)(q^3-q)(q^3-q^2)} + \frac{q^2-1}{(q-1)^2q^3} + \frac{q-1}{(q-1)q^2},$$

which may be directly checked. This together with the similar relation for $r = 1$ discussed in 4.1.1 leads to

$$\gamma_{E,3}(0) = N_1 \cdot \gamma_{E,3}(\mathcal{O}_E) = N_1 \cdot \beta_{E,2}(\mathcal{O}_E) = N_1 \cdot \frac{q+3}{q^2-1}.$$

So we are left to study $\beta_{E,3}(d)$, $d=0,1,2$. Easily, we have

$$\beta_{E,3}(1)=\beta_{E,3}(2)=N_1\cdot\frac{1}{q-1}$$

since here all semi-stable bundles become stable. Thus we are lead to consider only $\beta_{E,3}(0)$. So it suffices to give $\beta_{E,3}(\lambda)$ for any $\lambda\neq\mathcal{O}_E$. (Despite the fact that $\beta_{E,r}(\lambda)=\beta_{E,r}(\mathcal{O}_E)$ for any $\lambda\in\mathrm{Pic}^0(E)$, in practice, the calculation of $\beta_{E,r}(\lambda)$ with $\lambda\neq\mathcal{O}_E$ is easier than that for $\beta_{E,r}(\mathcal{O}_E)$.)

Now

$$\mathcal{M}_{E,3}(\lambda)=\Big((W_{E,3}^{2;1}(\lambda)^0)\cup W_{E,3}^{1;2}(\lambda)^0\cup W_{E,3}^{1;1,1}(\lambda)^0\Big)\cup W_{E,3}^{0;3}(\lambda)^0$$
$$\cup W_{E,3}^{0;2,1}(\lambda)^0\cup W_{E,3}^{0;1,1,1}(\lambda)^0.$$

Moreover, we have

(1) $W_{E,3}^{2;1}(\lambda)^0$ consists a single class [V], i.e., the one with $\mathrm{gr}(V)=\mathcal{O}_E^2\oplus\lambda$, which contains two vector bundles, i.e., $\mathcal{O}_E^2\oplus\lambda$ and $I_2\oplus\lambda$;

(2) $W_{E,3}^{2;1}(\lambda)^0\cup W_{E,3}^{1;2}(\lambda)^0\cup W_{E,3}^{1;1,1}(\lambda)^0\simeq\mathbf{P}^1$ with $W^{1;2}(\lambda)^0$ consists of 4 classes $[V]$, i.e., these such that $\mathrm{gr}(V)=\mathcal{O}_E\oplus\big(\lambda^{\frac12}\big)^{(2)}$, where $\lambda^{\frac12}$ denotes any of the four square roots of λ. Clearly then in each class $[V]$, there are also two vector bundles $\mathcal{O}_E\oplus\big(\lambda^{\frac12}\big)^{(2)}$ and $\mathcal{O}_E\oplus I_2\otimes\lambda^{\frac12}$;

(3) $W_{E,3}^{0;3}(\lambda)^0\cup W_{E,3}^{0;2,1}(\lambda)^0\cup W_{E,3}^{0;1,1,1}(\lambda)^0=\mathbf{P}^2\backslash\mathbf{P}^1$.

(3.a) $W_{E,3}^{0;3}(\lambda)^0$ consists of 9 classes $[V]$, i.e., these $[V]$ with $\mathrm{gr}(V)=\big(\lambda^{\frac13}\big)^{(3)}$ where $\lambda^{\frac13}$ denotes any of the 9 triple roots of λ. Moreover, in each $[V]$, there are three bundles, i.e., $\big(\lambda^{\frac13}\big)^{(3)}$, $\lambda^{\frac13}\oplus I_2\otimes\lambda^{\frac13}$ and $I_3\otimes\lambda^{\frac13}$.

(3.b) $\Big(W_{E,3}^{2;1}(\lambda)^0\cup W_{E,3}^{1;2}(\lambda)^0\Big)\cup\Big(W_{E,3}^{0;3}(\lambda)^0\cup W_{E,3}^{0;2,1}(\lambda)^0\Big)$ is isomorphic to E. Moreover, each class $[V]$ in $W^{0;2,1}(\lambda)^0$ consists of two bundles, i.e., $L^{(2)}\oplus\lambda\otimes L^{-2}$ and $I_2\otimes L\oplus\lambda\otimes L^{-2}$ when $\mathrm{gr}(V)=L^{(2)}\oplus\lambda\otimes L^{-2}$.

(One checks that in fact the refined Brill-Noether loci $\mathbf{P}^1$ and E appeared above are embedded in $\mathbf{P}^2$ as degree 2 and 3 regular curves. And hence the intersection should be 6: The intersection points are at $[V]$ with $\mathrm{gr}(V)=\mathcal{O}_E^{(2)}\oplus\lambda$ with multiplicity 2, and $\mathcal{O}_E\oplus\big(\lambda^{\frac12}\big)^{(2)}$ corresponding to four square roots of λ with multiplicity one. That is to say, the intersection actually are supported on $W_{E,3}^{2;1}(\lambda)^0\cup W_{E,3}^{1;2}(\lambda)^0$. So it would be very interesting in general to study the intersections of the refined Brill-Noether loci as well.)

From this analysis, we conclude that

$$\begin{aligned}\beta_{E,3}(\lambda) =&\Big(\frac{1}{(q^2-1)(q^2-q)(q-1)}+\frac{1}{(q-1)q(q-1)}\Big)\\&+4\Big(\frac{1}{(q^2-1)(q^2-q)(q-1)}+\frac{1}{(q-1)q(q-1)}\Big)+(q-4)\cdot\Big(\frac{1}{(q-1)^3}\Big)\\&+9\Big(\frac{1}{(q^3-1)(q^3-q)(q^3-q^2)}+\frac{1}{(q-1)^2q^3}+\frac{1}{(q-1)q^2}\Big)\\&+\big(N_1-(9+4+1)\big)\cdot\Big(\frac{1}{(q-1)(q^2-1)(q^2-q)}+\frac{1}{(q-1)(q-1)q}\Big)\\&+\big(q^2-(N_1-4-1)\big)\cdot\frac{1}{(q-1)^3}.\end{aligned}$$

As a direct consequence, we have the following

Proposition. *With the same notation as above, we have*

$$\begin{aligned}P_{E,3,\mathbf{F}_q}(t) =&1+\frac{q^2-1}{q+3}\cdot(t+q^2t^5)+\frac{(q^2-1)(q+1)}{q+3}\cdot(t^2+qt^4)+q^3t^6\\&-\frac{1}{(q+3)(q-1)}\cdot\Big[(q-1)(5q^3-3q^2-4q+4)+(q^2+q+1)N_1\Big]\cdot t^3.\end{aligned}$$

Also we would reminder the reader that here in fact 2- and 3- torsion points are involved naturally in the calculation.

4.2. Why Use only Semi-Stable Bundles

4.2.1. Degree 0

At the first glance, Theorem 2.1.2 and Proposition 3.2.2 suggest that in the definition of non-abelian zeta functions we should consider all vector bundles, just as what happens in the theory of automorphic L-functions. However, we here use an example with $r=2$ to indicate the opposite.

Thus we first introduce a new zeta function $\zeta_{E,r}^{\mathrm{all}}(s)$ by

$$\zeta_{E,r}^{\mathrm{all}}(s):=\sum_{V:\mathrm{rank}(V)=2}\frac{q^{h^0(V)}-1}{\#\mathrm{Aut}(V)}\cdot q^{-sd(V)}.$$

Then by our discussion on the non-abelian zeta functions associated to semi-stable bundle, we only need to consider the contribution of rank 2 bundles which are not semi-stable.

We start with a discussion on extension of bundles. Assume that V is not semi-stable of rank 2. Let L_2 be the line subbundle of V with maximal degree, then V is obtained from the extension of $L_2 := V/L_1$ by L_1

$$0 \to L_1 \to V \to L_2 \to 0.$$

But V is not semi-stable implies that all such extensions are trivial. Thus $V = L_1 \oplus L_2$. For later use, set d_i to be the degree of $L_i, i = 1, 2$. Then $d_1 + d_2 = d$ the degree of V, and

$$\#\mathrm{Aut}(V) = (q-1)^2 \cdot q^{h^0(E, L_1 \otimes L_2^{-1})} = (q-1)^2 \cdot q^{d_1 - d_2}.$$

Next we study the contribution of degree 0 vector bundles of rank 2 which are not semi-stable. Note that the support of the summation should have non-vanishing h^0. Thus $V = L_1 \oplus L_2$ where $L_1 \in \mathrm{Pic}^{d_1}(E)$ with $d_1 > 0$. So the contributions of these bundles are given by

$$\begin{aligned}\zeta_{E,2}^{=0}(s) =& Z_{E,2}^{=0}(t) \\ =& \sum_{d=1}^{\infty} \sum_{L_1 \in \mathrm{Pic}^d(E), L_2 \in \mathrm{Pic}^{-d}(E)} \frac{q^{h^0(L_1)} - 1}{(q-1)^2 q^{h^0(L_1 \otimes L_2^{\otimes -1})}} = \frac{N_1^2}{(q-1)^2} \cdot \sum_{d=1}^{\infty} \frac{q^d - 1}{q^{2d}} \\ =& \frac{q N_1^2}{(q^2-1)(q-1)^2}.\end{aligned}$$

4.2.2. Degree > 0

Now we consider all degree strictly positive rank 2 vector bundles which are not semi-stable. From above we see that $V = L_1 \oplus L_2$ with $d_1 > d_2$. Thus for $h^0(E, V)$, there are three cases:

(i) $d_2 > 0$, clearly then $h^0(E, V) = d$;

(ii) $d_2 = 0$. Here there are two subcases, namely, (a) if $L_2 = \mathcal{O}_E$, then $h^0(E, V) = d_1 + 1$; (b) If $L_2 \neq \mathcal{O}_E$, then $h^0(E, V) = d_1$;

(iii) $d_2 < 0$. Then $h^0(E, V) = d_1$.

Therefore, all in all the contribution of strictly positive degree rank 2 bundles which are not semi-stable to the zeta function $\zeta_{E,r}^{\mathrm{all}}(s)$ is given by

$$\zeta_{E,2}^{>0}(s) = Z_{E,2}^{>0}(t) = \Big(\sum_{(i)} + \sum_{(ii.a)} + \sum_{(ii.b)} + \sum_{(iii)}\Big) \frac{q^{h^0(V)} - 1}{\#\mathrm{Aut}(V)} t^d$$

where $\sum_{(*)}$ means the summation is taken for all vector bundles in case (*).

Hence, we have

$$\sum_{(i)} \frac{q^{h^0(V)}-1}{\#\mathrm{Aut}(V)} = N_1^2 \cdot \sum_{d=1}^{\infty} \sum_{d_1+d_2=d, d_1>d_2>0} \frac{q^d-1}{(q-1)^2 q^{d_1-d_2}} t^d,$$

$$\sum_{(ii.a)} \frac{q^{h^0(V)}-1}{\#\mathrm{Aut}(V)} = N_1 \cdot \sum_{d=1}^{\infty} \frac{q^{d+1}-1}{(q-1)^2 q^d} t^d,$$

$$\sum_{(ii.a)} \frac{q^{h^0(V)}-1}{\#\mathrm{Aut}(V)} = N_1(N_1-1) \cdot \sum_{d=1}^{\infty} \frac{q^d-1}{(q-1)^2 q^d} t^d,$$

$$\sum_{(iii)} \frac{q^{h^0(V)}-1}{\#\mathrm{Aut}(V)} = N_1^2 \cdot \sum_{d=1}^{\infty} \sum_{d_1+d_2=d, d_1>0>d_2} \frac{q^{d_1}-1}{(q-1)^2 q^{d_1-d_2}} t^d.$$

By a direct calculation, we find that

$$\sum_{(i)} \frac{q^{h^0(V)}-1}{\#\mathrm{Aut}(V)} = \frac{N_1^2 t^3}{q-1} \cdot \frac{q^2+q+1+q^2 t}{(1-t^2)(1-q^2t^2)(q-t)},$$

$$\sum_{(ii.a)} \frac{q^{h^0(V)}-1}{\#\mathrm{Aut}(V)} = \frac{N_1 t}{q-1} \cdot \frac{q+1-t}{(q-t)(1-t)},$$

$$\sum_{(ii.a)} \frac{q^{h^0(V)}-1}{\#\mathrm{Aut}(V)} = \frac{N_1(N_1-1)}{q-1} \cdot \frac{t}{(q-t)(1-t)},$$

$$\sum_{(iii)} \frac{q^{h^0(V)}-1}{\#\mathrm{Aut}(V)} = \frac{N_1^2 t}{(q-1)^2(q^2-1)} \cdot \frac{q^2+q-1-qt}{(1-t)(q-t)}.$$

4.2.3. Degree < 0

Finally we consider the contribution of bundles with strictly negative degree. First we have the following classification according to $h^0(E,V)$.

(i) $d_1 > 0 > d_2$. Then $h^0(E,V) = d_1$;

(ii) $d_1 = 0 > d_2$. Here two subcases. (a) $L_1 = \mathcal{O}_E$, then $h^0(E,V) = 1$; (b) $L_1 \neq \mathcal{O}_E$, then $h^0(V) = 0$;

(iii) $0 > d_1 > d_2$. Here $h^0(V) = 0$.

Thus note that the support of $h^0(E,V)$ is only on the cases (i) and (ii.a), we see that similarly as before, the contribution of strictly positive degree rank 2 bundles which are not semi-stable to the zeta function is given by

$$\zeta_{E,2}^{<0}(s) = Z_{E,2}^{<0}(t) = \Big(\sum_{(i)} + \sum_{(ii.a)}\Big) \frac{q^{h^0(V)}-1}{\#\mathrm{Aut}(V)} t^d.$$

Hence, we have

$$\begin{aligned}\zeta_{E,2}^{<0}(s) = Z_{E,2}^{<0}(t) =& N_1^2 \sum_{d=-1}^{-\infty} \sum_{d_1>0>d_2, d_1+d_2=d} \frac{q^{d_1}-1}{(q-1)^2 q^{d_1-d_2}} t^d \\ &+ N_1 \cdot \sum_{d=-1}^{-\infty} \frac{q-1}{(q-1)^2 q^{-d}} t^d \\ =& \frac{N_1^2}{(q-1)^2} \cdot \frac{q}{(qt-1)(q^2-1)} + \frac{N_1}{q-1} \cdot \frac{1}{qt-1}.\end{aligned}$$

I hope now the reader is fully convinced that our definition of non-abelian zeta function by using moduli space of semi-stable bundles is much better than that of others: Not only our semi-stable zeta functions have much neat structure, we also have well-behavior geometric and hence arithmetic spaces ready to use. In a certain sense, these non-abelian zeta functions are quite similar to the so-called new forms: Only after removing these not-semi-stable contributions, we can see the intrinsic beautiful structures. For more details, please see [We2].

REFERENCES

[A] E. Artin, Quadratische Körper im Gebiete der höheren Kongruenzen, I,II, *Math. Zeit*, **19** 153-246 (1924) (See also *Collected Papers*, pp. 1-94, Addison-Wesley 1965)

[At] M. Atiyah, Vector Bundles over an elliptic curve, *Proc. LMS, VII*, 414-452 (1957) (See also *Collected Works*, Vol. 1, pp. 105-143, Oxford Science Publications, 1988)

[DR] U.V. Desale & S. Ramanan, Poincaré polynomials of the variety of stable bundles, Math. Ann **216**, 233-244 (1975)

[HN] G. Harder & M.S. Narasimhan, On the cohomology groups of moduli spaces of vector bundles over curves, Math Ann. **212**, (1975) 215-248

[H] H. Hasse, *Mathematische Abhandlungen*, Walter de Gruyter, Berlin-New York, 1975.

[Mu] D. Mumford, *Geometric Invariant Theory*, Springer-Verlag, Berlin (1965)

[W] A. Weil, *Sur les courbes algébriques et les variétés qui s'en déduisent*, Herman, Paris (1948)

[We1] L. Weng, Stability, and New Zeta Functions for Number Fields, to appear

[We2] L. Weng, A Program for Geometric Arithmetic, preprint, 2001

LIST OF PARTICIPANTS

Ambro, Florin
Aoki, Masao
Ashikaga, Tadashi
Chen, Jungkai Alfred
Choi, Jaeyoo
Choi, Youngook
Fujisawa, Taro
Fujita, Takao
Goto, Yoshihiro
Hanamura, Masaki
Harui, Takeshi
Hashimoto, Toshiyuki
Hwang, Jun-Muk
Ikeda, Atsushi
Inaba, Michiaki
Ishii, Akira
Ishii, Shihoko
Ito, Yukari
Izumi, Shuzo
Kawaguchi, Shu
Kawahara, Tomokazu
Kawanoe, Hiraku
Keem, Changho
Kei, Miura
Keum, JongHae
Kim, Sung-Ock
Kitagawa, Shinnya
Kobayashi, Masanori
Kojima, Hideo
Konno, Kazuhiro
Lee, Seunghun
Lee, Yongnam
Lu, Steven Shin-Yi
Miura, Kei
Miyaoka, Yoichi
Moriwaki, Atsushi
Mukai, Shigeru
Murakami, Masaaki
Nakayama, Hiroto
Nakayama, Noboru
Namikawa, Yoshinori
NguyenKhac Viet
Nishiyama, Ken-ichi
Oguiso, Keiji
Ohbuchi, Akira
Pearlstein, Gregory J
Shimada, Ichiro
Sumihiro, Hideyasu
Sun, Xiaotao
Takahashi, Takeshi
Takeda, Yoshifumi
Tokunaga, Hiroo
Tsuda, Noriko
Usui, Sampei
Weng, Lin
Yamada, Shin-ichiro
Yanase, Daisuke
Yoshihara, Hisao
Yoshioka, Kota
Zhang, De-Qi